Normal Accidents

NORMAL ACCIDENTS

Living with High-Risk Technologies

CHARLES PERROW

BasicBooks
A Division of HarperCollins*Publishers*

Library of Congress Cataloging in Publication Data

Perrow, Charles.
Normal accidents.

References: p. 366
Includes index.
1. Industrial accidents. 2. Accidents. 3. Risk.
I. Title.
T54.P47 1984 363.1 83–45256
ISBN 0–465–05143–X (cloth)
ISBN 0–465–05142–1 (paper)

Printed in the United States of America
Designed by Vincent Torre

95 96 97 98 99 HC 12 11 10 9

Contents

ABNORMAL BLESSINGS

I started this book, without knowing I was starting a book, in August of 1979. Since then I have lived with it at three institutions, under three research grants, and have worked with a variety of secretaries, research assistants, colleagues and some technical critics. In this acknowledgments section I wish not only to list a few of the numerous people who helped me, some of whom will be unknown and perhaps a matter of indifference to most readers, but also to give some sense of the extraordinary good fortune and flow of resources that can take place in academic life. Most readers have only a dim sense of how a book comes to be written. In a book on normal accidents, I think it is appropriate to acknowledge my abnormal blessings. Just as complex systems threaten to bring us down, as I will argue in this book, so do complex systems bring us unimagined and probably undeserved bounty. Here is an accounting of my bounty.

When Professor Cora Marrett was appointed to the President's Commission on the Accident at Three Mile Island, she met with David Sills of the Social Science Research Council, of which she was a board member, to discuss plans for some social science input into what threatened to be an entirely engineering-oriented investigation. They asked a number of people to write ten page reports for the Commission. I turned down my assignment, that of "reliability in industry," but offered to do an organizational analysis if I could think of one. They gave me transcripts of the hearings of May, June, and July, 1979 and a three-week deadline. Since I was in the mountains I asked graduate students, notably Lee Clarke and Mitchel Abalofia to send me books and articles on accidents in nuclear power plants and in other locations. With their excellent critiques I produced a forty-page paper on time, and in it were the essential ideas for this book. If only books came as fast as ideas!

Producing the book took another three and a half years with delays and cost overruns to match those of the nuclear industry itself. First the University Grants Committee of the State University of New York provided a few dollars to hire some students to do library work, and then the Sociology Program of the National Science Foundation funded an ambitious grant proposal to investigate accidents in high-risk systems. In addition to funds to hire graduate research assistants and Mary Luyster (the first of a string of remarkable secretaries), it gave me some time free from teaching duties and helped make it possible to accept an invitation to spend a year at a "think tank"—the Center for Advanced Study in the Behavioral Sciences, at Stanford, California. You and I would not be meeting this way, Dear Reader, were it not for these two beneficent institutions, The National Science Foundation and the Behavioral Science Center.

The grant allowed me to put together a toxic and corrosive group of graduate research assistants who argued with me and each other for a year at my university in Stony Brook, before I went to California. Abalofia and Clarke were joined by Leo Tasca, Kevin McHale, and others in intense research and discussions which made us the gloomiest group on campus, known for our gallows humor. At our Monday meetings one of us would say, "It was a great weekend for the project," and rattle off the latest disasters. Even quiet Mary Luyster was occasionally seen to be reading disaster books during her lunch hour.

At the Center for Advanced Study in the Behavioral Sciences, I was very fortunate in recruiting a graduate student in political science, Jeff Stewart, who did the major work on the weapons and space programs, and brought his own expertise as a former producer for public television to the project. Meanwhile, Mitch Abalofia had single handedly produced an excellent draft of the DNA material and gained command of that field in an incredibly short time. Lee Clarke was off studying a disaster that is not covered in this book, the dioxin contamination of the Binghamton Office Building. Leo Tasca went from marine accidents to a study of the political economy of the shipping industry. All of our lives were changed by Three Mile Island. At the time of the accident I was researching the emergence of organizations in the nineteenth century. I still hope to return to that comparatively sane world.

While still at Stony Brook my initial paper caught the attention of a member of the National Academy of Sciences, and I was asked to serve on their newly formed Committee on Human Factors Research. This fortuitous appointment gave me access to people and documents that I never knew existed. Richard Pew, the Chairman, and Tom Sheridan led me to the work of Jens Rassmussen in Denmark and to other engineering literature. Baruch Fischhoff put me in touch with Dr. John Gardenier of the Coast Guard, whose original papers and critical commentary proved valuable. The original Social Science Research Council Panel had already put me in touch with Paul Slovic and Fischhoff, and despite my extremely negative and off-the-cuff criticisms of their paper for that panel, they sent me and have continued to supply me with invaluable material and a crash course in cognitive psychology. It is an example of the non-defensive "invisible college" at its best. I left the Committee on Human Factors Research because of policy differences, but Pew, Fischhoff, Sheridan, and others introduced me to unknown worlds. A paper that I did for that committee came to the attention of the sponsors of the committee, the Office of Naval Research, and helped me get a summer grant.

That grant brought me into contact with some unusual research in the Naval Personnel Research Station in San Diego, an exemplary "human factors" program used on the Boeing 767, and into contact with Rex Hardy and his colleagues at the Ames Research Center of the National Aeronautics and Space Administration. Hardy provided invaluable data, great encouragement, and put me in touch with Harry Orlady and his staff who commented generously and Jerry Lederer of the Flight Safety Foundation who set me straight on a vast number of points but could not convince me on others.

The center in California allows you to stay for only one year, and I have never regretted a leave taking as much as that one. The intellectual, library and secretarial resources were unparalleled at that institution. Director Gardner Lindzey put scholarship first and red tape last. Ron Rice carefully critiqued some drafts. John Ferejohn introduced me to personal computers and wrote a simple word processing program for my IBM PC until, fortuitously, a remarkable gentleman in Santa Barbara named Camilo Wilson wrote his excellent, user-friendly "Volkswriter" program. The computer and that program brought this book out at least six months earlier than I had expected.

Leaving Palo Alto behind, I proceeded to cannibalize Yale in a desperate effort to finish that book. Three students were willing victims; Becky Friedkin, John Mohr, and Gary Ransom. Friedkin in particular had an extraordinary eye for inconsistencies, sloppiness, conceptual confusion, and generally had ways to solve these problems. Mohr brought the DNA material up-to-date and he and Ransom provided additional critiques. Beverly Apothaker and Mary Fasano completed an extraordinary run of resourceful and good humored secretaries. Finally, Yale's Institute for

Social and Policy Studies, the Behavioral Science Center, and Stony Brook, had the faculty seminar audiences that could critique my work. Outside of these institutions, however, were people like Dale Bridenbaugh, John Meyer, Marshall Meyer, John Scholz, and Todd LaPorte who read parts of the manuscript and gave willing and helpful criticisms. Then there are numerous technical people, chemists, geologists, biologists, and engineers that I have discussed the work with at dinner parties, military installations, seminars, and even on airplanes. Capitol Airlines, in addition, allowed me to ride in the cockpit on a long flight and the crew were most helpful.

Steve Fraser of Basic Books was a most perceptive and encouraging editor; editors are important, as any author will tell you. Authors also always tell you how dear, long-suffering, and supportive their families are. Mine hardly pays much attention anymore; they have been through it before and somewhat cheerfully cope. But it was impossible to write this book without two members of the next generation continually in mind. Nick and Lisa are inheriting our radioactive, toxic, and explosive systems, and I am aware that we are passing on a planet more degraded than we inherited. So I dedicate the book to them. I hope they can do more than Edith and I have been able to do.

Normal Accidents

INTRODUCTION

Welcome to the world of high-risk technologies. You may have noticed that they seem to be multiplying, and it is true. As our technology expands, as our wars multiply, and as we invade more and more of nature, we create systems—organizations, and the organization of organizations—that increase the risks for the operators, passengers, innocent bystanders, and for future generations. In this book we will review some of these systems—nuclear power plants, chemical plants, aircraft and air traffic control, ships, dams, nuclear weapons, space missions, and genetic engineering. Most of these risky enterprises have catastrophic potential, the ability to take the lives of hundreds of people in one blow, or to shorten or cripple the lives of thousands or millions more. Every year there are more such systems. That is the bad news.

The good news is that if we can understand the nature of risky enterprises better, we may be able to reduce or even remove these dangers. I have to present a lot of the bad news here in order to reach the good, but it is the possibility of managing high-risk technologies better than we are doing now that motivates this inquiry. There are many improvements we can make that I will not dwell on, because they are fairly obvious—such as better operator training, safer designs, more quality control, and more effective regulation. Experts are working on these solutions in both government and industry. I am not too sanguine about these efforts, since the risks seem to appear faster than the reduction of risks, but that is not the topic of this book.

Rather, I will dwell upon characteristics of high-risk technologies that suggest that no matter how effective conventional safety devices are, there is a form of accident that is inevitable. This is not good news for systems that have high catastrophic potential, such as nuclear power plants, nuclear weapons systems, recombinant DNA production, or even ships carrying highly toxic or explosive cargoes. It suggests, for example, that the probability of a nuclear plant meltdown with dispersion of radioactive materials to the atmosphere is not one chance in a million a year, but more like one chance in the next decade.

Most high-risk systems have some special characteristics, beyond their toxic or explosive or genetic dangers, that make accidents in them inevitable, even "normal." This has to do with the way failures can interact and the way the system is tied together. It is possible to analyze these special characteristics and in doing so gain a much better understanding of why accidents occur in these systems, and why they always will. If we know that, then we are in a better position to argue that certain technologies should be abandoned, and others, which we cannot abandon because we have built much of our society around them, should be modified. Risk will never be eliminated from high-risk systems, and we will never eliminate more than a few systems at best. At the very least, however, we might stop blaming the wrong people and the wrong factors, and stop trying to fix the systems in ways that only make them riskier.

The argument is basically very simple. We start with a plant, airplane, ship, biology laboratory, or other setting with a lot of components (parts, procedures, operators). Then we need two or more failures among components that interact in some unexpected way. No one dreamed that when X failed, Y would also be out of order and the two failures would interact so as to both start a fire and silence the fire alarm. Furthermore, no one can figure out the interaction at the time and thus know what to do. The problem is just something that never occurred to the designers. Next time they will put in an extra alarm system and a fire suppressor, but who knows, that might just allow three more unexpected interactions among inevitable failures. This interacting tendency is a characteristic of a system, not of a part or an operator; we will call it the "interactive complexity" of the system.

For some systems that have this kind of complexity, such as universities or research and development labs, the accident will not spread and be serious because there is a lot of slack available, and time to spare, and other ways to get things done. But suppose the system is also "tightly coupled," that is, processes happen very fast and can't be turned off, the failed parts cannot be isolated from other parts, or there is no other way to keep the production going safely. Then recovery from the initial disturbance is not possible; it will spread quickly and irretrievably for at least some time. Indeed, operator action or the safety systems may make it worse, since for a time it is not known what the problem really is.

Probably many production processes started out this way—complexly interactive and tightly coupled. But with experience, better designs, equipment, and procedures appeared, and the unsuspected interactions were avoided and the tight coupling reduced. This appears to have happened in the case of air traffic control, where interactive complexity and tight coupling have been reduced by better organization and "technological fixes." We will also see how the interconnection between dams and earthquakes is beginning to be understood. We now know that it involves a larger system than we originally thought when we just closed off a canyon and let it fill with water. But for most of the systems we shall consider in this book, neither better organization nor technological innovations appear to make them any less prone to system accidents. In fact, these systems require organizational structures that have large internal contradictions, and technological fixes that only increase interactive complexity and tighten the coupling; they become still more prone to certain kinds of accidents.

If interactive complexity and tight coupling—system characteristics—inevitably will produce an accident, I believe we are justified in calling it a *normal accident,* or a *system accident.* The odd term *normal accident* is meant to signal that, given the system characteristics, multiple and unexpected interactions of failures are inevitable. This is an expression of an integral characteristic of the system, not a statement of frequency. It is normal for us to die, but we only do it once. System accidents are uncommon, even rare; yet this is not all that reassuring, if they can produce catastrophes.

The best way to introduce the idea of a normal accident or a system accident is to give a hypothetical example from a homey, everyday experience. It should be familiar to all of us; it is one of those days when everything seems to go wrong.

A Day in the Life

You stay home from work or school because you have an important job interview downtown this morning that you have finally negotiated. Your friend or spouse has already left when you make breakfast, but unfortunately he or she has left the glass coffeepot on the stove with the light on. The coffee has boiled dry and the glass pot has cracked. Coffee is an addiction for you, so you rummage about in the closet until you find an old drip coffeemaker. Then you wait for the water to boil, watching the clock, and after a quick cup dash out the door. When you get to your car you find that in your haste you have left your car keys (and the apartment keys) in the apartment. That's okay, because there is a spare apartment key hidden in the hallway for just such emergencies. (This is a safety device, a *redundancy,* incidentally.) But then you remember that you gave a friend the key the other night because he had some books to pick up, and, planning ahead, you knew you would not be home when he came. (That finishes that *redundant pathway,* as engineers call it.)

Well, it is getting late, but there is always the neighbor's car. The neighbor is a nice old gent who drives his car about once a month and keeps it in good condition. You knock on the door, your tale ready. But he tells you that it just so happened that the generator went out last week and the man is coming this afternoon to pick it up and fix it. Another "backup" system has failed you, this time through no connection with your behavior at all (*uncoupled* or independent events, in this case, since the key and the generator are rarely connected). Well, there is always the bus. But not always. The nice old gent has been listening to the radio and tells you the threatened lock-out of the drivers by the bus company has indeed occurred. The drivers refuse to drive what they claim are unsafe buses, and incidentally want more money as well. (A safety system has foiled you, of all things.) You call a cab from your neighbor's apartment, but none can be had because of the bus strike. (These two events, the bus strike and the lack of cabs, are tightly connected, dependent events, or *tightly coupled* events, as we shall call them, since one triggers the other.)

You call the interviewer's secretary and say, "It's just too crazy to try to explain, but all sorts of things happened this morning and I can't make the interview with Mrs. Thompson. Can we reschedule it?" And you say to yourself, next week I am going to line up two cars and a cab and make the morning coffee myself. The secretary answers "Sure," but says to himself, "This person is obviously unreliable; now this after pushing for weeks for an interview with Thompson." He makes a note to that effect on the record and searches for the most inconvenient time imaginable for next week, one that Mrs. Thompson might have to cancel.

Now I would like you to answer a brief questionnaire about this event. Which was the primary cause of this "accident" or foul-up?

1. Human error (such as leaving the heat on under the coffee, or forgetting the keys in the rush)? Yes_____ No_____ Unsure_____
2. Mechanical failure (the generator on the neighbor's car)? Yes_____ No_____ Unsure_____
3. The environment (bus strike and taxi overload)? Yes_____ No_____ Unsure_____
4. Design of the system (in which you can lock yourself out of the apartment rather than having to use a door key to set the lock; a lack of emergency capacity in the taxi fleet)? Yes_____ No_____ Unsure_____
5. Procedures used (such as warming up coffee in a glass pot; allowing only normal time to get out on this morning)? Yes_____ No_____ Unsure_____

If you answered "not sure" or "no" to all of the above, I am with you. If you answered "yes" to the first, human error, you are taking a stand on multiple failure accidents that resembles that of the President's Commission to Investigate the Accident at Three Mile Island. The Commission blamed everyone, but primarily the operators.[1] The builders of the equipment, Babcock and Wilcox, blamed *only* the operators. If you answered "yes" to the second choice, mechanical error, you can join the Metropolitan Edison officials who run the Three Mile Island plant. They said the accident was caused by the faulty valve, and then sued the vendor, Babcock and Wilcox. If you answered "yes" to the fourth, design of the system, you can join the experts of the Essex Corporation, who did a study for the Nuclear Regulatory Commission of the control room.[2]

The best answer is not "all of the above" or any one of the choices, but rather "none of the above." (Of course I did not give you this as an option.) The cause of the accident is to be found in the complexity of the system. That is, each of the failures—design, equipment, operators, procedures, or environment—was trivial by itself. Such failures are expected to occur since nothing is perfect, and we normally take little notice of them. The bus strike would not affect you if you had your car key or the neighbor's car. The neighbor's generator failure would be of little consequence if taxis were available. If it were not an important appointment, the absence of cars, buses, and taxis would not matter. On any other morning the broken coffeepot would have been an annoyance (an *incident,* we will call it), but would not have added to your anxiety and caused you to dash out without your keys.

Though the failures were trivial in themselves, and each one had a backup system, or redundant path to tread if the main one were blocked, the failures became serious when they interacted. It is the *interaction* of the multiple failures that explains the accident. We expect bus strikes occasionally, we expect to forget our keys with that kind of apartment lock (why else hide a redundant key?), we occasionally loan the extra key to someone rather than disclose its hiding place. What we don't expect is for all of these events to come together at once. That is why we told the secretary that it was a crazy morning, too complex to explain, and invoked Murphy's law to ourselves (if anything can go wrong, it will).

That accident had its cause in the interactive nature of the world for us that morning and in its tight coupling—not in the discrete failures, which are to be expected and which are guarded against with backup systems. Most of the time we don't notice the inherent coupling in our world, because most of the time there are no failures, or the failures that occur do not interact. But all of a sudden, things that we did not realize could be linked (buses and generators, coffee and a loaned key) became linked. The system is suddenly more tightly coupled than we had realized. When we have interactive systems that are also tightly coupled, it is "normal" for them to have this kind of an accident, even though it is infrequent. It is normal not in the sense of being frequent or being expected—indeed, neither is true, which is why we were so baffled by what went wrong. It is normal in the sense that it is an inherent property of the system to occasionally experience this interaction. Three Mile Island was such a normal or system accident, and so were countless others that we shall examine in this book. We have such accidents because we have built an industrial society that has some parts, like industrial plants or military adventures, that have highly interactive and tightly coupled units. Unfortunately, some of these have high potential for catastrophic accidents.

Our "day in the life" example introduced some useful terms. Accidents can be the result of *multiple failures.* Our example illustrated failures in five components: in design, equipment, procedures, operators, and environment. To apply this concept to accidents in general, we will need to add a sixth area—supplies and materials. All six will be abbreviated as the DEPOSE components (for design, equipment, procedures, operators, supplies and materials, and environment). The example showed how different parts of the system can be quite dependent upon one another, as when the bus strike created a shortage of taxis. This dependence is known as *tight coupling.* On the other hand, events in a system can occur independently as we noted with the failure of the generator and forgetting the keys. These are *loosely coupled* events, because although at this time they were both involved in the same production sequence, one was not caused by the other.

One final point which our example cannot illustrate. It isn't the best case of a normal accident or system accident, as we shall use these terms, because the interdependence of the events was comprehensible for the person or "operator." She or he could not do much about the events singly or in their interdependence, but she or he could understand the interactions. In complex industrial, space, and military systems, the normal accident generally (not always) means that the interactions are not only unexpected, but are *incomprehensible* for some critical period of time. In part this is because in these human-machine systems the interactions literally cannot be seen. In part it is because, even if they are seen, they are not believed. As we shall find out and as Robert Jervis and Karl Weick have noted,[3] seeing is not necessarily believing; sometimes, we must believe before we can see.

Variations on the Theme

While basically simple, the idea that guides this book has some quite radical ramifications. For example, virtually every system we will examine places "operator error" high on its list of causal factors—generally about 60 to 80 percent of accidents are attributed to this factor. But if, as we shall see time and time again, the operator is confronted by unexpected and usually mysterious interactions among failures, saying that he or she should have zigged instead of zagged is possible only after the fact. Before the accident no one could know what was going on and what should have been done. Sometimes the errors are bizarre. We will encounter "noncollision course collisions," for example, where ships that were about to pass in the night suddenly turn and ram each other. But careful inquiry suggests that the mariners had quite reasonable explanations for their actions; it is just that the interaction of small failures led them to construct quite erroneous worlds in their minds, and in this case these conflicting images led to collision.

Another ramification is that great events have small beginnings. Running through the book are accidents that start with trivial kitchen mishaps; we will find them on aircraft and ships and in nuclear plants, having to do with making coffee or washing up. Small failures abound in big systems; accidents are not often caused by massive pipe breaks, wings coming off, or motors running amok. Patient accident reconstruction reveals the banality and triviality behind most catastrophes.

Small beginnings all too often cause great events when the system uses a "transformation" process rather than an additive or fabricating one.

Where chemical reactions, high temperature and pressure, or air, vapor, or water turbulence is involved, we cannot see what is going on or even, at times, understand the principles. In many transformation systems we generally know what works, but sometimes do not know why. These systems are particularly vulnerable to small failures that "propagate" unexpectedly, because of complexity and tight coupling. We will examine other systems where there is less transformation and more fabrication or assembly, systems that process raw materials rather than change them. Here there is an opportunity to learn from accidents and greatly reduce complexity and coupling. These systems can still have accidents—all systems can. But they are more likely to stem from major failures whose dynamics are obvious, rather than the trivial ones that are hidden from understanding.

Another ramification is the role of organizations and management in preventing failures—or causing them. Organizations are at the center of our inquiry, even though we will often talk about hardware and pressure and temperature and the like. High-risk systems have a double penalty: because normal accidents stem from the mysterious interaction of failures, those closest to the system, the operators, have to be able to take independent and sometimes quite creative action. But because these systems are so tightly coupled, control of operators must be centralized because there is little time to check everything out and be aware of what another part of the system is doing. An operator can't just do her own thing; tight coupling means tightly prescribed steps and invariant sequences that cannot be changed. But systems cannot be both decentralized and centralized at the same time; they are organizational Pushmepullyous, straight out of Dr. Doolittle stories, trying to go in opposite directions at once. So we must add organizational contradictions to our list of problems.

Even aside from these inherent contradictions, the role of organizations is important in other respects for our story. Time and time again warnings are ignored, unnecessary risks taken, sloppy work done, deception and downright lying practiced. As an organizational theorist I am reasonably unshaken by this; it occurs in all organizations, and it is a part of the human condition. But when it comes to systems with radioactive, toxic, or explosive materials, or those operating in an unforgiving, hostile environment in the air, at sea, or under the ground, these routine sins of organizations have very nonroutine consequences. Our ability to organize does not match the inherent hazards of some of our organized activities. Better organization will always help any endeavor. But the best is not good enough for some that we have decided to pursue.

Nor can better technology always do the job. Besides being a book about organizations (but painlessly, without the jargon and the sacred texts), this is a book about technology. You will probably learn more than you ever wanted to about condensate polishers, buffet boundaries, reboilers, and slat retraction systems. But that is in passing (and even while passing you are allowed a considerable measure of incomprehension). What is not in passing but is essential here is an evaluation of technology and its "fixes." As the saying goes, man's reach has always exceeded his grasp (and of course that goes for women too). It should be so. But we might begin to learn that of all the glorious possibilities out there to reach for, some are going to be beyond our grasp in catastrophic ways. There is no technological imperative that says we *must* have power or weapons from nuclear fission or fusion, or that we *must* create and loose upon the earth organisms that will devour our oil spills. We could reach for, and grasp, solar power or safe coal-fired plants, and the safe ship designs and industry controls that would virtually eliminate oil spills. No catastrophic potential flows from these.

It is particularly important to evaluate technological fixes in the systems that we cannot or will not do without. Fixes, including safety devices, sometimes create new accidents, and quite often merely allow those in charge to run the system faster, or in worse weather, or with bigger explosives. Some technological fixes are error-reducing—the jet engine is simpler and safer than the piston engine; fathometers are better than lead lines; three engines are better than two on an airplane; computers are more reliable than pneumatic controls. But other technological fixes are excuses for poor organization or an attempt to compensate for poor system design. The attention of authorities in some of these systems, unfortunately, is hard to get when safety is involved.

When we add complexity and coupling to catastrophe, we have something that is fairly new in the world. Catastrophes have always been with us. In the distant past, the natural ones easily exceeded the human-made ones. Human-made catastrophes appear to have increased with industrialization as we built devices that could crash, sink, burn, or explode. In the last fifty years, however, and particularly in the last twenty-five, to the usual cause of accidents—some component failure, which could be prevented in the future—was added a new cause: interactive complexity in the presence of tight coupling, producing a system accident. We have produced designs so complicated that we cannot anticipate all the possible interactions of the inevitable failures; we add safety devices that are deceived or avoided or defeated by hidden paths in the systems. The systems have become more complicated because either they are dealing with more deadly substances, or we demand they function in ever more hostile environments or with ever greater speed and volume. And still new systems keep appearing, such as gene splicing, and others grow ever more complex and tightly tied together. In the past, designers could learn from the collapse of a medieval cathedral under construction, or the explosion of boilers on steamboats, or the collision of railroad trains on a single track. But we seem to be unable to learn from chemical plant explosions or nuclear plant accidents. We may have reached a plateau where our learning curve is nearly flat. It is true that I should be wary of that supposition. Reviewing the wearisome Cassandras in history who prophesied that we had reached our limit with the reciprocating steam engine or the coal-fired railroad engine reminds us that predicting the course of technology in history is perilous. Some well-placed warnings will not harm us, however.

One last warning before outlining the chapters to come. The new risks have produced a new breed of shamans, called risk assessors. As with the shamans and the physicians of old, it might be more dangerous to go to them for advice than to suffer unattended. In our last chapter we will examine the dangers of this new alchemy where body counting replaces social and cultural values and excludes us from participating in decisions about the risks that a few have decided the many cannot do without. The issue is not risk, but power.

Fast Forward

Chapter 1 will examine the accident at Three Mile Island (TMI) where there were four independent failures, all small, none of which the operators could be aware of. The system caused that accident, not the operators. Chapter 2 raises the question of why, if these plants are so complex and tightly coupled, we have not had more TMIs. A review of the nuclear power industry and some of its trivial and its serious accidents will suggest that we have not given large plants of the size of TMI time to express themselves. The record of the industry and the Nuclear Regulatory Commission is frightening, but not because it is all that different from the records of other industries and regulatory agencies. It isn't. It is frightening because of the catastrophic potential of this industry; it has to have a perfect performance record, and it is far from achieving that.

We can go a fair distance with some loosely defined concepts such as complexity, coupling, and catastrophe, but in order to venture further into the world of high-risk systems we need better definitions, and a better model of systems and accidents and their consequences. This is the work of Chapter 3, where terms are defined and amply illustrated with still more accident stories. In this chapter we explore the advantages of loose coupling, map the industrial, service, and voluntary organizational world according to complexity and coupling, and add a definition of types of catastrophes. Chapter 4 applies our complexity, coupling, and catastrophe theories to the chemical industry. I wish to make it clear that normal accidents or, as we will generally call them, system accidents, are not limited to the nuclear industry. Some of the most interesting and bizarre examples of the unanticipated interaction of failures appear in this chapter—and we are now talking about a quite well-run industry with ample riches to spend on safety, training, and high-technology solutions.

Yet chemical plants mostly just sit there, though occasionally they will send a several hundred pound missile a mile away into a community or incinerate a low flying airplane. In Chapter 5 we move out into the environment and examine aircraft and flying, and air traffic control and the airports and airways. Flying is in part a transformation system, but largely just very complex and tightly coupled. Technological fixes are made continuously here, but designers and airlines just keep pushing up against the limits with each new advance. Flying is risky, and always will be. With the airways system, on the other hand, we will examine the actual reduction of complexity and coupling through organizational changes and technological developments; this system has become very safe, as safety goes in inherently risky systems. An examination of the John Wayne International Airport in Orange County, California, will remind us of the inherent risks.

With marine transport, in Chapter 6, the opposite problem is identified. No reduction in complexity or coupling has been achieved. Horrendous tales are told, three of which we will detail, about the needless perils of this system. We will analyze it as one that induces errors through its very structure, examining insurance, shipbuilders, shippers, captains and crews, collision avoidance systems, and the international anarchy that prevents effective regulation and encourages cowboys and hot rodders at sea. One would not think that ships could pile up as if they were on the Long Island Expressway, but they do.

Chapter 7 might seem to be a diversion since dams, lakes, and mines are not prone to system accidents. But it will support our point because they are also linear, rather than complex systems, and the accidents there are foreseeable and avoidable. However, when we move away from the individual dam or mine and take into account the larger system in which they exist, we find the "eco-system accident," an interaction of systems that were thought to be independent but are not because of the larger ecology. Once we realize this we can prevent future accidents of this type; in linear systems we can learn from our mistakes. Dams, lakes, and mines also simply provide tales worth telling. Do dams sink or float when they fail? Could we forestall a colossal earthquake in California by a series of mammoth chiropractic spinal adjustments? How could we lose a whole lake and barges and tugs in a matter of hours? (By inadvertently creating an eco-system accident.)

Chapter 8 deals with far more esoteric systems. Space missions are very complex and tightly coupled, but the catastrophic potential was small and now is smaller. More important, this system allows us to examine the role of the operator (in this case, extraordinarily well-trained astronauts) whom the omniscient designers and managers tried to treat like chimpanzees. It is a cautionary tale for all high-technology systems. Accidents with nuclear weapons, from dropping them to firing them by mistake, will illustrate a system so complicated and error-prone that the fate of the earth may be decided more by inadvertence than anger. The prospects are, I am afraid, terrifying. Equally frightening is the section in this chapter on gene splicing, or recombinant DNA. In this case, in the unseemly haste for prizes and profits, we have abandoned even the most elementary safeguards, and may loose upon the world a rude beast whose time need not have come.

In the last chapter we shall examine the new shamans, the risk assessors, and their inadvertent allies, the cognitive psychologists. Naturally, as a sociologist, I will have a few sharp words to say about the latter, but point out that their research has really provided the grounds for a public role in high-risk decision making, one the risk assessors do not envisage. Finally, we will add up the credits and deficits of the systems we examined, and I will make a few modest suggestions for complicating the lives of some systems—and shutting others down completely.

CHAPTER 1

Normal Accident at Three Mile Island

Our first example of the accident potential of complex systems is the accident at the Three Mile Island Unit 2 nuclear plant near Harrisburg, Pennsylvania, on March 28, 1979. I have simplified the technical details a great deal and have not tried to define all of the terms. It is not necessary to understand the technology in any depth. What I wish to convey is the interconnectedness of the system, and the occasions for baffling interactions. This will be the most demanding technological account in the book, but even a general sense of the complexity will suffice if one wishes to merely follow the drama rather than the technical evolution of the accident.*

TMI is clearly our most serious nuclear power plant accident to date. The high drama of the event gripped the nation for a fortnight, as reassurance gave way to near panic, and we learned of a massive hydrogen bubble and releases that sent pregnant women and others fleeing the area. The President of the United States toured the plant while two feeble pumps, designed for quite other duties, labored to keep the core from

*This account draws from many sources, and I have not cited each point individually. See the references from the first part of the bibilography.

melting further. (One of them soon failed, but fortunately by the time the second pump failed the system had cooled sufficiently to allow for natural circulation.) The subsequent investigations and law suits disclosed a seemingly endless story of incompetence, dishonesty, and cover-ups before, during, and after the event; indeed, new disclosures were appearing as this book went to press. Yet, as we shall see in chapter 2 when we examine other accidents, the performance of all concerned—utility, manufacturer, regulatory agency, and industry—was about average. Rather sizeable bits and pieces of the TMI disaster can be found elsewhere in the industry; they had just never been put together so dramatically before.

Unit 2 at Three Mile Island (TMI) had a hard time getting underway at the end of 1978. Nuclear plants are always plagued with start-up problems because the system is so complex, and the technology so new. Many processes are still not well understood, and the tolerances are frightfully small for some components. A nuclear plant is also a hybrid creation—the reactor itself being complex and new and carefully engineered by one company, while the system for drawing off the heat and using it to turn turbines is a rather conventional, old, and comparatively unsophisticated system built by another company. Unit 2 may have had more than the usual problems. The maintenance force was overworked at the time of the accident and had been reduced in size during an economizing drive. There were many shutdowns, and a variety of things turned out, in retrospect, to be out of order. But one suspects that it was not all that different from other plants; after a plant sustains an accident, a thorough investigation will turn up numerous problems that would have gone unnoticed or undocumented had the accident been avoided. Indeed, in the 1982 court case where the utility, Metropolitan Edison, sued the builder of the reactor, Babcock and Wilcox, the utility charged the builder with an embarrassing number of errors and failures, and the vendor returned the favor by charging that the utility was incompetent to run their machine.[1] But Metropolitan Edison runs other machines, and Babcock and Wilson have built many reactors that have not had such a serious accident. We know so much about the problems of Unit 2 only because the accident at Three Mile Island made it a subject for intense study; it is probably the most well-documented examination of organizational performance in the public record. At last count I found ten published technical volumes or books on the accident alone, perhaps one hundred articles, and many volumes of testimony.

The accident started in the cooling system. There are two cooling systems. The primary cooling system contains water under high pressure and at high temperature that circulates through the core where the nuclear reaction is taking place. This water goes into a steam generator, where it bathes small tubes circulating water in a quite separate system, the secondary cooling system, and heats this water in the secondary system. This transfer of heat from the primary to the secondary system keeps the core from overheating, and uses the heat to make steam. Water in the secondary system is also under high pressure until it is called upon to turn into steam, which drives the turbines that generate the electric power. The accident started in the secondary cooling system.

The water in the secondary system is not radioactive (as is the water in the primary system), but it must be very pure because its steam drives the finely precisioned turbine blades. Resins get into the water and have to be removed by the condensate polisher system, which removes particles that are precipitated out.

The polisher is a balky system, and it had failed three times in the few months the new unit had been in operation. After about eleven hours of work on the system, at 4:00 A.M. on March 28, 1979, the turbine tripped (stopped). Though the operators did not know why at the time, it is believed that some water leaked out of the polisher system—perhaps a cupful—through a leaky seal.

Seals are always in danger of leaking, but normally it is not a problem. In this case, however, the moisture got into the instrument air system of the plant. This is a pneumatic system that drives some of the instruments. The moisture interrupted the air pressure applied to two valves on two feedwater pumps. This interruption "told" the pumps that something was amiss (though it wasn't) and that they should stop. They did. Without the pumps, the cold water was no longer flowing into the steam generator, where the heat of the primary system could be transferred to the cool water in the secondary system. When this flow is interrupted, the turbine shuts down, automatically—an automatic safety device, or ASD.

But stopping the turbine is not enough to render the plant safe. Somehow, the heat in the core, which makes the primary cooling system water so hot, has to be removed. If you take a whistling tea kettle off the stove and plug its opening, the heat in the metal and water will continue to produce steam, and if it cannot get out, it may explode. Therefore, the emergency feedwater pumps came on (they are at H in Figure 1.1; the regular feedwater pumps which just stopped are above them in the figure). They are designed to pull water from an emergency storage tank and run it through the secondary cooling system, compensating for the water in that system that will boil off now that it is not circulating. (It is

FIGURE 1.1

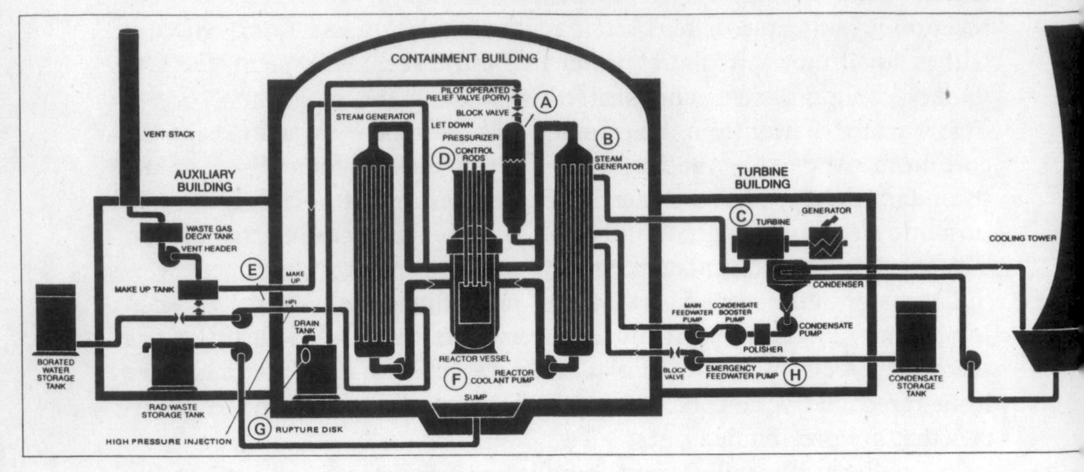

TMI Unit 2 March 28, 1978

Failure #1	Clogged condensate polisher line Moisture in instrument air line False signal to turbine
ASD*	Turbine stops
ASD	Feedwater pumps stop
ASD	Emergency feedwater pumps start
Failure #2	Flow blocked; valves closed instead of open
	No heat removal from primary coolant
	Rise in core temperature and pressure
ASD	Reactor scrams
	Reactor continues to heat, "decay heat"
	Pressure and temperature rise
ASD	Pilot Operated Relief Valve (PORV) opens
ASD	PORV told to close
Failure #3	PORV sticks open
Failure #4	PORV position indicator signifies it has shut
ASD	Reactor coolant pumps come on
	Primary coolant pressure down, temperature up
	Steam voids form in coolant pipes and core, restricting flow forced by coolant pumps, creating uneven pressures in system
ASD	Hi Pressure Injection (HPI) starts, to reduce temperature
	Pressurizer fills with coolant as it seeks outlet through PORV
"Operator error"	Operators reduce HPI to save pressurizer, per procedures
	Temperature and pressure in core continue to rise because of lack of heat removal, decay heat generation, steam voids, hydrogen generation from the zirconium-water reaction, and uncovering of core. Reactor coolant pumps cavitate and must be shut off, further restricting circulation.

*ASD (automatic safety device)

SOURCE: Kemeny, John, et al. Report of the President's Commission on the Accident at Three Mile Island. Washington, D.C.: Government Printing Office, 1979.

like pouring cold water over your plugged tea kettle.) However, these two pipes were unfortunately blocked; a valve in each pipe had been accidently left in a closed position after maintenance two days before. The pumps came on and the operator verified that they did, but he did not know that they were pumping water into a closed pipe.

The President's Commission on the Accident at Three Mile Island (the Kemeny Commission) spent a lot of time trying to find out just who was responsible for leaving the valves closed, but they were unsuccessful. Three operators testified that it was a mystery to them how the valves had gotten closed, because they distinctly remembered opening them after the testing. You probably have had the same problem with closing the freezer door or locking the front door; you are sure you did, because you have done it many times. Operators testified at the Commission's hearings that with hundreds of valves being opened or closed in a nuclear plant, it is not unusual to find some in the wrong position—even when locks are put on them and a "lock sheet" is maintained so the operators can make an entry every time a special valve is opened or closed.

Accidents often involve such mysteries. A safety hatch on a Mercury spacecraft prematurely blew open (it had an explosive charge for opening it) as the recovery helicopter was about to pick it up out of the water after splashdown. Gus Grissom, the astronaut, insisted afterwards that he hadn't fired it prematurely or hit it accidentally. It just blew by itself. (He almost drowned.) It is the old war between operators and the equipment others have designed and built. The operators say it wasn't their fault; the designers say it wasn't the fault of the equipment or design. Ironically, the astronauts had insisted upon the escape hatch being put in as a safety device in case they had to exit rapidly; it is not the only example we shall uncover of safety devices increasing the chances of accidents. The Three Mile Island operators finally had to concede reluctantly that large valves do not close by themselves, so someone must have goofed.

There were two indicators on TMI's gigantic control panel that showed that the valves were closed instead of open. One was obscured by a repair tag hanging on the switch above it. But at this point the operators were unaware of any problem with emergency feedwater and had no occasion to make sure those valves, which are always open except during tests, were indeed open. Eight minutes later, when they were baffled by the performance of the plant, they discovered it. By then much of the initial damage had been done. Apparently our knowledge of these plants is quite incomplete, for while some experts thought the closed valves constituted an important operator error, other experts held that it did not make much difference whether the valves were closed or not, since the supply of emergency feedwater is limited and worse problems were appearing anyway.

With no circulation of coolant in the secondary system, a number of complications were bound to occur. The steam generator boiled dry. Since no heat was being removed from the core, the reactor "scrammed." In a scram the graphite control rods, 80 percent silver, drop into the core and absorb the neutrons, stopping the chain reaction. (In the first experiments with chain reactions, the procedure was the same—"drop the rods and scram"; thus the graphic term *scram* for stopping the chain reaction.) But that isn't enough. The decaying radioactive materials still produce some heat, enough to generate electricity for 18,000 homes. The "decay heat" in this 40-foot-high stainless steel vessel, taller than a three-story building, builds up enormous temperature and pressure. Normally there are thousands of gallons of water in the primary and secondary cooling systems to draw off the intense heat of the reactor core. In a few days this cooling system should cool down the core. But the cooling system was not working.

There are, of course, ASDs to handle the problem. The first ASD is the pilot-operated relief valve (PORV), which will relieve the pressure in the core by channeling the water from the core through a big vessel called a pressurizer, and out the top of it into a drain pipe (called the "hot leg"), and down into a sump. It is radioactive water and is very hot, so the valve is a nuisance. Also, it should only be open long enough to relieve the pressure; if too much water comes through it, the pressure will drop so much that the water can flash into steam, creating bubbles of steam, called steam voids, in the core and the primary cooling pipes. These bubbles will restrict the flow of coolant, and allow certain spots to get much hotter than others—in particular, spots by the uranium rods, allowing them to start fissioning again.

The PORV is also known by its Dresser Industries' trade name of "electromatic relief valve." (Dresser Industries is the firm that sponsored ads shortly after the accident saying that actress Jane Fonda was more dangerous than nuclear plants. She was starring in the *China Syndrome,* a popular movie playing at the time that depicted a near meltdown in a nuclear plant.) It is expected to fail once in every fifty usages, but on the other hand, it is seldom needed. The President's Commission turned up at least eleven instances of it failing in other nuclear plants (to the surprise of the Nuclear Regulatory Commission and the builder of the reactor, Babcock and Wilcox, who only knew of four) and there had been two earlier failures in the short life of TMI-Unit 2. Unfortunately, it just so happened that this time, with the block valves closed and one indicator hidden, and with the condensate pumps out of order, the PORV failed to reseat, or close, after the core had relieved itself sufficiently of pressure.

This meant that the reactor core, where the heat was building up because the coolant was not moving, had a sizeable hole in it—the stuck-open relief valve. The coolant in the core, the primary coolant system, was under high pressure, and was ejecting out through the stuck valve into a long curved pipe, the "hot leg," which went down to a drain tank. Thirty-two thousand gallons, one third of the capacity of the core, would eventually stream out. This was no small pipe break someplace as the operators originally thought; the thing was simply uncorked, relieving itself when it shouldn't.

Since there had been problems with this relief valve before (and it is a difficult engineering job to make a highly reliable valve under the conditions in which it must operate), an indicator had recently been added to the valve to warn operators if it did not reseat. The watchword is "safety" in nuclear plants. But, since nothing is perfect, it just so happened that this time the indicator itself failed, probably because of a faulty solenoid, a kind of electromagnetic toggle switch. Actually, it wasn't much of an indicator, and the utility and supplier would have been better off to have had none at all. Safety systems, such as warning lights, are necessary, but they have the potential for deception. If there had been no light assuring them the valve had closed, the operators would have taken other steps to check the status of the valve, as operators did in a similar accident at another plant a year and a half before. But if you can't believe the lights on your control panel, an army of operators would be necessary to check every part of the system that might be relevant. And one of the lessons of complex systems and TMI is that *any* part of the system might be interacting with other parts in unanticipated ways.

The indicator sent a signal to the control board that the valve had received the impulse to shut down. (It was not an indication that the valve had actually shut down; that would be much harder to provide.) So the operators noted that all was fine with the PORV, and waited for reactor pressure to rise again, since it had dropped quickly when the valve opened for a second. The cork stayed off the vessel for two hours and twenty minutes before a new shift supervisor, taking a fresh look at the problems, discovered it.

We are now, incredibly enough, only thirteen seconds into the "transient," as engineers call it. (It is not a perversely optimistic term meaning something quite temporary or transient, but rather it means a rapid change in some parameter, in this case, temperature.) In these few seconds there was a false signal causing the condensate pumps to fail, two valves for emergency cooling out of position and the indicator obscured, a PORV that failed to reseat, and a failed indicator of its position. *The operators could have been aware of none of these.*

Moreover, while all these parts are highly interdependent, so that one affects the other, they are *not* in direct operational sequence. Direct operational sequence is a sequence of stages as in a production line, or an engineered safety sequence. The operator knows that a block in the condensate line will cause the condensate pump to trip, which will stop water from going to the steam generator and then going to the turbine as steam to drive it, so the turbine will shut down because it will have no source of power to turn it. This is quite comprehensible. But connected to this sequence, although not a part of its production role, is another system, the primary cooling system, which regulates the amount of water in the core. The water level in the core was judged to have fallen, which it had, because of the drop in the pressure and temperature in the primary cooling system. But for the operators there was no obvious connection between this drop and a turbine "trip" (shutdown). Unknown to them, there was an intimate connection because of the interactive complexity of the system. The connection is through the PORV, but that also has no production sequence or safety sequence connection to the trip of the turbines, or to the failure of the condensate polisher system, even had the operators been able to ascertain that this was the cause of the turbine trip. The PORV is expected to operate on the basis of core pressure, regardless of the functioning of the turbine, the secondary cooling system (feedwater to the steam generators and turbine), or the emergency core cooling pumps.

Even if there is a part of the system that is in direct operational sequence, an information failure in any part of that sequence can render the connection opaque, if not invisible. For example, the PORV is connected in a direct sequence to a drain pipe, then to a drain tank, and when that overflows, to a sump. A couple of readings of excessive radioactive water will appear along the way. But for the operators, this was water from an "unknown origin," since they were assured, by the signal light, that the PORV was closed. Since they assumed a pipe break somewhere and since the piping system in the plant is so complex that a member of the Presidential Commission had to use a magnifying glass to try to follow it on the drawings, there was reason to believe that the water could have come from any number of places. Indeed, later in the accident, they found that radioactive water was not traveling to the tank they intended, but because of complex flow and pressure interactions, was going to a different, wrong tank, which also overflowed, this time in the auxiliary building.

Here we have the essence of the normal accident: the interaction of multiple failures that are not in a direct operational sequence. You could underline this definition, but there is one other ingredient we have not explored in detail—incomprehensibility. In contrast to our appointment-car-key accident, which was quite comprehensible, most normal accidents have a significant degree of incomprehensibility. Let us go back to the TMI story to examine this incomprehensibility, which is the main reason why answer number one to the quiz, operator error, is so wrong for normal accidents.

The PORV was now open and would be for two hours and twenty minutes, and coolant from the core of the reactor was squirting out at a great rate down the hot leg to the drain tank, so pressure in the reactor dropped. This is dangerous unless the temperature is also going down rapidly, because without pressure on the superheated water (over 2,000° F.), it will become steam, which does not cool as well and creates bubbles that block the flow of coolant. So one of two reactor coolant pumps (another emergency system) started up automatically and another was started by the operators (thirteen seconds into the accident; check it out on your watch). For two or three minutes things looked fine; the coolant in the core appeared stable. But it wasn't. For a variety of reasons that can only be matters of conjecture, it appears that voids or steam bubbles formed in such a way as to give the appearance of stabilization after the two reactor coolant pumps came on. The operators were not aware that the steam generators were not getting water. When they boiled dry, the reactor coolant heated up again because the secondary coolant system was not removing heat from the primary one, which removes it from the core. Since the core was losing water, pressure in the coolant system dropped sharply.

At this point, two minutes into the accident, another emergency device came on—high-pressure injection, or HPI, which forces water into the core at a rapid rate. Now came the high drama, the action that has been called the major source of the accident and the key operator error. After letting HPI run full tilt for about two minutes, they reduced it drastically, thus not replacing the water that was boiling out through the PORV. This meant that the core was steadily being uncovered—the most fearsome danger in a nuclear plant, for it will then melt the vessel and perhaps loose radiation on the world.

Probing this action by the operators, investigating committees were led back to an earlier accident at an Ohio plant, memos from a TVA engineer, memos in the files of Babcock and Wilcox (the firm that built the reactor), and an accident in Belgium in a Westinghouse reactor. All of these warnings occurred well before TMI. A bureaucratic tale worthy of Franz Kafka came out of the investigation of TMI and the warnings, which we shall forego telling so we can stick with the villains of the piece, according to most reports: the hapless operators.

High-pressure injection involves the injection of cold water at a very high pressure into the reactor core in order to lower reactor temperatures. It goes in at about 1,000 gallons a minute, and could fill a swimming pool in twenty minutes. It is a risky business. The cold water may "shock" the core, producing hairline cracks in equipment in the core, or conceivably in the vessel itself (but probably only if it had been in operation for several years). The high pressure may also cause damage as the core fills up, putting a pressure strain on it. Most experts discount these dangers, but not all. As an indication of how little we understand nuclear systems, I should note that shortly after the accident, some even argued that it was fortunate that the operators cut back on HPI, although this was not the majority view.

Two years later, however, the Nuclear Regulatory Commission issued a report which gave substance to this danger. It disclosed that thirteen reactors, some of them only three to four years old, showed degrees of core vessel brittleness, because of the intense radioactive bombardment that were greater than predicted.[2] This raised serious safety concerns. Certainly the high-pressure injection of cold water into a brittle vessel could crack the vessel, leading to a meltdown and all its consequences. Fortunately, the TMI core had only been in operation at full power for about forty days.

Another problem with HPI is a matter of lively dispute. It may increase the pressure in the pressurizer by flooding it with water. The pressurizer is a kind of huge shock absorber and stabilizer. It is a large tank with, under normal circumstances, 800 cubic feet of water in the bottom and 700 cubic feet of steam above it. By using heaters in the tank, the pressure of the steam at the top can be raised or lowered, and this controls the pressure of the water cooling the core. If HPI sends too much water into the core, it will flood the pressurizer. (This is called "going solid"—solid water and no steam.) If there is a substantial pressure surge in the core, the cushion provided by the steam in the pressurizer would be lost and the coolant pipes could burst (one source of a LOCA, or loss of coolant accident), perhaps causing a meltdown. Even if the safety valves prevented a pipe burst, a full pressurizer still presents a serious situation. It is a first-line emergency safety device (ESD), and should not be disabled.

Operators were assiduously trained to avoid going solid in the pressurizer by both the vendor, Babcock and Wilcox, and the user, Metropolitan Edison, which operates TMI. There was no hint in the training manual or the the procedure manual that under some circumstances it might be preferable to go solid in the pressurizer rather than cut back on HPI. Such a directive was considered after an earlier accident at another plant, but was rejected by Babock and Wilcox. At this point, some two minutes into the accident, there was a circumstance in which HPI was needed more than a resilient pressurizer. The core was about to be uncovered.

After HPI came on, the operators were looking primarily at two dials, close to one another. One indicated that the pressure in the reactor was still falling, which was mysterious because the other indicated that pressure in the pressurizer was rising—indeed, it was dangerously high. But they should move together, and always had. They are connected by pipes, and the pressurizer is supposed to control the pressure in the coolant system; that is what it is there for. If pressure is up in the pressurizer, and it is connected to the core, it should be up in the core.

Perhaps the dials were wrong. It sometimes happens. But which one? If the reactor dial was correct, and pressure was falling in the reactor, there must be some large anomaly, because there was plenty of water going into the core through the reactor coolant pumps, which were still running, and through the high-pressure injection that had just started. Even if there were a small pipe break somewhere, the reactor coolant pumps would ensure that the core would remain covered even without HPI. With all this water going in, how could the pressure fall? On the other hand, since the operators knew that the emergency feedwater pumps came on (but not that they had nothing to pump because of the closed valves), they thought that the secondary cooling system should be cooling the core, so pressure in the core would be falling. But if it were, why did HPI come on? Perhaps the reactor pressure dial was wrong.

The other dial was a serious source of concern. The high pressure in the pressurizer eliminated a safety margin, and all instructions said the pressurizer should not be flooded. It stood between the operators and the possibility of a loss of coolant accident, a LOCA; because if there were no steam at the top, a pressure surge could lead to a pipe break. They could see the connection between HPI and the high pressure reading in the pressurizer. High-pressure injection was flooding the core and sending water up to flood the pressurizer. So they cut back on it drastically (they "throttled back on the makeup valves"). Pressure in the pressurizer, sure enough, came down, relieving the danger of going solid.

What they didn't know, and couldn't know, was that with the PORV open and the two feedwater valves blocked, preventing the removal of residual heat, they already had a LOCA, but not from a pipe break. The rise in pressure in the pressurizer was probably due to the steam voids rapidly forming because the core was close to becoming uncovered. They thought they were avoiding a LOCA when they were *in* one and were making it worse. With the PORV stuck open, the danger of going solid in the pressurizer was reduced because the open valve would provide some relief. But no one knew it was open.

The Kemeny Commission thought the operators should have known, and berated them in its report—they were "oblivious" to the danger; two readings "should have clearly alerted" them to the LOCA; "the major cause of the accident was due to inappropriate actions by those who were operating the plant," they said in their final report.[3] Babcock and Wilcox agreed; this was the sole cause of the accident, they argued in a press conference. The British Secretary of State for Energy was less diplomatic—the accident was caused by "stupid errors," he said.[4]

Actually, there were three readings that should have indicated a LOCA to the operator, and it is a lesson in the fate of warnings to examine them. First, we should note that a LOCA is the most feared of the probable accidents in a plant, for it means the core can melt, and in what are called worst-case analyses could cause a steam explosion and rupture the vessel, spewing radioactivity. Even without a steam explosion, the extreme heat of uncontrolled fissioning could breach containment. LOCA will occur when the water level drops below the level of the fuel rods, and they overheat. But there is no direct measure of water level in the core in the Babcock and Wilcox reactors. One could be put on, said a Babcock and Wilcox official during a press conference, but it would be hard to provide and would create other complications.[5] One hesitates to penetrate the core more than needed, and it would be hard to measure surging water under high pressure, about to flash into steam. So, let's examine the indirect measures.

One device measured drain tank pressures. But it is not considered a particularly vital indicator by the designers, and is located on the back side of a 7-foot high control panel, near the bottom. Not suspecting they were in a LOCA, no one bothered to examine it (though the record is vague on this question). Another indicator showed the temperature of the drain tank; with hundreds of gallons of hot coolant spewing out and going to the drain tank, that temperature reading should be way up. It was indeed up. But they had been having trouble with a leaky PORV for some weeks, meaning that there was always some coolant going through it, so it was usual for it to be higher than normal. It did shoot up at one point, they noted, but that was shortly after the PORV opened, and when it didn't come down fast that was comprehensible, because the pipe heats up and stays hot. "That hot?" a commissioner interrogating an operator asked, in effect. The operator replied, in effect, "Yes; if it were a LOCA I would expect it to be much higher." It was not the LOCA they were trained for on the simulators that are used for training sessions, since it had some coolant coming in through an emergency system, and some coming in through HPI, which was only throttled back, not stopped. Their training never imagined a multiple accident with a stuck PORV, and blocked valves. Well, what about the drop in pressure in the core itself; surely this would indicate that the coolant was getting out somehow. But the operators discounted that indicator as erroneous or simply mysterious because it contradicted the one next to it, the pressurizer indicator, which was rising. A supervisor testified:

> I think we knew we were experiencing something different, but I think each time we made a decision it was based on something we knew about. For instance: pressure was low, but they had opened the feed valves quickly in the steam generator, and they thought that might have been "shrink." There was logic at that time for most of the actions, even though today you can look back and say, well, that wasn't the cause of that, or, that shouldn't have been that long.[6]

We will encounter this man's dilemma a few more times in this book; it goes to the core of a common organizational problem. In the face of uncertainty, we must, of course, make a judgment, even if only a tentative and temporary one. Making a judgment means we create a "mental model" or an expected universe.

Suppose you get an ambiguous order from your boss. You don't know if you should do A or B because the order could mean either. Alternative A would be correct if something were terribly wrong or if the situation were quite unusual. B would be correct if it were a situation that had occurred a few times before and was not all that serious. You decide she must have meant B. This alternative has been used before, and is easy to carry out. To do it you perform steps 1, 2, and 3. Still uncertain, you check the consequences of each. After step 1, certain things should happen, and they do. The same with steps 2 and 3. Despite the fact that this is no proper test of the appropriateness of alternative B rather than A, it serves to "confirm" your decision. In so believing, you are actually creating a world that is congruent with your interpretation, even though it may be the wrong world. It may be too late before you find that out.

The operators at TMI were faced with this dilemma. Alternative A, believing in the core pressure indicator, would mean that the core was being uncovered. Uncovering the core is unheard of; it had never happened in a large (over 750 megawatts) commercial light water reactor in the 380 or so "reactor years" of large commercial light water reactor operation. (A reactor year measure adds together the number of years each reactor has operated. For reactors of around 1,000 Mws, the more appropriate comparison, there were only about thirty-five reactor years of operating experience). Believing the B gauge rather than the A one (or attributing A to some temporary phenomenon) was soon confirmed— pressure dropped in the pressurizer after HPI was cut back. The other anomolies were accounted for in rapid fashion. Since the light showed the PORV had shut, the pressure decline in the core could be due to "cold shock" (from the two-minute burst of HPI fluid), or it could be a faulty reading. There had been faulty readings in the past; the drain tank temperature was one example.

Besides, about this time—just four or five minutes into the accident— another more pressing problem arose. The reactor coolant pumps that had turned on started thumping and shaking. They could be heard and felt from far away in the control room. Would they withstand the violence they were exposed to? Or should they be shut off? A hasty conference was called, and they were shut off. (It could have been, perhaps should have been, a sign that there were further dangers ahead, since they were "cavitating"—not getting enough emergency coolant going through them to function properly.)

In the control room there were three audible alarms sounding, and many of the 1,600 lights (on-off lights and rectangular displays with some code numbers and letters on them) were on or blinking. The operators did not turn off the main audible alarm because it would cancel some of the annunciator lights. The computer was beginning to run far behind schedule; in fact it took some hours before its message that something might be wrong with the PORV finally got its chance to be printed. Radiation alarms were coming on. The control room was filling with experts; later in the day there were about forty people there. The phones were ringing constantly, demanding information the operators did not have.

Two hours and twenty minutes after the start of the accident, a new shift came on. The record is unclear, but either the new shift supervisor decided to check the PORV, or an expert talking with a supervisor over the telephone questioned its status, and the operators discovered the stuck valve, and closed a block valve to shut off the flow to the PORV. The operator testified at the Kemeny Commission hearings that it was more of an act of desperation to shut the block valve than an act of understanding. After all, he said, you do not casually block off a safety system. It was fortunate that it occurred when it did; incredible damage had been done, with substantial parts of the core melting, but had it remained open for another thirty minutes or so, and HPI remained throttled back, there would probably have been a complete meltdown, with the fissioning material threatening to breach containment.

But the accident was far from over. New dangers appeared every few hours. Thirty-three hours into the accident another unexpected and mysterious interaction occurred. Confusion still reigned when the first sign of the famous hydrogen bubble appeared; the bubble threatened the integrity of the plant for the next few days. Again we have a lesson in the meaning of warnings, and in the difficulty that even experts have in understanding such a complex human-made system as a nuclear plant. Here is the background:

The fuel rods—36,816 of them—contain enriched uranium in little pills, all stacked within a thin liner, like the cigarette paper around tobacco, only about 12 feet long. Water circulates through the stacks of rods and cools the cladding so it won't melt. When they get too hot, though, the liner, or "cladding," can react with the water in a zirconium-water reaction. This consumes oxygen, thus freeing hydrogen, making hydrogen bubbles, which then can make pockets of hydrogen gas if there is room for them, and a dandy explosion if there is also a bit of oxygen and a spark.

It is not a well-understood aspect of nuclear engineering, I take it. Three years before the accident, when a nuclear physicist from the University of Pittsburgh mentioned the danger in the Bulletin of the Atomic Scientist,[7] a nuclear physicist from Pennsylvania State University wrote a scoffing rebuttal, saying the matter had been well studied and there was no danger.[8] We might put this quarrel down to the traditional rivalry between these two universities and treat it as insignificant, except that the latter, scoffing, scientist turned out to be the advisor on nuclear power production to Governor Thornburg of Pennsylvania, and was in the thick of the expert advice at TMI. After TMI President Reagan appointed him Chairman of the Nuclear Regulatory Commission. Contrary to industry pronouncements, there is still a good bit of mystery about atomic power plants, and this was an unfortunate case, since it was hours or days (depending upon whose testimony you wish to believe) before the bubble was conceived by the experts. Thus, the operators might be forgiven for ignoring yet another signal that something was drastically wrong, the "spike."

Here is how the warning occurred. At 1:00 P.M. Wednesday, thirty-three hours into the accident, there was a soft but distinct bang heard in the control room. This is not the kind of thing you expect or like to hear. A quick glance showed that the reading of the amount of pressure in the containment building—the building that holds the core vessel itself, and the pressurizer, drain tank, sump pump, pipes, electrical connections, et cetera—had jumped suddenly. In fact, the pressure spike had reached half the design limit for the building (if the pressure had been twice as high, the building might have cracked). Here the story gets murky. The operator, interviewed by the President's Commission, said, "We kind of wrote it off at the time as possibly instrument malfunction of some sort." This was not an unreasonable conclusion, since instruments were malfunctioning. "We did not have a firm conclusion regarding the spike," he went on, "since it appeared and went away with such rapidity."[9]

But another story has it that someone on the floor—there were perhaps twenty people there—knew that there had been a hydrogen explosion. Fearing another pocket of gas might appear and be ignited by a spark, he asked another operator not to restart a failed pump. The operator replied, "I already have." (Pumps have motors; they are big and make sparks.) That means, the first fellow said, that we don't have more hydrogen.[10] That is, he knew there had already been one hydrogen "burn." If this story is true, a lot of people went through the rest of the day ignorant of a vital piece of information.

Why worry? Because with more hydrogen being produced, the gas might find other ways to be vented from the core—whose condition was unknown to the personnel—and collect in the containment building. With pumps starting and stopping and other activity, a spark could easily be available, and the containment building had oxygen in it. If the hydrogen managed to collect in a spot near a lot of equipment and explode there, the pressure force could send missiles flying. (Indeed, three years later they found the huge crane required to lift off the top of the reactor vessel had been damaged by missiles from the explosion; two engineers protested the crane was not safe enough to use and were fired.[11]) Even a small explosion might pierce a cable or two and cause a short circuit, shutting off the emergency cooling, or rupture a pipe, causing a more rapid LOCA, and so on, though the design does take into account the possibility of "guillotine" accidents where pipes enter the containment wall. Even after the PORV was closed, the build-up of hydrogen in the core vessel itself is extremely dangerous, because its bubble can prevent the flow needed for cooling. The hydrogen will not explode there, but it need not explode to be dangerous.

Of such complexities is the normal accident made. For all but one operator, presumably, and for all the experts, the pressure spike and the hydrogen bubble were incomprehensible. To understand the accident, they would have had to know that the core was seriously uncovered, and that a zirconium-water reaction was likely (a possibility disputed by an expert), and would have had to recall that the PORV had been open, allowing the hydrogen to get out of the core vessel into the building that contained it. These are not expected sequences in a production or safety system; they are multiple failures that interacted in an incomprehensible manner—for all but at least one person, who, incredibly enough, wasn't talking, or didn't examine the implications of his hunch. A warning such as the spike is only effective if it fits into our mental model of what is going on. As with the "warnings" of Pearl Harbor, it can get swamped by the multitude of signals that fit our expectations, and thus be discounted as "noise" in the system.

That's enough on the accident for now. We will return to Harrisburg a few more times. But first we should pose a question that may have been bothering you: If this is typical of a nuclear plant, why have we had only one TMI? Or is this just a bad apple in the nuclear barrel? In the next chapter I will try to show that TMI isn't unusual, and yet indicate why there has been only one TMI. In Chapter 3, we will have to examine our language and define major terms such as complexity, coupling, and catastrophes. Thus equipped, we will be ready to journey through other systems in subsequent chapters, exploring ways to prevent such threatening accidents as the near meltdown at TMI. For example, wouldn't better organization help, or more money and resources for better people and equipment? Not much, I shall argue.

CHAPTER 2

Nuclear Power as a High-Risk System: Why We Have Not Had More TMIs–But Will Soon

Why haven't we had more Three Mile Islands? If nuclear power is so risky, why has no one been killed by radiation exposure as a result of a nuclear power plant accident? If the safety systems have worked so far, nearly twenty years into the nuclear power age, why call this a high-risk system? One answer is that the "defense in depth" safety systems have worked, limiting the course of accidents. We shall examine these safety systems briefly. But a more accurate and less reassuring answer is that we simply have not given the nuclear power system a reasonable amount of time to disclose its potential. We do not really have twenty years of experience, but very little—too little, by most industrial standards, to make a reasonable assessment of the risks.

The nuclear industry does not agree that it lacks experience. Therefore, we must journey into the heart of industry experience, taking a close look at some serious accidents, some trivial ones, problems of reliability and management, and above all, the special characteristics of the nuclear power system. This will give us the necessary tools, in the form of ideas or concepts, to enter, in later chapters, the world of other high-risk systems that someone has decided we cannot live without.

Operating Experience

We have not given the nuclear power generation system enough time to express itself; and we are only just beginning to uncover the potential dangers that make any prediction of risk very uncertain. We are about twenty years into the era of commercial plant operation, but our experience is not all with one type or size of plant. Indeed, the oldest plant in operation in 1982 was a 430 megawatt (Mw) reactor operating more or less continuously since 1967. We do not build this size any more, so its sixteen years of operating experience is of somewhat limited value.

The small plants of around 400 Mws are different in many respects from the larger ones of around 1,000 Mws; changes in scale produce surprising results. For example, the larger plants appear to be less reliable; there is more downtime after the first two or three years. In addition to size, there are two different types of U.S. reactors, the pressurized water reactors (PWR) and the boiling water reactors (BWR). Experience accumulated in one does not necessarily enable us to judge the reliability of the other; some aspects are similar, some different. In addition to size and type, there are four different U.S. manufacturers. General Electric builds only BWRs, while Westinghouse, Babcock and Wilcox, and Combustion Engineering all build PWRs. The designs differ, of course, limiting the accumulation of experience to some degree.

Thus, to say, as proponents of nuclear power often do, that we have 500 "reactor years" of experience with commercial plants (summing up the number of plants times the number of years each has been operating) is quite misleading. There is no consensus on what would be adequate experience for such a complex and novel transformation process as controlled nuclear fission creating steam that drives turbines; there are thousands of years operating experience with large turbines, but very little with nuclear fission. The condensate polisher problem on the turbine side of the plant at TMI would have been trivial in a coal fired plant but was not in a nuclear plant. We have been building large pressure vessels since the late nineteenth century, but are only beginning to learn the problems with welded stainless steel vessels 40 feet high that are bombarded with neutrons. Every few months new problems appear in nuclear plants, including the failure of supposedly failure-proof emergency scram systems. At the time of TMI we had only thirty-five years experience with reactors the size of Unit 2; that is infancy for a system of this size and complexity.

The first order for a commercial plant that was not in part a demonstration project was placed in 1963. Before that plant was even operating, the boom was on. By the end of 1967 there were seventy-five plants on order; in 1966–67 alone, forty-nine firm orders were placed.[1] More important, by 1968 the utilities were taking orders for plants six times larger than the largest in operation. This extrapolation, from the size of a plant one has some experience with to another six times larger, is very unusual for large, complex installations.

Bupp and Derain, who give the history of commercial reactor development, note that "electric power generation was an industry which had previously operated on the belief that extrapolations of two to one over operating experience were at the outer boundary of acceptable risk."[2] By 1967, cumulative operating capacity, a measure of experience, was only 3.5 percent of ordered capacity, rather than two to one. In short, no one knew if the seventy-five plants on order would ever work. They also did not know what the capital costs of building them would be. The plants completed in 1975 were about three times the cost per kilowatt produced, in constant dollars, of those completed only five years earlier. "The learning that usually lowers initial costs has not generally occurred in the nuclear power business. Contrary to the industry's own oft-repeated claims that reactor costs were 'soon going to stabilize' and that 'learning by doing' would soon produce cost decreases, just the opposite happened."[3]

The technical learning curve with these plants (sometimes called "light water" plants) also failed to materialize, according to the study just quoted. "After more than a decade of experience with large light water nuclear power plants, important engineering and design changes were still being made. This is contrary to experience with most other complex industrial products."[4] After a decade the major problems of well-designed systems should be far behind, but not in this case. The reason, the authors believe, is the haste with which untried designs were ordered, and the stubborn refusal "to face up to the sheer technical complexity of the job that remained even after the first prototype nuclear power plants had been built in the mid- and late 1950s."[5] Nor was this due to a stodgy industry "boiler business" mentality. The utility industry had been one of the great growth areas in the postwar American economy. Energy production was doubling every nine or ten years, and operating costs were declining steadily, largely as a result of technological progress. Generating costs declined as fossil plant sizes increased and as improvements in operating efficiency continued. It was not a technologically stagnant industry. But it was unprepared for the technological complexities of controlled fission. New complexities are now being realized (and publicized) almost monthly. In time, it seems, the problems for which TMI was an early precursor will unfold in more TMIs.

For example, steam generators are a problem with all power plants; the pipes rust. Special care and materials are used in nuclear plants, but in 1981 it appeared that seventeen reactors, some only five or six years old, had serious rusting problems. The repairs on two plants owned by the Virginia Electric Power Company cost a total of $112 million. Rusting is a special problem in nuclear plants since the thin tubes in the generators are immersed in water continuously, and leaks will allow radioactive water to get in the secondary (nonradioactive) cooling system. Various steps were taken to reduce the rust, but apparently without success in some plants.[6] The point is, in a nuclear plant leaks in the generator are failures that can interact with other failures, and thus be a source of system accidents; repairs to such a system can be enormously expensive (in contrast to a conventionally fueled power plant); and there was no way to anticipate these problems in a new technology with such large design and construction lead times.

More serious is the problem of core embrittlement. The bombardment of the containment vessel by the nuclear reaction going on within it has had a greater impact than anticipated. The 40-foot stainless steel vessel is designed to last 40 years, but there are already potential brittleness problems in forty-seven plants, the Nuclear Regulatory Commission (NRC) announced in 1981, and of these, thirteen have serious problems. One of these is only three years old; three others, four years old. The problem is that the core is very hot—about 550° F.—and if you have an emergency and must force thousands of gallons of cold water into the core, the inside of the 8-inch-thick vessel will shrink faster than the outside, creating cracks. In an accident, the pressure must be kept high, further straining the core. These problems apply to PWR systems only, but PWRs account for two-thirds of the operating reactors in the U.S.[7]

These are technical reasons why we have not had sufficient time to have a truly serious nuclear accident—the system is quite new and has not been given a chance to reveal its full potential for danger. Unknown potential cannot be corrected, except by running the plants and taking the risks; without experience, we cannot be sure of the potential for damage inherent in the system's characteristics.

The Construction Problem

There are other problems that are not so directly related to the technological nature of the system, but rather to the nature of the utility and construction industries. Several weeks after TMI, the NRC reported on a continuing study of earthquake protection measures at operating plants. At that point they had identified thirty-five plants with "significant differences" between the way they were designed and the way they were built. This raised questions about "the whole procedure for checking plants," one NRC official said.[8] Since there is only about one engineer from the NRC to watch over each plant in construction, there is "almost complete reliance on the utility and its contractor to monitor themselves and report on deviations from acceptable standards," said a General Accounting Office report of the previous year.[9] One would think that reliance on the utility would be adequate, since it is the utility that owns the plant, not the government. But enough stories have appeared to question whether it is possible to rely on anyone to build safe nuclear plants.

For example, at the Marble Hill nuclear project in Madison, Indiana, it took affidavits filed by workers and former workers to alert the NRC to the fact that, as John Emshwiller puts it in his *Wall Street Journal* article, "the builders can't seem to get the hang of pouring concrete."[10] So far, 500 voids (some up to 180 cubic feet in size!), had been found in the concrete structures. Workers were ordered to do cosmetic patching jobs in order to get them past inspection. At another plant, the Brown and Root construction firm was accused of intimidating federal inspectors, in one case putting the inspector into the hospital for two days. On the other hand, engineers have resigned in protest from the NRC, charging coverups and intimidation by the NRC itself. The NRC was informed of falsified documents regarding the inspection of a safety system at one midwestern plant, but, according to the NRC administrator, he ignored them. Three months later two employees went public with the documents, and the NRC promised to investigate.[11]

Perhaps the most striking testimony on unsafe construction in this business is the Diablo Canyon case. Diablo Canyon, in central California, has been waiting for several years to be allowed to operate. After construction was underway, an earthquake fault was discovered a short distance from the site and extensive earthquake protection was required. A little more than a week before the plant was scheduled to open (after some dramatic protests from anti-nuclear groups and local residents and 1,600 arrests), a diffident, 25-year-old engineer for Pacific Gas and Electric Co., owners of Diablo Canyon, was staring at some drawings of a part of the plant. The drawings divided the floor of the containment building into five segments, and showed the location of some heavy equipment (fan coolers). Something about the drawings bothered him. "Just out of curiosity, I pulled some detailed cutaway engineering drawings out of the file—drawings that showed the actual placing of those coolers, and the two diagrams didn't match. It didn't make sense."[12] He insisted that he was not looking for flaws; his discovery was accidental.

What he found was that in 1977 the utility had mistakenly sent the wrong set of diagrams to its seismic engineering consultants, who were to provide seismic shock calculations to be used in strengthening the vulnerable parts of the plant. Instead they sent the diagrams for a second reactor, still under construction, which was the mirror image of the Diablo Canyon reactor about to be retrofitted. The work was performed, and many parts were needlessly reinforced, while others, which should have been strengthened, were left untouched.[13] Subsequent investigations turned up no fewer than 111 other flaws in the construction of this $2.5 billion reactor, and by the end of 1982 it was still not operating.

Shoddy construction and inadvertent errors, intimidation and actual deception—these are part and parcel of industrial life. No industry is without these problems, just as no valve can be made failure-proof. Normally, the consequences are not catastrophic. They may be, however, if you build systems with catastrophic potential. No less an authority than former reactor designer and former Dean of the Engineering College at Pennsylvania State University, Nunzio J. Palladino, appointed Chairman of the Nuclear Regulatory Commission in 1981, remarked in december of that year:

> During my first five months as NRC chairman a number of deficiencies at some plants have come to my attention which show a surprising lack of profes-

> sionalism in the construction and preparation for operation of nuclear facilities. The responsibility for such deficiencies rests squarely on the shoulders of management. . . . There have been lapses of many kinds—in design analyses resulting in built-in errors, in poor construction, in harassment of quality control personnel and inadequate training of reactor operators.[14]

Safer Designs?

If the plants are not built well, and we do not have enough operating experience to assure us the design and equipment are safe, could we turn to other, safer designs? *Are* there safer designs? Apparently there are, though it is well beyond my capacities and the argument of this book to be confident about this. The Canadian reactor, the CANDU, is said to be slower, more "forgiving," and less tightly coupled than our PWRs and BWRs. Operators have more time to take action, and can take more actions. This has not prevented Canada from having some nuclear accidents, but I gather they are less serious than those we have suffered. But the Canadian plants are also smaller and less efficient than ours.

Some engineers believe we missed the boat in not investing more heavily in the gas-cooled reactor, considered to be safer. A small commercial one was built, but has been shut down for some time, though the utility—Pennsylvania Electric—indicates it still wants to keep alive the possibility of developing gas-cooled reactors because of their increased margin for safety. A second, larger one recently began operation in another utility. A sodium-cooled breeder reactor is operating in France and a much larger one is being built there. These produce more fuel than they use, which is useful since the world supply of uranium is quite limited. But the technology of sodium-cooled breeder reactors is very new and some feel the dangers of radioactive sodium far exceed the dangers from light water reactors—PWRs and BWRs. We shall encounter this later in the chapter when we examine our experience with the Fermi breeder reactor. There are other designs, but there is no evidence that any nuclear reactor designs are significantly less complex and interactive, or significantly less tightly coupled than the light water ones we have been concerned with.

There is a good reason why our dominant design, the pressurized light water reactor (PWR), was adopted, even though heavy water reactors (CANDU), gas-cooled reactors, and perhaps other designs might be better. In the 1950s the U.S. government was very anxious to find peaceful uses for atomic energy, and, in particular, to develop atomic power production. The reasons for the government's haste are in dispute at this writing, but it certainly was not an expected power shortage or increase in energy costs. In fact, cheap oil and gas was driving out small hydroelectric dams in the northeast and the popular solar hot water heaters found in the south. The government had to offer large incentives to private utilities, and when that did not work, to threaten them with the prospect of socialized power—federal atomic power plants on the TVA model across the country—before the utilities would build them. The government had on hand a design for a reactor; it was being built for submarines. Such a reactor is very compact, very responsive, and can easily be refueled once a year when the submarine returns to port and does not need the power.

None of these characteristics were appropriate to utility installations; indeed, for these, the size, responsiveness, and refueling cycle of submarine reactors are counterproductive. A company does not want to have to shut down its plant each year for refueling, because replacement power has to be bought, and since it generally comes from the least efficient generating sources that are maintained only for peak loads (gas and oil plants with small output), it is very expensive. Compactness is not a requirement at a plant site. Responsiveness is not necessary since these are "base-load" installations, designed to handle the bulk of demand on a steady basis, rather than requiring fluctuations, and they do not need to come up to power or cool down quickly. Nevertheless, the firms that built and sold nuclear plants took over the designs for the submarine systems and modified and greatly enlarged them. There appears to have been a rush to get into the business. Indeed, the first "turnkey" plants—the vendor builds it and "turns" the key over to the utility—were sold at substantial losses in order to get established in the industry. It is a good example of a technological "push" rather than a demand "pull." This unseemly haste has left us with a particularly complex and tightly coupled design, and a design that was assumed to be capable of being scaled up in size without any serious complications.

Even if there were a technological breakthrough, and a much safer design were available, it is very unlikely that one would be built in the United States in the next decade or two. We have about seventy operating reactors now, and perhaps fifty more that might begin operating in the next few years (unless the rate at which they are being cancelled increases), according to an NRC commissioner.[15] Even the most enthusiastic proponents do not anticipate more than 120 reactors operating in the next five or so years. A new design would not attract much interest in the financial community; utilities generally find themselves with excess capacity because the rise in demand for electricity, for decades a stable 7 percent, has dropped steadily since 1974 to 1.7 percent in 1981. Further, it would take over ten years to design and build a new facility, even if it were significantly less complex than those we have now. Thus, we will have to live with the plants we have, safe or not; new, dramatically safer ones do not appear to be in the offing, and probably will not be built for a long time to come. Note that I am not saying there could never be a nuclear plant that was *not* highly interactive and tightly coupled (though I suspect the nature of the transformation process involved in this kind of energy production makes that impossible) but only that we shall not see one for many years. And we shall continue to see our existing and nearly ready plants for a longer time—perhaps forty years, if they live up to industry predictions.

Defense in Depth

There is yet a quite different answer to the initial question posed in this chapter: Why have there not been more accidents resembling TMI if these systems are all that dangerous? So far I have argued that we have not given them time. The design and construction flaws will not appear immediately nor in every reactor. But is it not possible that the "defense in depth" is working—that containment buildings do contain; that emergency core cooling systems do cool; that even if some unanticipated radioactivity escapes, the plants are sufficiently far from highly populated centers to reduce the risk to negligible proportions? Yes, but the situation, while reassuring, is not wholly so, because the possibilities for system accidents that evade these defenses still exist. Let us look at each of the defenses.

We can be glad that we have containment buildings. These are concrete shells that cover the reactor vessel and other key pieces of equipment, and are maintained at negative pressures—that is, at a lower air pressure than the atmosphere outside of them—so that if a leak occurs, clean air will flow in rather than radioactive air flowing out. The Soviet Union, which did not begin a large nuclear generating program until about 1970, is far less concerned about the chance of large accidents, so they did not build containment structures for their early reactors, nor do they yet require emergency core cooling systems. Had the accident at

Three Mile Island taken place in one of the plants near Moscow, it would have exposed the operators to potentially lethal doses, and irradiated a large population.

At TMI, the hydrogen explosion (or "burn") that took place in the containment building generated a pressure surge equal to one-half that which the building was designed to handle. The building was built this strong only because the state of Pennsylvania insisted that it meet the criterion of being able to withstand a direct hit from a jet airliner (it is close to the Harrisburg airport). The initial plans did not call for this. Even if the building were not reinforced, it is unlikely, I am told, that the hydrogen burn would have breached containment and allowed the radioactive particles to escape. However, such a disaster might occur in a plant with all those flaws in the concrete we heard of; the explosion might have taken place thirty minutes later when there would have been much more hydrogen to burn; and it could have happened in a part of the building where more missiles would have been created, which could have ruptured the many penetrations required in the building for controls and pipes. While containment is absolutely necessary, it may not be sufficient. It *can* be ruptured.

We almost had a good test of the ability of the concrete containment structure to withstand an airplane crash in 1971. A B-52 bomber was flying a routine practice flight near Charlevoix, Michigan, on the shores of Lake Michigan. Bombers and fighter-bombers from a nearby Strategic Air Command base routinely flew low-level (1,000 foot) sorties directly over the plant, despite Air Force instructions to stay clear. This time the plane was heading directly toward the reactor when it crashed, skipping off the surface of the water, and raising a fireball 200 to 600 feet in the air. A Grumman aerospace official suggested that it might have flown into radioactive gases from the plant's stack, which could interfere with the plane's electronics. The plane was two miles, or about twenty seconds, short of crashing into the plant and testing containment.[16]

Fortunately, we tend to build our plants in sparsely populated areas, though they are generally near big cities. The ideal spot for a nuclear plant cannot exist. It should be far from any population concentration in case of an accident, but close to one because of transmission economies; it has to be near a large supply of water, but that is also where people like to live; it should be far from any earthquake faults, but these tend to be near coastlines or rivers or other desirable features; it should be far from agricultural activities, but that also puts it far from the places that need its power. The result has been that most of our plants are near population concentrations, but in farming or resort areas just outside of them.

The Indian Point nuclear stations, for example, are on the Hudson River, but just thirty-five miles upwind of Manhattan. The owner of one of the plants there, Consolidated Edison, once proposed building a nuclear plant in the middle of Queens—truly one of the most densely populated areas in the United States. Some plants are built on earthquake-prone coastlines, others on rivers that supply fresh water for large cities and for irrigation. Some people have suggested isolated reactor parks, where several nuclear plants will be built, with long transmission lines to populated areas. But an accident in one of the plants might require the abandonment of the adjacent plants in the park (and thus possible additional accidents).

Despite all these problems, semi-remote siting has no doubt increased the safety of nuclear plants. Many have had small emissions of radioactive materials as a result of accidents. Were they located in Queens, the long-term dangers would be higher. Furthermore, though it is said to be minuscule by almost all experts, the plants do release radioactive materials to the environment in the course of normal operation. The farther you are from that, the better.

Finally, there is the Emergency Core Cooling System (ECCS). Should there be a danger of a core melt, this system will flood the core with water, cooling it. It is in the nature of the beast that we cannot use full-scale testing to see how effectively ECCS will work. In a series of tests with a 9-inch model reactor core, all tests failed.[17] Some critics, such as the Union of Concerned Scientists, believe that as presently constituted, ECCS is an inadequate safeguard. At the Browns Ferry nuclear station, the fire that shut down two reactors and burned out of control for several hours rendered the ECCS system inoperative. Fortunately, other means were used to prevent massive fuel melting. The assessment of the ECCS made in the most ambitious safety study commissioned and carried out by the Atomic Energy Committee (forerunner of the NRC), the *Reactor Safety Study* (RSS or WASH 1400, or Rasmussen Report as it is variously referred to), failed to consider that anything else might be wrong in a plant when there was an emergency that required ECCS. That is, the study ignored the possibility that there could be a variety of failures that in themselves would defeat this safety device. For example, steam generators are a continuous problem with nuclear plants; should many of the tubes in them fail in an accident, so would ECCS. There are also problems with other major subsystems. The integrity of the reactor vessel itself has been questioned, drawing upon industrial experience with vessels in nonnuclear systems.[18] Finally, in 1981 half the Browns Ferry control rods failed to drop on command, and in 1983 the automatic shutdown system at the Salem plant in South Jersey failed twice. Both events were assumed to have extremely low probability; both could easily defeat the ECCS.

It is true we can be glad that containment, siting, and major emergency systems exist to reduce the dangers. No doubt there would have been more severe accidents without them. But they are unlikely to prevent all future disasters. Siting is not remote enough; containment is vulnerable to hydrogen explosions, missiles, and faulty construction; and the "defense in depth" major emergency systems such as ECCS are defenses with perhaps not that much depth.

Trivial Events in Nontrivial Systems

Nothing is perfect; every part of every system, industrial or not, is liable to failure. Common, run-of-the-mill industrial plants have a steady run of unremarked failures. The more complicated, highly engineered continuous processing plants, such as chemical, pharmaceutical, and some steel processing plants, are no exception. The more complicated or tightly coupled the plant, the more attention is paid to reducing the occasion for failures, but as I shall argue in the next chapter, this can never be enough. If we add catastrophic potential, as we must with nuclear plants, the everyday failures should not go unremarked. They now become significant. What I will be reporting in this section would not even make a news story in the plant paper, let alone the *New York Times* and the like, if it did not occur in a nuclear plant. In fact, not until after Three Mile Island would most of these incidents even be picked up by the daily paper.

Utilities are quite sensitive to this unwanted and "unjustified" scrutiny, but we *should* be sensitive to trivial events in nontrivial systems. I will start with some trivia, to show the course of consequences in these expensive systems, and then proceed to a few of the famous accidents. Keep in mind that these types of mishaps go on all the time in most organizations; we are being unrealistic if we are surprised that they go on in nuclear plants.

Let's start with a trivial event like the ones that plague us all. In 1980 a worker in the North Anna Number 1 plant of the Virginia Electric and Power Company (VEPCO) was cleaning the floor in an auxiliary building. His shirt caught on a 3-inch handle of a circuit breaker protruding from a wall. He pulled it free, and apparently was unaware that in doing so he activated the breaker. This shut off the current to the control rod mechanism, and the reactor scrammed (shut off) automatically. This trivial event caused a four-day shutdown, which cost consumers several hundred thousand dollars. Fortunately, the weather was mild, so demand was low. The executive vice president of VEPCO termed the accident embarrassing, but suggested there was a fortunate lesson for us all: The incident "clearly demonstrates the sensitivity of nuclear station systems to the slightest deviation from normal and the ability of these systems to perform safely as designed in immediately stopping the unit."[19] Shutting off current to a major safety system is hardly a slight deviation from normal, and that it can be done so casually suggests an undue degree of sensitivity.

Piping is always a problem in any plant. In a nuclear plant this problem is a bit more severe. During the TMI accident, operators sent radioactive water to the wrong places because the plumbing was so complex and pressures could cause reverse flows. At one plant a small error sent radioactive waste water into the drinking water system that went to the fountains!

Clams are another problem. The filters used on cooling water intake systems from rivers and bays do not keep out the clam larvae, which then lodge in the cooling pipes in the plants and begin reproducing. Eventually, the pipes become clogged with thousands of clams. A report on one plant in Arkansas suggested a week-long shutdown to remove them. Clams foul non-nuclear plants too, but stopping and starting them is not as dangerous.

Even changing light bulbs has its dangers in these highly engineered, complex systems. In 1978 a worker changing a light bulb in a control panel at the Rancho Secco 1 reactor in Clay Station, California, dropped the bulb. It created a short circuit in some sensors and controls. Fortunately, the reactor scram controls were not among those affected, and the reactor automatically scrammed. But the loss of some sensors meant the operators could not determine the condition of the plant, and there was a rapid cooling of the core. As we have already noted, normally the inside temperature of the reactor vessel is at 550° F. Within an hour it had dropped to 280°. The colder, internal walls tried to shrink but the hotter, external ones would not allow shrinkage. This put strong internal stresses on the core. Meanwhile, to prevent a meltdown of the fuel rods, the internal pressure must remain high—2,200 pounds per square inch—while the temperature must drop. At the lower temperature of 280°, the strength of the vessel is reduced, but the pressure remains high. This rapid cooling, which can occur with high pressure injection, or with a loss of instrumentation and control, did not in this case damage the core. But this is probably only because the plant had been operating at full power for less than three years. A spokesman for the NRC said: "If it had been 10 to 15 full power years, instead of two to three, which it was, that vessel might have cracked."[20] A cracked vessel would result in a loss of coolant and a meltdown; no emergency system would be available to cool the core.

Knowledge of such problems, after Three Mile Island, should lead to extra surveillance; we should learn from experience. The record has not been encouraging, however. The Indian Point Number 2 nuclear plant, thirty-five miles upwind of New York City, run by Consolidated Edison (Con Ed), had been having problems with leaks in the fan cooling unit service for some time. The leaks occurred in the containment building. Early in October, 1980, a light went on, warning of high water in sumps in the building, and remained on for several days. The indicator light itself was apparently considered to be malfunctioning. But water was actually leaking into the building from the fan cooling unit; eventually 100,000 gallons would collect, covering the first 9 feet of the reactor vessel in salty, brackish, cold Hudson River water. A safety device, involving two moisture-level indicators, failed to detect the water, because the indicators were designed to detect hot, not cold water.

The leak might have gone undetected for hours or days more were it not for an operator error. A warning signal came on, indicating a fluctuation in reactor power. It presumably was not related to the water leak. Operators reduced power and checked; nothing seemed wrong, so they supposed it was a faulty signal (it possibly was; they are common). But to go up to full power again, an adjustment was necessary on a governor. It was made too quickly (the operator error), and the entire reactor shut down automatically. The technicians had to enter the containment building before starting up again. They then discovered areas flooded with over 9 feet of water. The two sump pumps, which should have removed the water, were both inoperative. In one the fuses were blown, in the other the float mechanism was stuck.

We should not be aghast; these are just the routine problems of industrial equipment. But this case occurred in a building that is inconvenient to enter, and is not visited except for maintenance, unless there is trouble. The supervisor then restarted the reactor twice, without considering whether having the bottom 9 feet immersed in cold water for hours or days might have led to thermal cracking or other problems. Fortunately, another supervisor, who just happened by on his day off, recognized the danger and shut the reactor down.

All this took place on a Friday, October 17, 1981, at 11:00 A.M. Contrary to an agreement with federal and state officials to notify them immediately of any trouble at the problem-plagued plant, nothing was done until 3:20 that afternoon. A plant official called the NRC's resident inspector on the site. But it was Friday afternoon and he was away. The plant official did leave a message on the answering system, but it was a message simply to call him, not one saying, "We have just found 100,000 gallons of water in the containment building." Nor did the Con Ed official call the emergency number that is given on the answering service tape. The resident inspector came back to work on Monday to find the plant shut down. He waited until 4:20 that afternoon to inform the NRC regional office of the problem. Con Ed waited another day—five all told—before it informed local officials and the public of the leak.

So it goes with organizational safeguards designed to save us from technological failures. The NRC proposed a fine of $210,000 for the utility, which Con Ed of course protested. To replace the power during the long shutdown (they eventually concluded there was no damage to the reactor vessel), expensive oil-fired power plants had to be used, at a cost of $800,000 a day to Con Ed's customers.[21] Stuck floats, light bulbs, and shirt tails are just a few of the trivialities to which this system is vulnerable.

Learning from Our Mistakes

Those were simple failures or shutdowns. It is time to get more deeply into run-of-the-mill accidents. The NRC puts out a journal called *Nuclear Safety*. One of its regular features is a compilation of safety-related occurrences, selected by the editor and briefly described. Though technical, they provide endless, numbing fascination as they describe all the things that can go wrong in these awesome plants. Here is one brief account, not particularly remarkable, but it will give you the flavor. Don't try to follow it too closely; just note the failures of the equipment, operators, and design, before we turn to the journal's editorial comment that the incident shows how the industry has managed to achieve its excellent safety record.

A small, early BWR reactor at Humboldt Bay, California, (Pacific Gas and Electric) lost its offsite power source on July 17, 1970, and scrammed, as designed. The emergency power supply came on, but it was not designed to provide power to the particular sensors that turned out to be needed. Reactor pressure rose, but the emergency condensor, which would reduce it, did not come on because the gate on the switch stuck in the guides, probably as a result of a poor setting on a valve. The operators knew the emergency condensor did not operate, but assumed that a safety valve had opened to reduce pressure. Instead, a different safety valve opened, and, due to coolant shrink from its discharge, a low-water level signal came on. This, combined with the loss of feedwater and an increase in dry-well pressure, opened the reactor vent system. Meanwhile, a pipe joint ruptured in the safety valve discharge line. The vent valves were open for four minutes before the operators discovered them. There was no indication of a rupture, so they closed them. Then the fire pumps started automatically, indicating excessive pressure in the reactor, low water level, high pressure in the dry well, and loss of power to some safety systems. The accident was successfully contained, but the pressure in the reactor had exceeded safety levels; 24,000 pounds of reactor water was "blown down" (forced out of the core), indicating that the top of the fuel rods in the core were in danger of being uncovered. This was not a particularly remarkable accident; many are far worse. What is interesting is the comment that precedes it, which I quote:

> The nuclear industry is not vastly different from other industries. Things do go wrong, as is attested to by these safety-related occurrences which are reported in each issue of *Nuclear Safety*. Even so, the nuclear industry has an excellent safety record. The items chosen for this article demonstrate how this record has been attained. For example, safety systems are designed with backups that take into account the possibility of failure, operations people watch for anomalies and investigate them quickly, and routine checks are run to assure that all is proceeding as planned.[22]

It is hard to believe the cheerful author read his own account of the Humboldt Bay accident—or any other accounts in *Nuclear Safety*. In the previous issue of the journal, a fuel meltdown was graphically described (though in a plant in France); in the issue with the above quote we find among others a report of another plant in which, even after a seven-month shutdown for repair of primary coolant piping, an important motor broke, and sixty-three valves malfunctioned—35 percent of those tested prior to start-up. Reassuringly, we are told that "the frequency of valve testing will be increased, and a better method of cleaning the air used for some valve operations will be studied."[23]

In the next issue of *Nuclear Safety,* after a discussion of some fires and other problems, we find the following: "A core-spray injection valve failed to close, and then it was discovered that the injection valves for the other core-spray system would not work either. Also, the valves on the low-pressure coolant-injection systems would not operate properly. While the problem of these valves was being pondered, one of four control valves on the main turbine unexpectedly closed completely with the reactor at full power."[24] And so *Nuclear Safety* goes on, issue after issue, editorializing how the "excellent safety record . . . has been maintained," in spite of the accounts presented in its pages.

But the following year a more plaintive note is struck. "Two-thirds of the problems discussed in this issue are strikingly similar to ones previously reported in *Nuclear Safety* in the hope and expectation that we will all be able to learn from the experience of others. . . . Operators should take particular note of these occurrences so that they can more readily avoid similar happenings in their own plants."[25]

The following is an unusually literate and straightforward account of an incident that has many parallels in marine tankers and chemical plants. Since gases are invisible, and the subtle interactions of pressure, temperature, and operator actions cannot be fully anticipated, these events are unavoidable in highly interactive systems. In this case, after the event, two additional valves and some additional procedures were added to the system to prevent its happening again, but then no one thought it could happen in the first place; it can also happen in a slightly different way in another location of the plant.

> During a shutdown, service personnel requested that demineralized water be made available in the containment in order to fill pails to be used for cleaning. The shift supervisor informed them that a valve lineup would have to be made by operations people before water would be available. When the service personnel entered the containment building, they tried the faucet to see if there was water yet. There was none, and so they closed the valve and waited a while. Then once again they tried the faucet, leaving it partially open while waiting for water, and called the main control room on the plant intercom to ask when the water would be available. Operations personnel said that a man was on his way into the containment building to align the demineralized water system. The service people closed the valve.
>
> Shortly thereafter the radiation monitors in the containment alarmed, and the control-room operators ordered an evacuation of the containment building. The increase in radiation levels in the containment building was traced to the gas escaping from another tank, the collecting tank, during the brief period the faucet was opened. Inasmuch as several tanks are interconnected by the same demineralized water supply, the valving in the system has to be properly aligned to prevent undesirable interaction. The premature manipulation of the

> sink valve before operations personnel could align the system resulted in the venting of the quench tank back through the primary water header to the open faucet. However, the radiation levels were low, and there were no overexposures.[26]

Dresden 2 is a nuclear plant outside of Chicago that is hardly a household name for most people, but for me it holds the distinction of providing the quintessential example of a system accident. It is owned and operated by Commonwealth Edison, reputedly one of the top two utilities in the country in terms of organization and management. This is important, for it indicates what can happen even in a well-run utility. The following description is much simplified, though you would hardly believe so in reading it. Do not try to understand the complex interactions, but let yourself be overwhelmed by the operators' frequent, uncomprehending attempts to cope with multiple equipment failures, false signals, and bewildering interactions.

> The steam valve began to malfunction and then closed. Fortunately, the reactor SCRAMmed automatically. The power dropped to the afterheat level, reducing the size of the steam bubbles in the core. This caused the water level in the reactor to drop, which caused the feedwater pumps to increase coolant flow into the reactor to avoid uncovering the core. As the water level rose, the operator noticed that the level indicator was reading a low level. Actually, however, the indicator was stuck and giving a false low-water-level reading. The operator reacted by manually increasing the feedwater flow still further, so that the water then filled the reactor and spilled over into the steam line. The feedwater-flow error was uncovered and corrected; but then the pressure began to rise, and two safety systems designed to cope with the problem and cool down the reactor were found inoperative. The operator then reduced pressure by opening a relief valve momentarily. At this point, water hammer occurred, produced by the water spill-over into the steam line, and this popped safety valves (pressure relief valves), which stuck open due to a design error. The relief valves then discharged reactor steam to the reactor containment atmosphere, which began to pressurize the containment. The loss of coolant through the stuck relief valves should have caused the ECCS to activate to inject replacement coolant; but one system was found inoperative, and the operators blocked the operation of the other system on the assumption that the loss-of-coolant problem was minor. However, they did not know the cause (stuck valve) and could not make a sound judgment (it could have been a leaky coolant pipe about to completely rupture). Meanwhile, the pressure in the containment rose beyond the range of the pressure gauge (5 psig). The containment is equipped with water sprays to quench the steam pressure whenever two psig pressure is exceeded, but the operators blocked this safety action because that would have cold-shocked some equipment and thereby damaged it. They did not, however, have sufficient knowledge of the events to justify their action. The containment reached 20 psig compared to 60 psig design pressure before the plant was finally brought under control.[27]

Fermi

Our final example is not, strictly speaking, of a system accident, but of a component failure accident, though the recovery effort involved some of the typical complexities of the system accident. (The distinction between system accidents and component failure accidents is fully developed in Chapter 3.) The account, based largely on a book by John Fuller, of the Fermi core meltdown will serve to illustrate dramatically the complexity for these systems, the pressures on operators, and the tremendous problems of clean-up. It also shows that attempts to make the system safer are sometimes ill-conceived and add danger; that completely novel accidents make the question of operator error irrelevant; and that the industry, instead of worrying about the disaster potential, only draws strength from the fact that it was not worse. The accident occurred in a demonstration reactor on Lake Erie in the small community of Lagoona Beach, near Monroe, Michigan—which is very near Detroit. A report by the Atomic Energy Committee completed before the accident (and promptly classified) predicted that, given a severe accident at Fermi with unfavorable wind conditions, 133,000 people would receive high doses of radiation, and one-half would quickly die. Another 181,000 could receive 150 rads.[28] As Fuller's account makes clear, it was a close call.[29]

The reactor was a sodium-cooled breeder reactor, large enough to produce substantial power as well as plutonium that could be used to fuel other conventional reactors. It was the first and only U.S. breeder reactor, and thus an untried design near Detroit's millions. In October of 1966 the operators were trying to achieve the first stage of a high power goal set by the company, slowly and carefully bringing up the temperature of the reactor. Delays and problems had been numerous in the past and continued this time. One of the steam generator valves malfunctioned, and six hours were spent correcting it. Then a boiler feedwater pump failed, but was quickly corrected. The operators once again increased the fissioning in the reactor. But the engineer on duty noticed some erratic changes in the neutron activity of the fission process, which could have been merely the electronic system picking up some "noise" or static. They paused, it disappeared, and they continued. Next, the engineer noted that for the amount of power the reactor was producing, the control rods, which shut off the fissioning when fully inserted, should have been raised only 6 inches, but they were 9 inches out of the core, and the neutron activity signal was again erratic. The reactor was put on hold, and the engineer went to check the instruments on the individual subassemblies of the fuel rods—some 30 feet away from the control board. The results were puzzling. The outlet temperature of one of the subassemblies was clearly too high, but they had been having trouble with that one. Indeed, since it appeared that the instrument was faulty, they had moved the instrument to a different part of the fuel bundle. But now a second subassembly showed high temperatures too, but none of the ones that were nearby and also instrumented were abnormal. Unfortunately, only one of every four subassemblies were instrumented, but if one overheated those near it that also had instruments should show overheating.

Then radiation alarms went off, the air horn began blasting twice every three seconds, and the public address system came on with a laconic, "How hear this. Now hear this. The containment building and the fission product detector building have been secured. There are high radiation readings, and they are sealed off. Do not attempt to enter. Stay out. Both buildings are isolated. This is a Class I emergency. Stand by for further instructions. Stand by for further instructions."[30] The operators first counted the crew to make sure that no one had been sealed in the containment building, with its high radiation readings. Everyone was safe. Next they had to pull down the power in the reactor. They were reluctant to scram the reactor immediately because of thermal shock from a sudden change in the temperature of the sodium coolant. One hypothesis that was quickly ruled out was that the radiation alarms were false; an engineer had been working on the fission product monitor and thought that he might have triggered a false alarm. But the temperature readings on the subassemblies indicated something real was going on.

Eleven minutes after they started to cool down the reactor, they decided to scram it manually. There was no way of knowing whether this was too late, or too soon, with this type of reactor. Indeed, there was no way of knowing what had happened inside the core. According to the design of the safety features, if there were any fuel melting, the reactor should scram automatically. It obviously hadn't scrammed, which suggested that the problem was not a fuel melt. Yet perhaps it was a fuel melt which just was not indicated. On the other hand, it might be an instrument problem; there had been problems with the instruments before. Still, they could not be sure if a fuel melt had been avoided or not. It was essential to find this out, because a fuel melt would block the flow of sodium coolant, and could lead to heat build-up and more melting, and thus a secondary accident. The assistant general manager took charge of the immediate efforts and announced, "We will go at this very, very slowly."[31] Fortunately, the Fermi engineers had time, since the core temperature continued to slowly decline. Since there were no procedures for such an emergency, they had to write ones out, and check them very carefully. They feared stirring up trouble inside the core. They soon found there had been both fuel melting and fuel redistribution. The redistribution could cause blockage and further fissioning.

The fuel melting conclusion was no doubt reluctantly arrived at for another reason—expert advice. Nobel Laureate physicist and nuclear power advocate Hans Bethe had confidently predicted a core meltdown could not happen with this reactor. Another expert was less certain, but he had predicted that at worst only one subassembly could melt. The evidence now was that two or more had melted. The second expert had also stated that the automatic safety devices would shut the reactor down if there were any melting; the devices did nothing of the sort. The Fermi engineers now talked of "hair-raising decisions" and "terrifying thoughts"; they were sitting on top of a volcano next to Detroit. They could not walk away and leave it there; they could not be sure there would not be a secondary accident, and in any case the melted uranium would eventually eat through the core and the concrete base of the building.

For a month the reactor sat there while the company let it cool and planned the next step. Then the engineers very carefully removed the top and hoped that none of the fuel subassemblies were stuck together in such a way as to produce "criticality" (the conditions for fissioning). If they could pull out the damaged subassemblies, it would be safe. It took three months to learn that four were damaged, and two stuck together. It took five more months to remove them. Special equipment was built; the deadly sodium had to be drained, and there was no provision for this in the reactor design. Almost a year from the accident, they were able to lower a periscope 40 feet down to the bottom of the core, where there was a conical flow guide—a safety device similar to a huge inverted ice-cream cone that was meant to widely distribute any uranium that might inconceivably melt and drop to the bottom of the vessel. Here they spied a crumpled bit of metal, for all the world looking like a crushed beer can, which could have blocked the flow of sodium coolant.

It wasn't a beer can, but the operators could not see clearly enough to identify it. The periscope had fifteen optical relay lenses, would cloud up and take a day to clean, was very hard to maneuver, and had to be operated from specially-built, locked-air chambers to avoid radiation. To turn the metal over to examine it required the use of another complex, snake-like tool operated 35 feet from the base of the reactor. The operators managed to get a grip on the metal, and after an hour and a half it was removed.

The crumpled bit of metal turned out to be one of five triangular pieces of zirconium that had been installed as a safety device at the insistence of the Advisory Reactor Safety Committee, a prestigious group of nuclear experts who advise the NRC. It wasn't even on the blueprints. The flow of sodium coolant had ripped it loose. Moving about, it soon took a position that blocked the flow of coolant, causing the melting of the fuel bundles.

During this time, and for many months afterwards, the reactor had to be constantly bathed in argon gas or nitrogen to make sure that the extremely volatile sodium coolant did not come into contact with any air or water; if it did, it would explode and could rupture the core. It was constantly monitored with Geiger counters by health physicists. Even loud noises had to be avoided. Though the reactor was subcritical, there was still a chance of a reactivity accident. Slowly the fuel assemblies were removed and cut into three pieces so they could be shipped out of the plant for burial. But first they had to be cooled off for months in spent-fuel pools—huge swimming pools of water, where the rods of uranium could not be placed too close to each other. Then they were placed in cylinders 9 feet in diameter weighing 18 tons each. These were designed to withstand a 30-foot fall and a 30-minute fire, so dangerous is the spent fuel. Leakage from the casks could kill children a half a mile away. It took three years to remove the poisonous materials from the plant and to seal the radioactive sodium up in steel drums for storage at the site (none of the six burial grounds in the country would take it) where it will have to be monitored for generations. The plant, incredibly enough, was re-commissioned some years later and operated at low power for a short time. It was finally permanently decommissioned after more troubles.

This account clearly illustrates some of the principles investigated in this book, as can be seen below:

1. The problem originated with a safety device. Indeed, installation of the device was prompted by concerns of a prestigious committee made up of nuclear scientists and engineers, many from the elite universities, responsible for advising the NRC on safety matters. They were worried about a fuel drop, and the sheets were part of the response.[32]

2. Poor design and negligent construction led to the accident. Though it did not start with diverse failures, it is hardly reassuring that the sheets were poorly secured and the force of the surging coolant not anticipated, and the addition left off the final drawings.

3. As in other accidents, some parties were to suggest operator error, when in fact there was no clear procedure to follow; nothing like this had been anticipated. R. L. Scott, in his account of the accident for *Nuclear*

Safety, hints that one of the major problems was a failure of the operators to scram the reactor immediately. But some of the technical papers that he cites in his article point out that there was insufficient information available for the operators to know what the danger was and what was going on.[33]

4. Finally, we should note once again that those attached to high-risk systems can be uncommonly cheerful about these system failures. Scott is pleased to point out in the NRC journal that the melted fuel resolidified only a short distance from the hot spot, and did not cause the melting of adjacent subassemblies. This should give us more confidence, presumably, in breeder reactors. He next tells us, "Much additional benefit was derived from the recovery operations . . . not the least of these was the experience gained by the personnel directly involved." We may be very happy that these personnel had their experience increased, but unhappy that most of Detroit had to be at risk to secure the gain. He goes on, "Many innovations are required to cope with the new and different problems that presented themselves."[34] As an example of positive thinking about this, he lists the number of changes subsequently made in the system, such as a provision for draining the radioactive sodium from the reactor vessel. One would hope that a serious accident would not be required to bring this matter to the attention of designers. Finally, he cheerfully concludes that "the Fermi fuel melting incident [sic] has been quite instructive, emphasizing the need for design provisions for inservice inspection and the desirability for a simple, rapid presentation of critical operating information to the operator, together with adequate procedures and precise criteria for operator action."

In our accident is our salvation.

The Fuel Cycle as a System

We have treated the plant as the unit of analysis, and generally will continue to do so as we investigate other high-risk systems. But nuclear power involves the whole "fuel cycle"—the sequence from mining uranium ore, processing it into fuel, burning it in reactors to boil water, and the disposing of the many kinds of wastes. All of these involve serious hazards. Indeed, while we will not discuss the waste problem in this book, it probably has a greater long-run catastrophic potential (if we include military wastes) than nuclear plant operation. Mining is probably responsible for more radiation-induced deaths than any other part of the cycle to date (for the waste problem will take much longer to reveal itself), although these deaths are generally not the result of system accidents. But system accidents do occur in the fuel processing stage. A quick look at this stage will suggest that the processing of dangerous materials is, as a rule, associated with system accidents. These accounts are included for another reason: to point out the trivial details that can have large consequences, and the lack of understanding still evident in a production process that is well beyond the research stage.

Thirteen accidents involving fabrication of fuel are described in a *Nuclear Safety* article.[35] Some appear to be due to carelessness or inadequate technology. For example, there is the spontaneous ignition of contaminated wastes that are unaccountably stored in cardboard cartons in a waste storage room. Part of the plutonium released from this fire was washed from the building by the fire hoses, contaminating the surrounding ground. In another case, plutonium-casting residues were placed in a plastic bag, and burned through. In another, a five-year-old filter "heavily loaded with plutonium dust" caught fire from the sparks of a welding torch.

Cleanup is difficult when radioactive materials are involved. In an explosion at the Oak Ridge National Laboratory on November 19, 1959, "buildings and nearby streets were contaminated by the air flow through open pipes and other cell wall penetrations." The streets had to be scraped up. But the author of the *Nuclear Safety* article is reassuring. He concludes that "in all plutonium incidents to date, only a small fraction of the plutonium involved was released."[36] That is like saying that in a war, only a small fraction of the bullets kill anyone.

A bit more revealing is another discussion of seven "criticality" accidents. If plutonium, which is exceedingly volatile and hard to machine or handle, experiences the proper conditions, it can attain a self-sustaining fission chain reaction. Criticality depends upon the quantity of the plutonium, the size, shape, and material of the vessel that holds it, the nature of any solvents or dilutants, and even adjacent material, which may reflect neutrons back into the plutonium. It is apparently hard to know when these conditions might be just right. In the seven critical accidents that occurred between 1958 and 1970, fifteen workers were reported as receiving significant degrees of irradiation (an average of 140 rems, while the current legal yearly maximum for nuclear personnel is 5 rems) and two more died within two days of an accident.[37]

The accidents reveal the highly interactive nature of the systems. In one case, two poorly working pumps were involved, along with a line that may have been plugged. In the attempt to free the line, a bubble of high pressure air was created, though no one knew it. This forced 40 liters of a solution up a 5-inch-diameter storage pipe and out into another vessel that just happened to have the proper dimensions for criticality, given this particular solution and its volume. In another case, a plug of uranium nitrate crystals was found in a line. Operators dissolved it with steam, but the liquid was then drained into some available bottles, which just happened to be identical with those used to store a much safer liquid. One of the bottles, now containing U-235, was poured into a makeup tank. After stirring was commenced, it blew up, knocking the operator to the floor. He managed to escape the building but died forty-nine horrible hours later. Two operators went in to drain the solution into safe containers, but in turning off the stirrer, apparently (and who can know with this technology) the change in geometry added enough reactivity to again produce criticality. The operators did not know this had happened because the alarm that would indicate the danger was still sounding from the first "excursion." These two men received dosages of from 60 to 100 rads. (A total of 50 rads is the exposure level needed to double the risks of genetic defects, and is the legal maximum accumulated dosage for nuclear workers over twenty-seven years of age.)

A government report, WASH-1192, documents 111 accidents involving unplanned release of radioactivity that exposed 317 people to excess radiation from a few to as many as 80,000 rads. These occurred between 1959 and 1970.[38] The *average* dose of workers at the West Valley Reprocessing Plant (now closed), near Buffalo, New York, run by the Getty Oil Company, was 6.7 rads in 1971 and 7.1 rads in 1972, well beyond the legal minimum dose.[39] These assaults upon personnel, most of which will not reveal their damage for two decades, are not considered in the statistics that show that nuclear power is "safe."

As we have just seen, the power generation phase of the nuclear cycle is not the only one prone to system accidents; the fuel processing and reprocessing systems are at risk also. Transformation processes in the nuclear fuel cycle seem to have an inherent degree of unpredictability. But two questions still remain. How frequent are system accidents, and could not management learn to at least prevent the minor failures that can occasionally come together to create a system accident? We can only guess at the first, but there is unmistakable evidence regarding the latter.

Can We Handle It?

I hope I have convinced you of the frequency of serious accidents or near accidents in nuclear plants, and the existence of system accidents in the above examples. How many system accidents there are is impossible to tell; the reports in *Nuclear Safety* are often not detailed enough to judge. One serious attempt to analyze accidents in terms of the multiplicity of failures and the variety of component failures supports the argument of this book. Morris and Engelken examined eight Loss of Coolant Accidents (LOCAs) in BWRs in a two-year period when there were only twenty-nine plants operating. They occurred in six different BWRs. The authors estimate there will be a LOCA for each two reactor years of operation.

They conclude that "No two of the incidents were initiated by a common system or component malfunction. . . . The reactor primary coolant was released during these transients through safety and relief valves that either operated prematurely or that operated correctly but failed to close."[40] So each accident was unique, and each involved, among other things, the failure of a key safety device. These could easily be system accidents. In their summary of the eight occurrences, they identify eight categories of failures (such as, valves lifting below the set point at which they are supposed to lift; valves failing to reseat; flooding of steam lines; isolation valves closing too soon; condensor malfunctions; violation of operating procedures). Each accident involved from two to four of these failures. In half of the eight accidents there were violations of operating procedures, but they always occurred in conjunction with at least two and as many as five other failures. Failures not only were spread over the eight categories, they were widespread for all vendors and manufacturers. In one sample of valves, 15 percent were thinner than the design specified. Deficient valves were found at twenty plants owned by fifteen different utilities and supplied by ten different suppliers.[41] It is from such minor failures that system accidents can grow; the uniqueness of the accidents and the multiplicity of failures in this survey of just one system suggests that system accidents are not all that rare.

Could not management prevent these failures? The authors' investigation of eight LOCAs led them to a broad indictment of management practices and to a further study. Frequently, "abnormal situations and incidents . . . have not been thoroughly investigated"; "minor abnormalities were often ignored or their implications not understood, and these sometimes led to more serious conditions." It is in minor abnormalities, we might interject, that the system accident is spawned. They note, "Of course, there is always a very strong incentive to keep the plant on line, i.e., producing power." Finding all this, they conducted a management appraisal program in the plants, and found that "even though a deliberate effort was not made to look for violations," there were seventy-five in only seven appraisals. Of these, eighteen were failures to test "vital safety equipment." "In summary, there has been a surprising lack of knowledge, understanding, and effort by some utility executives to discharge their own responsibilities and those imposed by the specific requirements of an Atomic Energy Commission license."[42]

Strong stuff. But that was in 1972, and the industry was young. Since then, there have been major accidents at Dresden, Browns Ferry, and TMI, and several critical reviews of performance. Yet in an NRC review of operating plants, conducted in 1980 as a result of the excoriating criticism of the NRC by the Kemeny Commission, little seems to have changed. In possibly the most dangerous industrial activity that humans have yet to engage in, the study described the twenty-one "below average" facilities in numbing, repetitious terms: inadequate technical staff, insufficient training, poor supervision, failure to follow procedures, radiation protection weaknesses, incomplete licensee event reports and failure to consider their implications, unmonitored and uncontrolled release of airborne radioactive material, noncompliance with quality assurance programs, inadequate control over liquid and solid radioactive waste, repetitive equipment problems, problems in management coordination and attention, inadequate fire protection, failure to meet commitments made to the NRC, "repetitive instances of system misalignments, impaired ECCS equipment operability and containment integrity," personnel overexposure, and longstanding and uncorrected design problems. Most of these items appeared several times.[43]

We are not told anything about the average plants; the above list refers to the 29 percent found to be below average. Let's take a plant studied by the NRC from May 1979 to May 1980 and rated as average—San Onofre, owned by Southern California Edison. It is a small plant, 436 megawatts, which has been in operation for thirteen years. (Actually, in those thirteen years it has operated at full power for only 8.8 years, or 68 percent of the time, which is above the industry average.) In 1980, it tied for first place in the newly established NRC category of having "especially significant mishaps" (serious incidents). It is one of the eight plants listed by the NRC as having he most serious weakening of steel, which could cause the core vessel to crack. Some of the recent problems: In November 1979, a nest of field mice (a sign of poor housekeeping) caused an electrical fire that shut the plant down for a week and cost $2 million.

From April 1980 to June 1981 it was shut down for steam generator repairs (a problem that plagues all power plants), costing $68 million; the repairs are good, at best, for five years. During the overhaul, seventy-three workers were overexposed to radiation (the NRC fined the utility $100,000 for that and $50,000 for additional violations and exposure to workers). Fifty truckloads of radioactive sand had to be removed from the ocean beach in front of the plant in May 1981. A fire in an auxiliary diesel generator shut the plant down for four weeks July 1981, costing $2.5 million in repairs. During this time there was an explosion in a radioactive gas holding tank with a release of 8.8 curies of radioactive Krypton gas to the atmosphere. In September of 1981 a failure in a voltage regulator was investigated and the company found inoperative valves in the ECCS—estimated, by the NRC, to have been inoperative since 1977. This resulted in a finding by the Commission of "deficiencies in management and procedure controls." The NRC estimated that the unit may become unsafe to operate by 1983 due to "embrittlement" problems that could crack the core vessel.

Much of this took place just after the evaluation that rated it as average, rather than before or during it. But even in 1980, while the evaluation was going on, there were thirty-seven safety-related failures they were required by law to report to the NRC and seven "especially significant mishaps."[44] If this is an average plant, those below average by the NRC's standards might give their neighbors cause for concern.

Well, if things have not improved much since the 1972 study of a few plants, judging from the NRC's 1980 evaluation, perhaps we shall do better with the plants being built and those about to come on-stream. The Diablo Canyon plant of Pacific Gas and Electric has been ready for a long time, and in 1980 the NRC also evaluated it, along with seventy-five others that were in various stages of construction. It rated Diablo Canyon as average (the highest rating given for those under construction). Yet the next year an engineer in the utility accidentally discovered that the required earthquake reinforcements of key equipment had been incorrectly installed, as we noted earlier, and then 111 other violations were found. Something similar had happened to a second unit being built at San Onofre, where the reactor was installed 180 degrees out of alignment, and it took Southern California Edison seven months to discover the error. To correct it they reversed the wiring in the control room; but it was not that simple at Diablo Canyon.

Conclusion

We have not had more serious accidents of the scope of Three Mile Island simply because we have not given them enough time to appear. But the ingredients for such accidents are there, and unless we are very lucky, one or more will appear in the next decade and breach containment. Large nuclear plants of 1,000 or so megawatts have not been operating very long—only about thirty-five to forty years of operating experience exists, and that constitutes "industrial infancy" for complicated, poorly understood transformation systems. There is ample evidence that problems abound in these large systems, and that they are different from the problems of the smaller units where we have a bit more experience. For all nuclear power plants, the steam generator and the core embrittlement problems are awesome. Small failures can interact and render inoperative the safety systems designed to prevent a steam generator failure from being catastrophic. Trivial events can place stress on the embrittled core in ways unimagined by designers. The sources of other errors and failures appear all too numerous, judging from the events covered in this chapter.

The catastrophic potential of nuclear plant accidents is acknowledged by all, but defense in depth is held by experts to reduce accident probabilities to nearly zero. Yet core containment, emergency cooling systems, and isolated siting all appear to be inadequate; all have been threatened. Nor can we have any confidence whatsoever that quality control in construction and maintenance is near the heroic levels necessary to make these dangerous systems safe. A long list of construction failures, coverups, threats, and sheer ineptitude plagues the industry. I have argued that construction problems are probably no worse than in most other industries, but that is no comfort; it has to be much better. Nor has the actual operation of nuclear plants appeared to be as far above normal industrial standards as would be required of such a dangerous undertaking. If anything, it is somewhat below industrial standards. These statements regarding construction, maintenance, and organizational management are based upon the reviews and statements of the Nuclear Regulatory Commission itself, including its chairman. Finally, a review of some of the serious accidents that have occurred reveals the complexity of the plants, the difficulty of recovery from minor accidents so that they will not become major ones, the unlikelihood that the industry will even learn from the accidents, and the sanguine and casual response of the industry and the NRC.

Nuclear Power as a High-Risk System

When the Kemeny Commission was writing its final report, its members debated two key issues at length: Are these plants different from other industrial plants, and thus need to be judged by different criteria; and if they are different, what kind of organization is required to run them safely? A group of pro-industry members of the panel argued first that the plants are not different, and the restrictions being considered were not necessary; they then argued that if there were unique dangers, the plants should be run on a paramilitary basis. This position frightened other Commission members and led them to ask if a peacetime economy really needed an authoritarian, dictatorial segment managing a system with such catastrophic potential. These are heady issues, going beyond shoddy construction, inept management, untried and hasty designs. We will discuss these issues at length after we have reviewed other high-risk systems such as nuclear weapons and DNA engineering.

Yet despite the glaring failures of the nuclear power industry, it is clear that its design, construction, and operating problems do not, in themselves, constitute the cause of system accidents. It is instead the potential for unexpected interactions of small failures in that system that makes it prone to the system accident. Some systems with catastrophic potential are not liable to these complex failures; their accidents have different, more mundane sources. Some highly interactive systems are without catastrophic potential. To tread our way through these complexities, we need some careful analysis and more precise terms and concepts. That is the task of the next chapter.

CHAPTER 3

Complexity, Coupling, and Catastrophe

To make a systematic examination of the world of high-risk systems, and to address problems of reorganization of systems, risk analysis, and public participation, we need to carefully define our terms. Not everything untoward that happens should be called an accident; to exclude many minor failures, we need an exact definition of an accident. Our key term, *system accident* or *normal accident,* needs to be defined as precisely as possible, and distinguished from more commonplace accidents. We will define it with the aid of two concepts used loosely so far, which now require definition and illustration: *complexity* and *coupling.* We also need to explain the terms *catastrophe* and *victim* as they are used in our analysis. Then we will be in a position to chart the world of organized systems, predicting which will be prone to system accidents and which will not. This chapter, drawing upon numerous examples, will construct an apparatus that will carry us through the systems discussed in the rest of the book. It constitutes a theory of systems, of their potential for failure and recovery from failure. As such, it is, I believe, unique in the literature on accidents and the literature on organizations.

Perhaps the most original aspect of the analysis is that it focuses on the properties of systems themselves, rather than on the errors that owners, designers, and operators make in running them. Conventional explanations for accidents use notions such as operator error; faulty design or equipment; lack of attention to safety features; lack of operating experience; inadequately trained personnel; failure to use the most advanced technology; systems that are too big, underfinanced, or poorly run. We have already encountered ample evidence of these problems causing accidents. But something more basic and important contributes to the failure of systems. The conventional explanations only speak of problems that are more or less inevitable, widespread, and common to all systems, and thus do not account for variations in the failure rate of different kinds of systems.

What is needed is an explanation based upon system characteristics. In Chapter 6, on marine transport, you will see how technological fixes at best make no difference, and can even make the situation worse. Chapter 8, on the space program, reveals how the best talent and organizational resources, while they certainly help, cannot overcome system accident potentials. Several chapters show how system-related production pressures defeat safety improvements. The accidents covered all chapters challenge the ready explanation of "operator error." This is not to say that the concepts and definitions of system characteristics presented here will solve all analysis problems. They are preliminary; definitional problems remain. This is a first attempt at a "structural" analysis of risky systems, but I believe it goes substantially beyond conventional "item" analyses, and yields, as we shall see in the last chapter, long-run strategies for handling risks—rather than short-run tactics of posting safety notices or levying trivial fines.

Defining Accidents

What do we mean by an *accident?* At the minimum, an accident is an unintended and untoward event. If you are driving home and take a wrong turn and lengthen your trip, it is both unintended and unfortunate. But we generally mean something more serious than this by the term *accident.* You would not, upon arriving home late, announce, "I had an accident on the way home." If you did, you would be worriedly asked, "Was anyone hurt?" or, "Was the car damaged?"

An accident, then, involves some damage to people, objects, or to both. But uncertainties remain. Suppose you scratched the paint of the car on a post while leaving a parking lot. You probably would not call it an accident (even though it was accidental) if it didn't interrupt the trip or impair the functioning of either the car or the post. We might, then, say that the damage to objects or people must be sufficient to disrupt the ongoing "task" or future tasks that will be demanded of the objects or people.

This introduces a further complication: tasks analysis. What task is involved? That depends on what we want to call the system. If I had planned to take the car to a rally the next day and show it off to other automobile buffs, I might very well call the scratch in the parking lot an accident. The system, from my point of view, involves my going to a rally to meet people, impress people, and show off my car, and it is interrupted. If the post was bent over so much that cars trying to enter the lot were endangered by what I had done, that too might qualify the event as an accident. Drivers would notify the maintenance people, "There has been an accident, the post is knocked over so far that I can hardly get into the lot." It is part of the parking system to have a post separating lanes; it is part of my recreational system to show off a beautifully maintained and polished car. The event has disturbed these systems.

But there are also degrees of disturbance to systems. The rally will not be disturbed in any perceptible way if I show up with a scratch on my '57 De Soto, or if, ashamed of the state of my car, I do not go at all. But *my* system might be greatly disturbed if I stayed home rather than meeting or impressing people. The degree of disturbance, then, is related to what we define as the system. If the rally is the system under analysis, there is no accident. If that part of my life concerned with custom cars is the system, there *is* an accident. A steam generator tube failure in a nuclear plant can hardly be anything other than an accident for that plant and for that utility. Yet it may or may not have an appreciable effect upon the nuclear power "system" in the United States.

So far we have defined an accident as unintended damage to people or objects that affects the functioning of the system we choose to analyze. But we can also affect system functioning by "damaging" symbols, communication patterns, legitimacy, or a number of factors that are not, strictly speaking, people or objects. This observation will become important if we are to consider such organizations as universities.

An accident, then, involves damage to a defined system that disrupts the ongoing or future output of that system. But not all such disruptions should be classified as accidents; the damage must be reasonably substantial. In a nuclear plant, the momentary failure of the off-site power supply (the electricity coming into the plant to run its machinery) usually means that the reactor must be shut down. Similarly, a valve failure may result in a shutdown. But we will not call these *accidents,* even though they were not desired and have unfortunate or untoward consequences—purchasing replacement power alone may cost a few hundred thousand dollars before the reactor is restarted and the plant placed back on-line. We will call them *incidents.* Though the reactor is shut down, it does not sustain any damage. We need, then, criteria, inevitably somewhat arbitrary and rough, to distinguish between "minor" events such as these and accidents. Furthermore, we need a scheme that can apply equally to a steam generator as a system, or to a plant, or to the nuclear power industry.

I propose that the system be divided into four levels. Disruption to the third or fourth levels will be called *accidents* and disruption to the first and second will be called *incidents.* It is a scheme that need not be applied carefully in all parts of this book; sometimes we will ignore it and speak quite loosely of accidents when the meaning is obvious. But it is important for selecting accidents for analysis, for understanding safety devices, and for risk analysis; at times it will be crucial.

Consider a nuclear plant as the system. A *part* will be the first level—say a valve. This is the smallest component of the system that is likely to be identified in analyzing an accident. A functionally related collection of parts, as, for example, those that make up the steam generator, will be called a *unit,* the second level. An array of units, such as the steam generator and the water return system that includes the condensate polishers and associated motors, pumps, and piping, will make up a *subsystem,* in this case the secondary cooling system. This is the third level. A nuclear plant has around two dozen subsystems under this rough scheme. They all come together in the fourth level, the nuclear plant or system. Beyond this is the environment.

With this scheme we reserve the term *accident* for serious matters, that is, those affecting the third or fourth levels; we use the term *incident* for disruptions at the first or second level. The transition between incidents and accidents is the nexus where most of the engineered safety features (ESFs) come into play—the redundant components that may be activated; the emergency shut-offs; the emergency suppressors, such as core spray; or emergency supplies, such as emergency feedwater pumps. The scheme has its ambiguities, since one could argue interminably over the dividing line between part, unit, and subsystem, but it is flexible and adequate for our purposes. It must be flexible because at times we might want to consider, for example, the Apollo rocket and modules as a system; at other times we might want to consider all moon shots as a system. What is an accident in the first might be a mere incident in the second.

We are now ready for a formal definition. An *accident* is a failure in a subsystem, or the system as a whole, that damages more than one unit and in doing so disrupts the ongoing or future output of the system. An *incident* involves damage that is limited to parts or a unit, whether the failure disrupts the system or not. By disrupt we mean the output ceases or decreases to the extent that prompt repairs will be required. Since we have drawn a dividing line between the unit and the subsystem, and since many of the ESFs are clustered around that dividing line, it will often mean that an ESF will be one of the components that fails.

Victims

Note that, in contrast to common usage, we have not explicitly made damage to humans a part of our definition. This is because we are primarily interested in systems and the way they work. Humans are part of all the systems considered in this book, and a group of humans (the flight crew on an airliner) or a single human (the astronaut in a space capsule) may constitute a subsystem. Damage to them will mean that we have an accident.

But it is important for analysis to treat humans in most systems as mere parts. We kill about 5,000 people a year outright in U.S. industry. The vast majority of these "accidents," however, are only "incidents" in our scheme, for no subsystem or system damage is entailed. Only a "part" has been destroyed.

I am aware that this sounds like a quite heartless scheme, but while the ultimate concern of this book is reducing actual and potential damage to humans, I believe it is the character of systems that cause that damage, and thus we need a definition of accidents that focuses upon system characteristics. The failure of what we have defined as parts and units plays a vital role in subsystem and system failures, but if the analysis is limited to these we will lose our focus upon the kinds of systems that business and government leaders decide should be built. Our analysis, then, must be of systems, and falling off a ladder becomes a mere incident.

More important, we are concerned about those systems that have catastrophic potential—can cause damage to a great many humans. Our concern here is not the blue-collar worker who backs into a grinding machine, or the scientist who accidentally creates the proper criticality conditions for the plutonium he is experimenting with and is fatally irradiated. More or less mundane precautions and training will reduce the frequency of such accidents; fairly minimal attention to mine safety greatly reduced the number of coal mining deaths in a few years. Our concerns are more cosmic. To bring them into focus, we will turn to a victim classification scheme that will be used throughout this book.

Most of the work concerned with safety and accidents deals, rightly enough, with what I call first-party victims, and to some extent second-party victims. But in this book we are concerned with third- and fourth-party victims. Briefly, first-party victims are the operators; second-party victims are nonoperating personnel or system users such as passengers on a ship; third-party victims are innocent bystanders; fourth-party victims are fetuses and future generations. Generally, as we move from operators to future generations, the number of persons involved rises geometrically, risky activities are less well compensated, and the risks taken are increasingly unknown ones. We will take a close look at each of the four classes.

First-party victims are the operators of the system. In this book operators will include not only those actually running the system (nuclear plant operators, pilots, ship officers) but others in attendance on regular shifts, such as first-level supervisors, maintenance personnel, low-level engineering personnel, and laborers and assisting personnel. Most industrial accidents involve operators; in fact, most involve only one operator. Most industrial accidents are attributed to "operator error" or "human error" by those who study and seek to prevent such accidents. There is a growing recognition, however, that this is a great oversimplification; worse, it involves blaming the victim. It also suggests an unwitting—or perhaps conscious—class bias; many jobs, for example, require that the operator ignore safety precautions if she is to produce enough to keep her job, but when she is killed or maimed, it is considered her fault. Some operators are compensated for the risks they run by receiving higher pay than the skill level would suggest, but in industry as a whole there is no clear relationship between risk and pay.[1] There is no impersonal, fair market that rewards those that risk their lives with higher wages. Indeed, "jumpers" or the "glow boys" in the nuclear industry, temporary help who dash into a radioactive area to make repairs, will be hired for two or three weeks' work, at only six dollars an hour, even though they can receive high doses of radiation in the few minutes they are near the core.[2] There is no evidence of compensation for the long-term effects of toxic chemicals or contaminated atmospheres. Textile workers are not compensated for brown lung disease, nor are chemical plant workers compensated for cancer showing up ten or twenty years after exposure.

Second-party victims are those associated with the system as suppliers or users, but without influence over it. They are not innocent bystanders (third-party victims), because they are aware (or could be informed) of their exposure, even though such exposure may not be entirely voluntary. The largest class of second-party victims are passengers on ships, trains, airlines, cars, and busses. To some extent they "choose" to participate in the system, and thus elect to share at least some of the risk. If I accept a ride home from a party with an inebriated driver, I accept the risk. The difference between first- and second-party victims with regard to voluntary exposure may be slight when we compare an unemployed person who has no choice but to accept risky work, and an employee who must travel extensively to keep his job. Neither has realistic options to participating in the system. But there are less ambiguous examples. Comparing second-party victims to innocent bystanders illustrates the slightly voluntary nature of the former: we feel differently about airline passengers killed in a crash than we do about innocent bystanders on the ground killed by the airliner's crash. The former accepted the risk of flying, the latter did not.

There are other types of second-party victims. For example, consider the office personnel hurt by a refinery explosion or the truck driver delivering goods for his company who just happens to be on the site when the explosion occurs. These are the voluntary actions of people who elect to participate in a system but have no influence over its operation. Nevertheless, without these second party or system-associated participants there would be no system. The refinery could not run. Like passengers, they are participants in the system and aware of some risks.

Third-party victims, innocent bystanders, have no such involvement in the system. I have heard some nuclear proponents argue, in a conference on establishing safety goals for nuclear power plants, that people can choose not to live near a nuclear plant (or choose to avoid going to events at a stadium that is in the flight path of an O'Hare Airport runway). But I think these arguments can be dismissed. Nuclear plants exist near all densely populated areas in the United States; there is simply no practical means to avoid being within 50 miles of one. Even if one could, a severe core meltdown with breach of containment under the proper weather conditions could contaminate areas as large as the whole Northeast. The plutonium that might have crashed on Madagascar during the *Apollo 13* emergency reentry of the atmosphere would have contaminated innocent bystanders in a distant part of the world that one would have thought safe from such high-technology disasters. People are aware of risks associated with flying; but I suspect that most people living downstream from large dams do not realize the dam is near enough to endanger their lives in the case of a failure.

Fourth-party victims are, for the most part, victims of radiation and toxic chemicals. They are fetuses being carried at the time of exposure; the would-be children that damaged parents will not be able to conceive; stillborn or deformed children conceived after exposure; and all those people who will be contaminated in the future by residual substances, including those substances that will become concentrated as they move up the food chain. Note that we do not include run-of-the-mill pollution here. Nuclear plants may regularly give off radiation in doses large enough to affect fetuses or future reproductive capacity, as a few scientists argue (though most deny); but these consequences are not the result of accidents. (Although one might argue that such radiation is a design failure that should, but doesn't, interrupt the output.) Nor are the pollutants from industrial plants of concern here, unless they are released in an accident. This routine and mostly conscious contamination of the planet probably has more serious long-term consequences than any accident we shall consider in this book, other than military accidents, but it is beyond the scope of our concern.

The importance of fourth-party victims in risk analysis is growing with the increasing concern and sophistication about the long-term consequences of some systems. Yet recognition is slow. A draft version of the NRC's "Safety Goals for Nuclear Power Plants" indicated that while the NRC was aware of "inter-generational" risks from a nuclear power plant accident through genetic effects or long-term contamination, "we cannot suggest a good way to handle the issue in a safety-goal context." So they ignored it.[3] If the risks of an accident are kept low enough, they said, there will be no problem with ignoring inter-generational effects. This conclusion answers the question about consequences of accidents by saying they will be trivial because there will be so few accidents.

Fourth-party victims potentially constitute the most serious class of victims. Chemical or radioactive contamination of land areas could have far-reaching effects upon the health of future generations. Genetic defects harm future generations in other ways, including adding the burden of lifetime care and treatment of victims. Future generations carry the burden; the present generation reaps whatever rewards there may be from the activity.

These issues are comparatively new—less than one generation old. Some influential scientists and academics, at the workshop that tried to formulate safety goals for the NRC, argued that the present generation is more important than future ones—for we need nuclear power to prevent economic and political crises—and who knows, there may be a technological fix that would mitigate the burden for the future generations of an accident in the present one. Thus at least some of the "experts" do not see the issue as a clear-cut one of our responsibility to future generations. Other experts, however, believe we have a responsibility to pass on to our offspring a world that is at least no more contaminated and degraded than the one we inherited.

Accident Definitions

We now have a definition of an accident, distinguishing it from incidents on the basis of levels in the system. The following list presents these and adds the definition of the two types of accidents: component failure accidents and system accidents, which we will now take up.

Systems are divided into four levels of increasing aggregation: units, parts, subsystems, and system.

Incidents involved damage to or failures of parts or a unit only, even though the failure may stop the output of the system or affect it to the extent that it must be stopped.

Accidents involve damage to subsystems or the system as a whole, stopping the intended output or affecting it to the extent that it must be halted promptly.

Component failure accidents involve one or more component failures (part, unit, or subsystem) that are linked in an anticipated sequence.

System accidents involve the unanticipated interaction of multiple failures.

Component failure accidents and system accidents are distinguished on the basis of whether any interaction of two or more failures is anticipated, expected, or comprehensible to the persons who designed the system, and those who are adequately trained to operate it. A system accident, in our definition, must have multiple failures, and they are likely to be in reasonably independent units or subsystems. But system accidents, as with all accidents, start with a component failure, most commonly the failure of a part, say a valve or an operator error. It is not the *source* of the accident that distinguishes the two types, since both start with component failures; it is the presence or not of multiple failures that interact in unanticipated ways.

The vast majority of component failure accidents involve a series of failures. If a valve fails there is good chance that the pump will overheat and fail, and if this happens, the boiler is quite likely to overheat because the coolant is insufficient. Designers know this, and so do operators—though they may not be able to prevent it, or intervene in the series of failures. There will be some accidents, however, where the initial failure is so drastic that it is not worth tracing out the subsequent sequence, if there is one. If the wing comes off an airplane in flight, or an earthquake shatters a dam, we hardly need further analysis. These might be called "final accidents"; there is no possible intervention by operators, and no point in detailing sequences. We will not deal with any examples of these in this book; for one thing they are fairly rare, but more important, analysis of the system itself does little to contribute to understanding these accidents.

Incidents are overwhelmingly the most common untoward system events. Accidents are far less frequent. Among accidents, component failure accidents are far more frequent than system accidents. I have no reliable way to estimate these frequencies. For the systems analyzed in this book the richest body of data comes from the safety-related failures that nuclear plants in the United States are required to report. Roughly 3,000 Licensee Event Reports are filed each year by the 70 or so plants. Based upon the literature discussing these reports, I estimate 300 of the 3,000 events might be called accidents; 15 to 30 of these might be system accidents. So far, at least as far as we know, all the system and component failure accidents in nuclear plants have had only first-party victims, and very few of these.

Complex and Linear Interactions

What kinds of systems are most prone to system accidents? In the last chapter we kept mentioning two concepts: interactiveness, which could confuse operators, and tight coupling, which could prevent speedy recovery from an incident. With a more precise definition of these two terms, we can classify systems and be forewarned about those that are most prone to system accidents. We will take up interactions first.

The notion of baffling interactions is increasingly familiar to all of us. It characterizes our social and political world as well as our technological and industrial world. As systems grow in size and in the number of diverse functions they serve, and are built to function in ever more hostile environments, increasing their ties to other systems, they experience more and more incomprehensible or unexpected interactions. They become more vulnerable to unavoidable system accidents.

Interactiveness per se, though, is not a useful concept. Almost any organization of any size, whether public or private, will have many parts that interact once we look closely at them. The existence of many parts is no great trouble for either system designers or system operators if their interactions are expected and obvious. If a part or a unit fails in an assembly line, it is quite clear what will happen to the parts and units "downstream" of the failure, and we know that the products "upstream" will start piling up fast. We can shut down the line and fix it, or bypass that work station and assemble the missing parts later on, or temporarily shunt the product into a storage bin. These are "linear" interactions: production is carried out through a series or sequence of steps laid out in a line. It doesn't matter much whether there are 1,000 or 1,000,000 parts in the line. It is easy to spot the failure and we know what its effect will be on the adjacent stations. There will be product accumulating upstream and incomplete product going out downstream of the failure point. Most of our planned life is organized that way.

But what if parts, or units, or subsystems (that is, components) serve multiple functions? For example, a heater might both heat the gas in tank A and also be used as a heat exchanger to absorb excess heat from a chemical reactor. If the heater fails, tank A will be too cool for the recombination of gas molecules expected, and at the same time, the chemical reactor will overheat as the excess heat fails to be absorbed. This is a good design for a heater, because it saves energy. But the interactions are no longer linear. The heater has what engineers call a "common-mode" function—it services two other components, and if it fails, both of those "modes" (heating the tank, cooling the reactor) fail. This begins to get more complex.

This source of complexity was slow to gain recognition in the nuclear power field. The first analytical model to consider common mode failures appeared only in 1967, according to a perceptive and disturbing article by one of the editors of *Nuclear Safety,* E. W. Hagen.[4]

The monumental reactor safety study issued in 1975, WASH-1400 or the Rasmussen Report, was criticized by a subsequent NRC study for giving insufficient and overly simplified attention to this problem. Hagen concludes that potential common-mode failures are "the result of adding complexity to system designs." Ironically, in many cases, the complexity is added to reduce common-mode failures. The addition of redundant components has been the main line of defense, but, as Hagen illustrates, also the main source of the failures. "To date, all proposed 'fixes' are for more of the same—more components and more complexity in system design."[5] The Rasmussen safety study relied upon a "PRA" (probabilistic risk analysis), finding that core melts and the like were virtually impossible. Hagen notes (p. 191) that PRAs, "using established techniques of reliability and statistical analysis," constitute the main source of public reassurance about such potential dangers. But, he notes, this involves a "narrow definition" of common-mode failures, and the analysts "are busily working in an area which has not been shown to be the main problem." The main problem is complexity itself, Hagen argues. With that we can agree.

Common-mode failures are just one indication of interactiveness in systems. Proximity and indirect information sources are two others. For a graphic illustration of complexity in the form of unanticipated interactions from these sources, let us go to a different system, marine transport. A tanker, the *Dauntless Colocotronis,* traveling up the Mississippi River near New Orleans, grazed the top of a submerged wreck. The wreck had been improperly located on some of the charts. Furthermore, it was closer to the surface than the charts indicated because its depth had been determined when the channel was deep, and the listing for mariners had not been corrected for those seasons when the river would be lower. The wreck, only a foot or so too high for the tanker, sliced a wound in the bottom of the tanker, and the oil began to seep out. Unfortunately, the gash occurred just at the point where the tank adjoined the pump room. Some of the oil seeped into the pump room. At first it was probably slow-moving, but the heat in the room made it less viscous, allowing it to flow more rapidly, drawing more oil into the pump room. When enough accumulated it reached a packing "gland" around a shaft near the floor that penetrated into the engine room next to it. The oil now leaked into the engine room. In the hot engine room, evaporation was rapid, creating an explosive gas. There is always a spark in an engine room, from motors and even engine parts striking one another. (In fact, nylon rope can create sparks sufficient to cause explosions in tanker holds.) When enough gas was produced through evaporation, it ignited, causing an explosion and fire.

In this example, an unanticipated connection between two independent, unrelated subsystems that happened to be in close proximity caused an interaction that was certainly not a planned, expected, linear one. The operators of the system had no way of knowing that the very slight jar to the ship made a gash that would supply flammable or explosive substances to the pump and engine rooms; nor of knowing, until several minutes had passed, that there was a fire in the engine room; or of the extensive amount of oil involved. They tried to put it out with water hoses, perhaps not realizing it was an oil fire, but the water merely spread the oil and broke it up into finer, more flammable particles.

This accident went on for several hours, because the crew made various mistakes (e.g., a key fire door had been tied open and was not closed by an escaping crew member, allowing the fire to spread), and they did not have a clear notion of the location of the many rooms and closets and passageways in that part of the ship. Late in the accident a trained fire crew boarded the ship equipped with protective devices and proper extinguishing equipment. But when they opened one door there was a series of explosions. They immediately closed the door and made no further attempt to put out the fire in that part of the ship, believing that the oil was producing explosive gases. Subsequently, however, it was determined that there were no explosive gases involved by this time; instead, three small empty gas tanks stored just inside the door, before being exchanged for full ones, had exploded when the heat expanded the residual amounts of freon, oxygen, and acetylene in them.[6]

Thus even in the recovery stage of the accident a nonlinear interaction intervened to mislead the recovery attempt. It was a sensible place to store the tanks; who could imagine that there might be a fire, that firefighters would not be certain that there were not explosive mixtures behind the closed doors, and that these tanks would go off just as the firemen were entering the passageway? A requirement that empty tanks be stored elsewhere would hardly make the ship safer; who could know where the next fire might be?

Recall the illustration that opened this book, linking three "subsystems": breakfast, getting to the appointment, and the job interview. In the world we plan out and think through, this seems like a very linear, straightforward problem—get some food and coffee, get the car, drive to the interview. One would expect the car keys to be linked to using the car, but one would not expect the failure of the coffeepot to be linked to using the car. One would also not expect that even if the car failed, the alternative of a taxi would be linked to a contract dispute, and the neighbor's car would be unavailable just that day. These represent interactions that were not in our original design of our world, and interactions that we as "operators" could not anticipate or reasonably guard against. What distinguishes these interactions is that they were not designed into the system by anybody; no one intended them to be linked. They baffle us because we acted in terms of our own designs of a world that we expected to exist—but thc world was different.

I will refer to these kinds of interactions as *complex interactions,* suggesting that there are branching paths, feedback loops, jumps from one linear sequence to another because of proximity and certain other features we will explore shortly. The connections are not only adjacent, serial ones, but can multiply as other parts or units or subsystems are reached.

The much more common interactions, the kind we intuitively try to construct because of their simplicity and comprehensibility, I will call *linear interactions.* Linear interactions overwhelmingly predominate in all systems. But even the most linear of systems will have at least one source of complex interactions, the environment, since it impinges upon many parts or units in the system. The environment alone can constitute a source of failure that is common for many components—a common-mode accident. But even the most complex systems of any size will be primarily made up of linear, planned, visible interactions.

Based upon a general knowledge of various systems, considerable experience with reading accident reports, and some site visits, I suggest that only 1 percent of all possible parts or units in a linear system are capable of producing "complex" interactions, while about 10 percent of those in a complex system will be capable of doing so. But that 10 percent represents more than a tenfold increase in the potential for system accidents. The potential interactions produced by, say, four parts or units that are interrelated in a complex, rather than linear way, is twelve. (There are twelve possible paths between the four parts.) Suppose this exists in a system where there are 400 parts or units. If 10 percent of the units had such characteristics, rather than only 1 percent, there would be forty such parts or units. The potential complex interactions of each one of these with the remaining 399 would be in the millions, since the possible paths increase exponentially. Of course, in a large system some parts are so remote from each other that the chances of their interacting in unexpected ways can be disregarded; but many will not be so remote. Thus, increasing the proportion of possible complex interactions from 1 percent to 10 percent will have an enormous impact upon the potential for system accidents—given the fact that, since nothing is perfect, component failures are inevitable.

The possibility of many unintended interactions is recognized by designers, so they introduce buffers and other safety devices to prevent some kinds of interactions. Imagine a chemical plant where gas is expected to flow from tank A to tank B, because the pressure is kept higher in tank A than in tank B. Various things go into tank A besides the basic gas—reagents, purifiers, "inerting" gases, and so on. Then the gas flows to B, where additional substances are piped in to alter that mixture even further. This is quite straightforward and linear.

But there is a danger that it might become nonlinear. A feedback loop could occur if pressure declined in tank A because of a failure there or elsewhere, and the now-modified gas in tank B flowed back into A. This could create problems; it might even be dangerous. The engineers are aware of this, so they design a butterfly valve to be placed between the two tanks to prevent reverse flow. An engineered safety device (ESD) is installed to keep the system as linear as possible (in this case, flowing in one direction). But valves can fail, especially rarely used ones. If the pressure in tank A were to fall and the butterfly valve to fail, the feedback loop would open again, and the operators might not expect the interaction. Actually, this example is so simple that most operators would take the unplanned interaction into account as a possibility.

For a more complicated example involving quite remote and independent units, recall the case in the last chapter of the demineralized water needed to clean up the containment room in the nuclear plant. A faucet (valve) was opened to see if the water had been sent to the men, and a complex series of backflows and pressure adjustments occurred that allowed radioactive materials to enter containment. There was nothing very linear about that, to the operators in containment, the operators in the control room, or the engineers who designed the system. Nor did anyone make an error or even a mistake in design. Subsequently they could figure out what had happened and introduce measures to prevent a recurrence, but that meant altering an unexpected and unplanned interaction between two normally independent subsystems, core maintenance and steam generation. As I suggested, there are probably a number of other complex interactions ready to be revealed by a casual event in that very complicated piping system.

Linear interactions are overwhelmingly those of an "expected production sequence"—this is the way the system is designed to run, and anyone operating it will know that. Complex interactions will generally be those not intended in the design. No designer of the *Dauntless Colocotronis* said, "Let's put the number 5 tank next to the pump room so they can interact." There is just no way to isolate all tanks in a tanker from a room that generates sparks. Or the nonlinear interactions may be intended but rarely activated, and thus operators or designers forget about them. It could have been foreseen by a designer that demineralized water might sometimes be needed in containment, so she could have made it possible to line up various valves to provide the water. But if it is rarely used or is usually lined up before a crew enters containment, the faucet creates no problems, it is not an expected production sequence but an infrequently used system possibility (in this case, for maintenance, not production). Thus, complex interactions may be unintended ones, or intended but unfamiliar ones.

While linear interactions occur overwhelmingly in an anticipated production sequence, there is another kind of interaction that does not occur in a production sequence but is nevertheless obvious, and thus can be defended against. This is a visible interaction, even though it is outside of the normal sequence. If the operator of a crane sees that the part of the crane that holds up the load (cable, ratchet, motor, hook) has failed and the load is about to drop on a boiler on the floor, he knows what this interaction will produce. There is nothing mysterious about the connection between these events, though they are certainly not in any expected production sequence. Knowing there is a remote possibility of a large load falling, the operator usually tries to avoid passing it over the boiler, but he cannot always do so. Similarly, the designer may have considered covering the boiler or moving it, but estimated that the problems involved were too great, given the remote possibility of a large load falling at just that point.

To summarize our work so far: *Linear interactions* are those interactions of one component in the DEPOSE system (Design, Equipment, Procedures, Operators, Supplies and materials, and Environment) with one or more components that precede or follow it immediately in the sequence of production. *Complex interactions* are those in which one component can interact with one or more other components outside of the normal production sequence, either by design or not by design.

To elaborate on the properties of these interactions as they affect the operators:

Linear interactions are those in expected and familiar production or maintenance sequence, and those that are quite visible even if unplanned.

Complex interactions are those of unfamiliar sequences, or unplanned and unexpected sequences, and either not visible or not immediately comprehensible.

A note on the terms "complex" and "linear" is in order. It is difficult to find precise terms that are also brief; I have opted for brevity. "Complex" should read "interactions in an unexpected sequence"; "linear" should read "interactions in an expected sequence." One problem with the terms *complex* and *linear* is that the opposite of complex is "simple," and the opposite of linear is "nonlinear." Linear interactions are "simple" in the sense that they are easily comprehensible, but "simple" implies unsophisticated processes and technologies, or systems with few parts and units or routine and uncomplicated operation and maintenance. But the production of, say, pharmaceuticals or F-16 fighter planes is anything but simple, though it is linear. On the other hand, "nonlinear" does not readily convey the notion of possible incomprehensibility, while the term *complex* does. I will occasionally slip in such phrases as "complexly interactive" and "linear sequences" to remind the reader that neither of the chosen terms is really satisfactory.

Another warning is in order. Since linear interactions predominate in all systems, and even the most linear systems can occasionally have complex interactions, systems must be characterized in terms of the degree of either quality. It is not a matter of dichotomies. Furthermore, systems are not linear or complex, strictly speaking, only their interactions are. Even here we must recall that linear systems have very few complex interactions, while complex ones have more than linear ones, but complex interactions are still few in number.

Finally, the reader should not equate the notion of a linear system with the physical layout of the plant or production process. It does not necessarily imply an assembly line, although these production systems tend to be linear. Nor does the designation "complex system" necessarily imply highly sophisticated technology, numerous components, or many stages of production. We will characterize universities as complex systems, but as most large-scale organizations go, they have few of the above characteristics.

In order to understand systems, we need to go beyond the distinction between the two types of interactions. By examining, in the next section, how systems cope with hidden interactions, we will explore the basic attributes of complex systems, and be able to characterize them more systematically.

Coping with Hidden Interactions

Linear systems also have interactions that are not visible, of course, but they occur within well-defined and segregated segments of the production or maintenance sequence. Controls, such as dials, warning lights, audible alarms, and switches read the presence of these interactions and inform the operators and allow them to intervene. In systems with a fair degree of complex interactions, however, well-defined and segregated segments do not necessarily exist. Instead, jiggling unit D may well affect not only the next unit, E, but A and H also. This increases the number of controls that must be installed and monitored. The control panel of a nuclear plant is considerably denser and larger than that of an oil or coal plant, in part because so many components are linked in branching paths and feedback loops.

Attempts are continually made to reduce the number of controls by automating the subsidiary interactions and leaving only the main parameters for the operators to worry about. But this decreases the system's flexibility; the operator loses the ability to correct a minor failure in a part rather than shutting down a whole unit or subsystem. The operator cannot exit from the high-level, summary controls to the low-level specific one required to deal with a single part. Larry Hirschhorn provides an instructive example in a perceptive article on the limits of "cybernetic" (self-correcting) systems, which I shall elaborate on considerably.[7]

Consider a piston engine and a fuel pump. If particles get into a cylinder, perhaps during a maintenance procedure, or from wear from some other part, they can cause the piston to move more sluggishly. The result is that there is less output from this part of the engine. A control device monitors the output (but not the many causes of reduced output, such as foreign particles, piston wear, low octane fuel, over- or under-heating).

Noting the reduced power, the automatic control calls for more fuel from the fuel pump. This does compensate for the new source of friction, and the machine functions adequately, though less efficiently. But suppose that a surge in power is called for from the engine. The fuel pump is asked to increase its output. Soon its limit is reached. A typical cybernetic monitor will now decide that the fuel pump has failed; it will not know that the requirements expected of a (now degraded) piston/cylinder part cannot be met. An engineered safety device comes on, sending fuel from another source to the cylinder.

Any number of things might happen at this point. If the first pump is left running (since it has not, in fact, failed at all) there will be too much fuel supplied, which might burst some fuel carriers, stall the engine with fuel flooding in, overpressurize the fuel supply so that it tries to flow backwards (an eventuality that was not conceived by designers, so no protection is provided), or cause a runaway engine that rapidly overheats and explodes.

If, instead of the above, power to the fuel pump is automatically shut off because it is believed to have failed, the pump may be shut off with the valves in the open position, rather than the closed position they would be in if the pump had indeed failed. This could create back-flow or other problems since the other systems have not been shut off. Designers might have anticipated that if an undamaged pump were shut off, it would be because of a power failure that would affect other units of the system, and thus no back-flow problems would occur. In this case, however, an undamaged pump loses its power but keeps its valve setting, while other related parts or units do not lose their power and thus keep operating.

After the failure, the pump would probably be replaced, since the monitor indicated that this is what had failed. But the problem would remain, since it resided in the piston; only the automatic cybernetic control system thought the problem was the pump. This elaborated example is fabricated, but it illustrates the difficulty control systems have in reading the nature and source of failures. In a system without this added complexity, an operator presented with this sequence would probably cut back on the power demands, reasoning that either the engine or the fuel supply was malfunctioning, check the pump by varying engine speed, and conclude that there was sluggishness in the engine, which in fact is the correct conclusion.

One limitation of this example is that there is no pressing reason to have automatic controls in such a simple system (though if the motor exists in connection with many other units or subsystems, it may be advisable). But in some systems automatic controls are necessary because there is simply not sufficient time for operator reaction. The nuclear reactors designed by Babcock and Wilcox are of this type. The reactor is extremely responsive, and this has important economic benefits. But some of its subsystems may be too responsive. With a loss of coolant, the operator may have only thirty to sixty seconds before the steam generator boils dry. Consider this accident at the Crystal River, Florida, plant in February 1980.

For an unknown reason a short circuit occurred in some of the controls in the control room. The utility said it could have been due to a bent connecting pin in the control panel, so sensitive are these devices, or the malfunction may have been caused by some maintenance work being done on an adjacent panel. The short circuit distorted some of the readings in the control system, in particular the important and very sensitive one of coolant temperature. The computer "thought" the coolant was growing too cold, so it speeded up the reaction in the core. (The Babcock and Wilcox reactor operates within a very efficient, but quite narrow temperature band.) The reactor overheated, the pressure in the core went up to the danger level, and then the reactor automatically shut down. The computer was apparently now puzzled, and correctly ordered the pressure relief valve (PORV) to open, but incorrectly ordered it to remain open until things settled down. This was an "error" on its part, because pressure dropped so quickly that it automatically caused high pressure injection to come on, and stay on, flooding the primary coolant loop—including the core, steam tubes, and pressurizer. A valve stuck and 43,000 gallons of radioactive water were dumped on the floor of the reactor building. Fortunately it was not worse; after several minutes an operator noticed the computer's error in keeping the relief valve open and closed the valves manually. Had the frequent injunction been followed that the computer knows best, and that the (dumb) operators should keep their hands off the system until its routines are carried out, the sump would have overflowed and we would indeed have had a wet reactor.[8]

Operators tend to resist the introduction of more general, high-level controls such as cathode ray tubes (CRTs), which produce a television screen display of the state of a number of units or subsystems, because they feel they cannot make selective interventions as easily. Only these high-level controls can be manipulated; the selective ones are more difficult to get to because it is assumed they will not be needed. They may be out of the way, or accessible only by a complicated series of steps that inactivate the more general controls. On the other hand, operators also complain about the arrangement of less automated systems that allow selective intervention, because they are confronted with 15-foot banks of identically designed switches with tiny numbers above them. These are not even grouped in any way that reflects operation, but only in the way that reflects ease of installation. One of the most common "operator errors" in nuclear plants is, understandably, operating the wrong switch. Surely the choice need not be reduced discretion versus endless, identical displays.[9]

Complicating the dilemma of too few vs. too many controls is the problem of uncertain "default" status. Default status is the normal status of a control; for example, you must choose to change a switch to "on"; by default it is "off." The position of a relief valve is normally closed. The temperature reading on a drain pipe is normally cool. The position of most block valves is normally open. One might decide to have the default status represented by a green light on the control panel—go ahead, there is no danger. But one also might decide to have green represent open valves or circuits that are open to current flow—they are "on," and on is the default status. One can readily see the problem: if something is normally closed, this status cannot be represented by the color green, since green also means it is on, or open. Nor can red be reserved for "danger"; if the reactor is supposed to be on, you may want a red signal to warn you that it *is* on, even though it is not dangerous, but normal under these circumstances. But then what would green mean? (I learned the hard way that on my personal computer closing a "dip switch" means to open it!)

Some switches and valves have to be on at certain times and off at others, so there is no default position. Some systems use colored lights to handle this. If operating mode A is in effect, then switch 1 should be on, and an amber light next to the switch indicates this. If operating mode B is in effect, the switch should be off. If the switch is in the wrong position for the operating mode, a red or blinking light appears. If that part of the system is not in operation, a green light may be on, regardless of the position of the switch, or no light may be on. But modes and switches are not always clearly linked; there may well be a mode A and a mode A1, and the correct switch positions for the two would be different. Operators sometimes disconnect these sophisticated indicators (or much more commonly, ignore them) because of such complications.

These problems exist in all industrial and transportation systems, but they are greatly magnified in systems with many complex interactions. This is because interactions, caused by proximity, common mode connections, or unfamiliar or unintended feedback loops, require many more probes of system conditions, and many more alterations of the conditions. Much more is simply invisible to the controller. The events go on inside vessels, or inside airplane wings, or in the space craft's service module, or inside computers. Complex systems tend to have elaborate control centers not because they make life easier for the operators, saving steps or time, nor because there is necessarily more machinery to control, but because components must interact in more than linear, sequential ways, and therefore may interact in unexpected ways.

In addition to there being many interactions to control, the information about the state of components or processes is more indirect and inferential in complex systems. The reactor core at Three Mile Island had no direct measure of coolant level. Given the high pressures and the flow of coolant in the core, such a measure would be very hard to provide, and would introduce more penetrations of the vessel, something to be avoided. So, operators had to estimate coolant level from indirect indicators. The presence of steam voids and hydrogen bubbles made even these unreliable.

To cite other examples, pilots in ships or airplanes can fix on the wrong star. Ships may find navigation beacons obscured by shore lights, distorted by refractions, or simply out. Routine fluctuations in pressure and temperature in chemical plants can mislead operators. Operators discount an abnormal reading, thinking it is part of the routine fluctuation. Generally they are correct; occasionally they are wrong. At TMI the operators knew the temperature on the hot leg of a drain pipe was abnormal because of a leak. During the accident, when the temperature exceeded even this abnormal reading, they treated it as a particularly wide fluctuation, whereas it really indicated an open relief valve.

In 1977 New York City experienced a massive and very costly blackout. One key contribution to the accident was an operator's expectation about the default reading for current flowing over a particular line. Normally that line carried little or no current. The operator did not know there had been two relay failures—one that would automatically lead to a high flow of current over that line; and a second that blocked the flow over the line. The operator treated the zero current reading as normal. In fact, it was abnormal, but only in this particular set of circumstances. This ambiguity led to a systematic, by-the-book sequence of actions to handle the problems that were showing up in other parts of the system, ending in the system being brought to a halt.[10] The only evidence the operator could really "see," in terms of sensory confirmation, was the lights going out in his control room.

In a typical (linear) manufacturing plant both errors and interactions are more visible. If an operator misunderstands an instruction the supervisor will probably know it soon enough because she will see the operator set up for the wrong task. I am told that one of the advantages of old-style steam valves, where the stem of the valve will rise when opened, is that a quick glance by an operator or supervisor over a huge room will show which valves are open, and which are shut—they just stick up when open. If an employee heads for the wrong valve after misunderstanding an order, that will be quite visible too. In complex systems, where not even a tip of an iceberg is visible, the communication must be exact, the dial correct, the switch position obvious, the reading direct and "on-line."

The problem of indirect or inferential information sources is compounded by the lack of redundancy available to complex systems. If we stopped to notice, we would observe that our daily life is full of missed or misunderstood signals and faulty information. A great deal of our speech is devoted to redundancy—saying the same thing over and over, or repeating it in a slightly different way. We know from experience that the person we are talking to may be in a different cognitive framework, framing our remarks to "hear" that which he expects to hear, not what he is being told. The listener suppresses such words as "not" or "no" because he doesn't expect to hear them. Indeed, he does not "hear" them, in the literal sense of processing in his brain the sounds that enter the ear. All sorts of trivial misunderstandings, and some quite serious ones, occur in normal conversation. We should not be surprised, then, if ambiguous or indirect information sources in complex systems are subject to misinterpretation.

Transformation Processes

As we gain more experience with systems, and design them more effectively, the high degree of interactiveness may be reduced. In Chapter 5 we shall examine the case of air traffic control, where exactly this appears to have happened. It is also true that a poorly trained or inexperienced operator may see a system as replete with unsuspected interactions or "traps," but after gaining experience may find it to be more linear. Finally, new technological breakthroughs may bring linearity into a once complexly interactive system, as when the jet engine replaced the piston engine, or the transistor replaced the vacuum tube. Though there is a tendency to then demand more of the redesigned system, and thus build in new interactions, as has been dramatically recounted in the case of a jet fighter,[11] we should emphasize that new or improved designs and more operator experience may reduce the possibilities of unexpected interactions. Some we shall examine in this book are aspects of marine transport, the air traffic control system, dams, and mining. But our major concern is with systems that appear to be irretrievably complex, or at least will remain so for the next few decades. Generally these are systems that transform their raw materials, rather than fabricate or assemble them.

Transformation processes exist in recombinant DNA technology, chemical plants, nuclear power production, nuclear weapons, and some aspects of space missions. Most of these are quite new, but it is significant that chemical processing is not. While experience has helped reduce accidents, accidents continue to plague transformation processes that are fifty years old. These are processes that can be described, but not really understood. They were often discovered through trial and error, and what passes for understanding is really only a description of something that works. Some of industrial chemical production is of this nature; much of iron and steel production was of this nature, although it has been greatly altered through scientific research.

The existence of transformation processes without full understanding certainly characterizes nuclear power; recall that the nuclear scientist who advised Governor Thornburg of Pennsylvania during the accident had asserted three years before in a scientific journal that there could be no problem with a zirconium-water reaction—the process was well understood. Yet precisely this problem produced the hydrogen bubble. Each space mission introduces new systems that may be unreliable because of lack of knowledge. Experience will increase understanding of some of these problems (our rockets do not blow up on the launch pad as readily as they once did), but the performance of space vehicles is still subject to much uncertainty. Recombinant DNA research is fraught with gaps in knowledge. Limited knowledge, then, allows unsuspected interactions, and requires many control parameters and indirect sources of information.

This completes our discussion of the attributes of systems with complex interactions. To summarize, complex systems are characterized by:

- Proximity of parts or units that are not in a production sequence;
- many common mode connections between components (parts, units, or subsystems) not in a production sequence;

- unfamiliar or unintended feedback loops;
- many control parameters with potential interactions;
- indirect or inferential information sources; and
- limited understanding of some processes.

Complex systems are not necessarily high-risk systems with catastrophic potential; universities, research and development firms, and some government bureaucracies are complex systems, as I will argue shortly. We can now briefly contrast complex and linear systems.

Linear Systems

The parts or units of linear systems that are not in a direct production sequence tend to be spatially spread out. Fabrication or assembly allow this, while transformation processes often have to be compact. Linear systems lack the common-mode connections that require proximity. It is also a design criterion to separate various stages of production for sheer ease of maintenance access or replacement of equipment. Linear systems not only have spatial segregation of separate phases of production, but within production sequences the links are few and sequential, allowing damaged components to be pulled out with minimal disturbance to the rest of the system. In complex systems, removing a component or shutting it down means temporarily severing numerous ties with consequent readjustments, capping, product storage, removal to get access, and reconfigurations because parts and units tend to be multiply linked. Linear systems also favor serial production—a series of linked but semi-independent production steps—rather than what organizational theorist James Thompson calls "pooled interdependence," where all components (including operators) must coordinate their input if the system is to function at all.[12]

In contrast to complex systems, there is minimal specialization of labor, materials, and pools of supply in linear systems. Equipment is specialized ("dedicated," engineers call it), but the people who run it tend to be generalists. Operators are trained on several tasks because they tend to rotate, bid on various jobs, or fill in for people. Maintenance people can operate equipment in an emergency, operators perform emergency maintenance. There are, of course, limits to such substitutions, including the demands set by unions. Later we shall see that substitutability is important for recovery from an accident—people can fill in and know something about the other person's job. Here, in discussing complexity and linearity, the emphasis is upon awareness of interdependencies. In complex systems, not only are unanticipated interdependencies more likely to emerge because of a failure of a part or unit, but those operating the system (or managing it) are less likely, because of specialized roles and knowledge, to predict, note, or be able to diagnose the interdependency before the incident escalates to an accident.

Consider some key jobs in complex systems and note how little substitutability there is between them: fighter pilots and maintenance; bombardiers and pilots and navigators; nuclear plant engineers and operators; operators and welders and other maintenance people rated to do specialized jobs; lab technicians and biochemists in a recombinant DNA firm; chemists with advanced degrees and lab technicians, in a chemical plant, or chemical engineers and control room operators; astronauts and ground control managers; navigators, radio operators, captain, and helmsmen on a ship.

Though I don't want to claim a vast difference between employees in complex and linear systems, the latter appear to have fewer specialized and esoteric skills, allowing more awareness of interdependencies if they appear. The welder in a nuclear plant is more specialized (and specially rated), and presumably more isolated from other personnel, than the welder in a fabrication plant. Specialized personnel tend not to bridge the wide range of possible interactions; generalists, rather than specialists, are perhaps more likely to see unexpected connections and cope with them.

What is true of labor is also true of materials and supplies. If materials and supplies are substitutable, more latitude of response is available, limiting failures to incidents and preventing failures in the first place. But complex systems appear to have more exacting requirements for materials and supplies; the fuel cannot be off-standard, nor one fuel substituted for another, whether we are dealing with nuclear plants, aircraft, spacecraft, or chemical production. Substitutions are more likely in linear systems.

Finally, linear systems have minimal feedback loops, and thus less opportunity to baffle designers or operators. There are fewer interactions of control parameters, because controls are more decentralized, and attached to special-purpose equipment. And the information used to run the system is more likely to be directly received, and to reflect actual operations.

The two systems are summarized in Table 3.1, with some summary terms listed that will be used in the rest of the book.

TABLE 3.1
Complex vs. Linear Systems

Complex Systems	*Linear Systems*
Tight spacing of equipment	Equipment spread out
Proximate production steps	Segregated production steps
Many common-mode connections of components not in production sequence	Common-mode connections limited to power supply and environment
Limited isolation of failed components	Easy isolation of failed components
Personnel specialization limits awareness of interdependencies	Less personnel specialization
Limited substitution of supplies and materials	Extensive substitution of supplies and materials
Unfamiliar or unintended feedback loops	Few unfamiliar or unintended feedback loops
Many control parameters with potential interactions	Control parameters few, direct, and segregated
Indirect or inferential information sources	Direct, on-line information sources
Limited understanding of some processes (associated with transformation processes)	Extensive understanding of all processes (typically fabrication or assembly processes)

Summary Terms

Complex Systems	*Linear Systems*
Proximity	Spacial segregation
Common-mode connections	Dedicated connections
Interconnected subsystems	Segregated subsystems
Limited substitutions	Easy substitutions
Feedback loops	Few feedback loops
Multiple and interacting controls	Single purpose, segregated controls
Indirect information	Direct information
Limited understanding	Extensive understanding

Which Is Best?

This litany of problems with complex systems and the advantages of linear systems might suggest the latter are much preferable and complex systems should be made linear. Unfortunately, this is not the case. Complex systems are more efficient (in the narrow terms of production efficiency, which neglects accident hazards) than linear systems. There is less slack, less underutilized space, less tolerance of low-quality performance, and more multifunction components. From this point of view, for design and hardware efficiency, complexity is desirable.

It also appears that we have comparatively little choice in the design of some of our systems. Some complex systems can be redesigned to be more linear, such as in air traffic control or with the substitution of jet engines for the very interactive piston engines. Nuclear plants could be made marginally less complex if the spent storage pool were removed from the premises. (Should a plant with a full load of fresh rods in its pool have to be evacuated and the pool left unattended or without power, the water could boil off in a few days. Without the water, they would commence to fission like sparklers on the Fourth of July.) This would eliminate some proximity and common-mode problems. Using one control room to run two different reactors also seems like an unnecessary source of common-mode problems. The steam generation and turbine systems could be separated from the nuclear system to reduce unintended feedback loops, though at considerable expense. And perhaps, as we noted in the last chapter, more "forgiving" designs exist. But by and large no extensive reductions in complexity seem possible in the nuclear power industry. The transformation system simply requires many nonlinear interactions. The same is true of chemical plants such as refineries; there is probably no efficient way to crack crude oil except in a highly interactive system.

On the whole, we have complex systems because we don't know how to produce the output through linear systems. If these complex systems also have catastrophic potential, then we had better consider alternative ways of getting the product, or abandoning the product entirely. Short of that there does not seem to be a choice possible between complexity and linearity for high-risk systems. Complexity is inherent in some forms of production. It is not intrinsically undesirable; we welcome complexity in some bureaucracies and resist the "rationalization" of our disorderly life because the unexpected interactions lead to innovations, amuse or interest us, or provide variety. If the system has catastrophic potential, however, and we cannot prevent the propagation of incidents and intervene before an accident has occurred, the issue is much graver. Let us see what the idea of tight and loose coupling has to say about this. It is our second major dimension in analyzing systems.

Tight and Loose Coupling

Engineers speak of tight and loose coupling in the normal course of their work, and in that context the terms are quite clear. It's not as simple as coupling a garden hose to the faucet, and doing it tightly if we don't want to get sprayed, but that is a fairly close analogy; *tight coupling* is a mechanical term meaning there is no slack or buffer or give between two items. What happens in one directly affects what happens in the other.

Sociologists and social psychologists took up the term in the mid-1970s to conceptualize a particular phenomenon.[13] Some public service organizations, public schools in particular, seemed to be characterized by an unusually large gap between official programs and actual behavior. In public schools researchers investigated the gap between new techniques and actual changes in the behavior of teachers, or between new programs and what the students actually learned. One might expect hierarchically organized bureaucracies to be capable of altering the behavior of subordinate personnel such as teachers and capable of producing outcomes in students that bear some relationship to the official goals. The explanation for the school's failure to achieve its goals was that while the programs and goals were real enough, they were only loosely connected to other matters with which the organization had to be preoccupied, such as political demands from the environment, the autonomy of professional teachers, and the ineffective mobilization of parental demands.

For example, there could be a tight connection (responsiveness) between a remedial program required by the school district and the school in that it is funded, staffed, and given space and a place in the curriculum. But the program might only be loosely coupled to other parts of the school. For example, two new teachers, presumably hired for the program, are actually the ones the school has long wanted to hire—a respected art teacher and a person who can teach a course in computers. They are used for these positions, and the two teachers who are the least respected by parents and students are assigned to the new program. The decisions might almost seem to have been made by themselves, so loosely coupled is the demand for the new program with the existing problems of the school.

The program budget is a concrete, discrete item, but its appearance does not result in purchasing the proper supplies and doing the expected remodeling. Instead, the supply orders are delayed, and the money used for gym equipment and refinishing the gym floor. There is no intent of fraud; the latter have long been pressing items, subject to an informal agreement with the district that the expenditures can be made just as soon as there is some slack in the system. The school principal reasons that the remedial program will be taken care of when the new budget passes. Space for the new program might turn out to be the most inconvenient space, on the incontrovertible grounds that all the convenient space is utilized, and the program should not disrupt successful existing programs. Existing programs are buffered, in this way, from the impact of the new program. However, the district's intent was to alter the school's existing priorities, which would require that the remedial program should have an impact on—be tightly coupled to, and change—the existing priorities.

Finally, the school could find it easy—almost inadvertent—to assign students to the program not on the basis of their reading difficulties, but because they are disruptive, impolite, or racially stereotyped. They all look as if they need "extra help." And the classroom materials, carefully designed by educators in elite universities, may bear little relationship to the needs of the stigmatized students.

In all this a number of connections are being made—student social status and program assignment; supply shortages and the power of certain departments; low performing teachers and periphery assignments. The system is not lacking in connections, but they are "loose" ones. Some are invoked for this program but perhaps not for the next. Some were invoked by accident—a chance discovery that the way the budget was drawn up it could cover costs of the gym floor. Others might have depended upon fleeting events, perhaps a recent controversy regarding the effectiveness of two teachers. Certainly the connection between the official mandate for a remedial program and the actual practice is quite loose. The characteristics of the system—its loosely coupled nature—make it possible for it to respond to the outside demand in such a loose fashion.

Loosely coupled systems tend to have ambiguous or perhaps flexible performance standards, and they may, as in the case of the school, have little consumer monitoring, so the absence of the intended connection can remain unobserved.

You might be tempted to call it a very inefficient system, or even a rip-off of state or federal funds, and call for "tight coupling." This would make the remedial program work as everyone outside the system thought it should work. But it would be a mistake to call it inefficient. The system is quite efficient for accomplishing many things that many participants desire; they are just not what the school district or the federal government, which supplies the money, had in mind. A bit of authoritarian leadership designed to put the official program in effect might well disable parts of the system that others value. If the poor teachers remained in the regular curriculum, parents would continue to object to them; if the college preparatory subjects were put in the temporary building (temporary since the end of World War II) they might also object; everyone uses the gym and it needs fixing; and what is one to do with disruptive students who reportedly interfere with the learning of the docile ones?

Loose coupling, then, allows certain parts of the system to express themselves according to their own logic or interests. Tight coupling restricts this. Loose coupling, however, is not the same as disorganization, unless we mean lack of centralized control by that term. Informally, the school is well organized, with a variety of coherent (though occasional shifting or episodic) interests, mechanisms for accommodative arrangements, slack resources to meet unexpected challenges, and stable interaction patterns. The degree of organization is independent of the degree of coupling.

We have not strayed as far from the engineer's usage of tight and loose coupling as it may seem. Elaborating the concept as used by organizational theorists will allow us to examine the responsiveness of systems to failures, or to shocks. Loosely coupled systems, whether for good or ill, can incorporate shocks and failures and pressures for change without destabilization. Tightly coupled systems will respond more quickly to these perturbations, but the response may be disastrous. Both types of systems have their virtues and vices.

For example, a continuous processing plant requires tight coupling. In some continuous processing plants where the technology is well understood and only linear interactions take place, such as pharmaceutical plants, bakeries, confectionaries, or ball-bearing plants, the characteristics of the product are altered frequently in response to market demands. Processes are also altered frequently to reflect changes in raw materials or operating conditions. Decisions to alter the process or shift to a slightly different product must proceed quickly through the organization, producing nearly unreflective changes in operator behavior. Resources are strictly allocated, schedules strictly followed, and reporting systems must be exact. Surveillance, both of processes and people, is continuous, though it is usually through remote and unobtrusive indicators—at least when higher-level personnel are monitored. Deviations from standards are quickly noted or reported, since a long product stream is affected. Responses to deviations must be standardized and immediate. The result is high operating efficiency. To loosely couple such a production process would be to invite disasters, as well as inefficiencies. If the system is linear, rather than complexly interactive, tight coupling appears to be the optimum mode of organization.

In contrast, consider a loosely coupled linear aircraft manufacturing and assembly plant. Fabrication of the tail section will be separate from the fabrication of the fuselage section it will be attached to. Different metals are involved, different tolerances, and different heat-treating processes. The constraints on the two sections are different, and their interdependence is minimal; the interdependence is taken care of in the design stage. Personnel practices may be different too, since particularly skilled personnel are needed for the tail construction, whereas fewer skills are needed for the fuselage, which involves more routine tasks than tail construction. This can lead to personnel decisions that resemble the allocation practices we discussed in the school, reflecting unit power and politics. Quality control, instead of being "built into" the system as in continuous processing, may be a "stand-alone" function, performing its own tests and inspections, physically isolated in the plant, and deliberately not integrated into the other functions because its independence must be maintained. But this buffering of quality control also means loose coupling, with the possibility of parochial interests developing, or unauthorized bargains with units.[14]

Characteristics of Coupling In Systems

We are now ready to more systematically lay out what is meant by tight and loose coupling in systems. These characteristics are reasonably independent of our other major dimensions—complex interactions and linearity.

1. Tightly coupled systems have more time-dependent processes: they cannot wait or stand by until attended to. Sometimes this is expressly for efficiency reasons, but generally it is because the production process does not allow for cooling and reheating, for forgetting and then relearning. Storage room may not be available, so products must move through continuously. Reactions, as in chemical plants, are almost instantaneous and cannot be delayed or extended. In loosely coupled systems, delays are possible; processes can remain in a standby mode; partially finished products (tail assemblies or students) will not change much while waiting.

2. The sequences in tightly coupled systems are more invariant. B must follow A, because that is the only way to make the product. Partially finished products cannot be rerouted to have Y done to them before X; Y depends upon X's having been performed. Though it may be expensive, in an aircraft assembly line a door or seat or the radio controls could be added later if there were a disruption. This is not the case for a nuclear or chemical plant, or a pharmaceutical production line. In a trade school the sequence of courses is carefully prescribed and more

tightly coupled than a university where it does not matter much when the science or math or language requirement is met.

3. In tightly coupled systems, not only are the specific sequences invariant, but the overall design of the process allows only one way to reach the production goal. A nuclear plant cannot produce electricity by shifting to oil or coal as a fuel; but oil plants can shift to coal and vice versa with little or no reconversion. While continuous processing plants (tightly coupled) can vary product mixes and production volume within limits, they cannot make significant changes in the way the product is made. However, manufacturing plants (generally loosely coupled) may, even on a temporary basis, eliminate heat-treating by substituting another metal; may shift from unit to batch production; contract out for subassemblies; install or remove robotic devices; substitute plastic for metal, and so on. In this sense, there are many ways to produce the item. But for tightly coupled systems such as dams, chemical plants, power grids, bakeries, and recombinant DNA technologies, there is little flexibility. Loosely coupled systems are said to have "equifinality"—many ways to skin the cat; tightly coupled ones have "unifinality."

4. Tightly coupled systems have little slack. Quantities must be precise; resources cannot be substituted for one another; wasted supplies may overload the process; failed equipment entails a shutdown because the temporary substitution of other equipment is not possible. No organization makes a virtue out of wasting supplies or equipment, but some can do so without bringing the system down or damaging it. In loosely coupled systems, supplies and equipment and humanpower can be wasted without great cost to the system. Something can be done twice if it is not correct the first time; one can temporarily get by with lower quality in supplies or products in the production line. The lower quality goods may have to be rejected in the end, but the technical system is not damaged in the meantime.

Recovery from Failure

Coupling is particularly germane to recovery from the inevitable component failures that occur. One important difference between tightly and loosely coupled systems deserves a more extended comment in this connection. In tightly coupled systems the buffers and redundancies and substitutions must be designed in; they must be thought of in advance.

In loosely coupled systems there is a better chance that expedient, spur-of-the-moment buffers and redundancies and substitutions can be found, even though they were not planned ahead of time.

Since failures occur in all systems, means to recovery are critical. One should be able to prevent an incident, a failure of a part or unit, from spreading. All systems design-in safety devices to this end. But in tightly coupled systems, the recovery aids are largely limited to deliberate, designed-in aids, such as engineered safety devices (in a nuclear plant, emergency coolant pumps and an emergency supply coolant) or engineered safety features (a more general category, which would include a buffering wall between the core and the source of coolant). While some jury-rigging is possible, such possibilities are limited, because of time-dependent sequences, invariant sequences, unifinality, and the absence of slack. In loosely coupled systems, in addition to ESDs and ESFs, fortuitous recovery aids are often possible. Failures can be patched more easily, a temporary rig can be set up, a crane moved over, a pancake landing made without power. There may, fortuitously, be plenty of space separating a burning subsystem from other systems; it may be possible and relatively harmless to flood an area to put out a fire, though no designer planned for that. Tightly coupled systems offer few such opportunities. Whether the interactions are complex or linear, they cannot be temporarily altered.

This does not mean that loosely coupled systems necessarily have sufficient designed-in safety devices; typically, designers perceive they have a safety margin in the form of fortuitous safety devices, and neglect to install even quite obvious ones. Most of the safety devices required after inspections by the Occupational Safety and Health Administration (OSHA) are fairly obvious items, such as railings, rough treads on stairs or passageways, emergency switches to shut off equipment, and requirements that the coupling for hoses be arranged so that an explosive gas such as hydrogen cannot be inadvertently supplied to an area that needs an inert gas. Even in loosely coupled systems there is still enormous room for improvement in safety features.

Tightly coupled systems are not completely devoid of unplanned safety devices. In two of the most famous nuclear plant accidents, Browns Ferry and TMI, imaginative jury-rigging was possible and operators were able to save the systems through fortuitous means. At TMI two pumps were put into service to keep the coolant circulating, even though neither was designed for core cooling. Subjected to intense radiation they were not designed to survive; one of them failed rather quickly, but the other kept going for days, until natural circulation could be established. Some-

TABLE 3.2
Tight and Loose Coupling Tendencies

Tight Coupling	*Loose Coupling*
Delays in processing not possible	Processing delays possible
Invariant sequences	Order of sequences can be changed
Only one method to achieve goal	Alternative methods available
Little slack possible in supplies, equipment, personnel	Slack in resources possible
Buffers and redundancies are designed-in, deliberate	Buffers and redundancies fortuitously available
Substitutions of supplies, equipment, personnel limited and designed-in	Substitutions fortuitously available

thing more complex but similar took place at Browns Ferry. The industry claimed that the recovery proved that the safety features worked; but the designed-in ones did not work. In both accidents the critical safety features were disabled.

What is true for buffers and redundancies is also true for substitutions of equipment, processes, and personnel. Tightly coupled systems offer few occasions for such fortuitous substitutions; loosely coupled ones offer many.

The characteristics of the two systems are summarized in Table 3.2. These are tendencies; no system has, for example, absolutely invariant sequences. Nor is a system likely to have all of the various characteristics.

The Organizational World According to Complexity and Coupling

Figure 3.1, the Interaction/Coupling Chart (I/C chart) puts interaction and coupling together in a two-variable array. The placement of systems is based entirely on subjective judgments on my part; at present there is no reliable way to measure these two variables, interaction and coupling. One may well quarrel with the placement in the figure. One good reason for a quarrel is that there is no precise specification of just what constitutes the system. Consider marine transport. On the high seas the system includes the ship, radio communication, the weather, and perhaps one other ship. But as the ship enters a crowded channel, we have not only the weather, but bank effects (the suction created as the ship passes close

FIGURE 3.1
Interaction/Coupling Chart

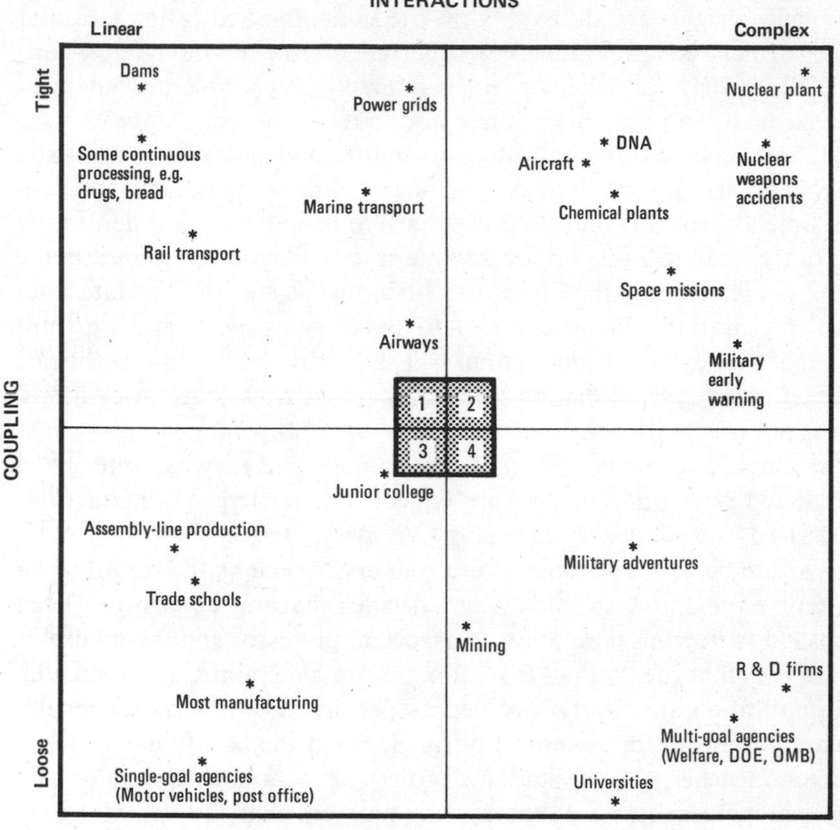

to the underwater bank of a channel), tides and current flow, wrecks and rocks, bridges, tows and other ships, a crowded radio channel, and navigation lights mixed in with the lights of highways, industrial plants, and distant towers. The high seas and the channel systems are very different. With "flying" we mix commercial airlines with recreational or sport flying. Mining includes strip mining, which should be closer to manufacturing, and underground mining. These ambiguities are a problem, but for the sake of a general argument I think the rough placement of vaguely defined systems is still useful.*

*One serious problem cannot be avoided, but should be mentioned: To some unknown extent it is quite possible that the degree of coupling and type of interactions have been

By combining our two variables in this way, a number of conclusions can be made. First, it is clear that the two variables are largely independent. Examine the top of the chart from left to right. Dams, power grids, and nuclear plants are all roughly on the same line, indicating a similar degree of tight coupling. But they differ greatly on the interaction variable. While there are few unexpected interactions possible in dams, and not that many in power grids, there are many in nuclear plants.

Or, looking across the bottom, universities and post offices are quite loosely coupled. If something goes wrong in either of these, there is plenty of time for recovery, nor do things have to be in a precise order. In the post office, mail can pile up for a while in a buffer stack without undue alarm; people tolerate the Christmas rush, just as students tolerate lines at fall registration. But in contrast to universities, post offices do not have many unexpected interactions—it is a fairly well laid out (linear) production sequence without a lot of branching paths or feedback loops. This is not true of universities. Here there are multiple functions—teaching, research, and public service, for example—and they can interact in unexpected ways. Indeed, they are expected to; synergistic interactions are desired, though "negative synergy" is also possible.

For example, let us suppose that a university reviews the record of an assistant professor of sociology, and decides that not enough has been published to warrant promotion to associate professor and the tenure or job security that goes with it. It is all quite straightforward, since research and publication are clearly specified as performance criteria for faculty and are outputs of the system. But imagine that the faculty member is a very good teacher, and the students protest the action. Imagine also that dismissal will threaten a public service program that she ran. Not only do the students protest, but some influential community members protest. To convince the students that the department is not indifferent to teaching quality, it has to institute a teacher evaluation program, and this now threatens the status of the senior faculty (associate or full professors with tenure) who have not had to pay much attention to this criterion before. Furthermore, the local churches send a delegation of church officials to protest the demise of the service program, and in the process,

---

inferred from a rough idea of the frequency of system accidents in the various systems, rather than derived from analysis of the properties of the systems independent of the nature of their failures. That is, if there are few accidents caused by air traffic control, that "must" mean it is not highly complex and tightly coupled, and then evidence for that conclusion is sought. Since the analytical scheme evolved from examination of many systems, there is no way to avoid this possible circularity. The scheme would have to be tested by examining systems not included here, as well as by collecting data based upon more rigorous concepts and verifying the placement of the systems that are included here.

demand more say in university personnel practices. The dean now has three problems on her hands: the department has to publish more (to justify firing the assistant professor), spend more resources on teaching to mollify the students, who have now made teaching a part of the formal personnel evaluation, and face a mobilized external group that demands some say in personnel decisions. This is an example of complex interactions; in contrast to the linear post office, the university is high on complexity.

But note that we are not likely to have a "system accident." Because universities are loosely coupled there is ample slack to limit the impact of this one personnel decision on other areas. Senior faculty members can be quietly reassured that student evaluations will be processed slowly, and evidence of the unreliability of evaluations will be discreetly passed about. (In one case that I witnessed, the university couldn't find the $2,000 needed to keypunch and process a student-initiated teacher evaluation program, the university's first.) Community members will be promised more access to the Student Activities Council, but realistically they will not be expected to attend often. The university research foundation is quietly told to favor sociology applications in the next cycle of grants, or joint appointments are handed out to make the sociology program look more productive. No interruption of output is experienced. The event remains an incident, with damage limited to the "part"—the teacher. While interactive, recovery is easy because of loose coupling.

The post office and the university, then, are similar with regard to coupling; both can recover from upsets because sequences are not that inevitable, there is slack in resources, substitutions are possible. But the post office is largely limited to linear interactions, while the university is full of many potentially complex ones that can reach unexpectedly into other parts of the system.

Educational systems are represented in three places in the I/C chart: universities are loosely coupled and complex, as we have seen; trade schools are somewhat loosely coupled but very linear; and junior colleges are near the middle of both dimensions. Trade schools have a well-defined curriculum and fairly explicit set of subjects students are expected to learn. This permits linear interactions to dominate and there are few surprises. However, the cafeteria style of offerings allows students to retake courses, making it somewhat more loosely coupled than junior colleges, and trade schools are also staffed by part-time, often temporary instructors who come and go with the changing student demands. Junior colleges are a bit more tightly coupled because of sequencing within programs (blocks of courses) and because of personnel inflexibility. They are less linear than trade schools because they undertake more functions, in particular, socialization and value impregnation (though not nearly as much as universities do). Neither of these institutions would be expected to have system accidents that result in stopping the output of subsystems; if such subsystems as courses or community programs do fail, it would be because of component failures—lack of students, instructors, or community support.

Conclusion

This completes our major analytic work in this book. We have accomplished quite a bit: defined accidents, distinguishing them from incidents; defined various classes of victims in a way that allows us to better assess catastrophic potential; defined system and component failure accidents; and defined our two key concepts, the types of interaction (complex and linear) and the types of coupling (loose and tight). These variables have been laid out so that we can locate organizations or activities that interest us and show how these two variables, interaction and coupling, can vary independently of each other.

We now have the basic tools to proceed through a variety of systems and examine them in terms of the type of accidents they have, and their catastrophic potential. We shall start with chemical plants, which resemble nuclear power plants in their system accidents, though not in their consequences, yet which illustrate a comparatively old and stable technology.

CHAPTER 4

Petrochemical Plants

Petrochemical plants produce far fewer headlines and much less controversy than nuclear power plants. They have been around for about one hundred years, so ample engineering experience exists, and the public is familiar with the sight of their gangly towers and squat storage tanks. Fires rage uncontrolled in them at times, and a few operators might be killed, but we do not have protest marches, a Chemical Regulatory Commission, scientific panels and conferences that are covered by the media, or a search for alternatives to our plastics and gasoline. It is a low-profile industry—deliberately, as we shall see. It is also quite safe; the fires and explosions kill few operators and few bystanders.

Yet the chemical industry is far from an unpromising subject for accident analysis; in fact, it provides some of the best examples of system accidents we shall come across. It is quite tightly coupled, and has many complexly interactive components. We are concerned with it primarily because it illustrates the presence of system accidents in a mature, well-run industry that has a substantial economic incentive to prevent accidents. None of these factors are true of the nuclear power industry; it is new, poorly run, and short of catastrophic accidents that would close all plants overnight, accidents by and large only generate costs that can be passed on to ratepayers. It takes something really large, such as TMI, to even affect stockholders.

In the chemical industry, with mature technologies, good management, and strong incentives to keep the plants from blowing up, the persistence of system accidents should suggest something intrinsic to the processes themselves, which is what our theory of system accidents would suggest. We will attempt, then, to demonstrate that the system is complexly interactive and tightly coupled enough so that the accidents that do occur are more likely to be system accidents than in other systems. We are also concerned with the catastrophic potential that is present in the chemical industry (although to a far lesser extent than in the nuclear power industry). It is true that there are few potential first- and second-party victims in these giant complexes, since the industry is very capital-intensive. Even large fires and explosions have rarely created third-party victims in surrounding communities. However, experts tell us that the plants are getting bigger, the communities closer, and the substances more dangerous.[1]

There are two very important aspects of the petrochemical industry that we shall not be considering here because of our concern with system accidents: the health effects of employment, and the problems of toxic wastes and toxic discharges that might affect the public. Were this a book about the health dangers of industry, these would have to be extensively discussed. The health hazards encountered by employees during their employment in the industry are by some accounts, severe.[2] But like the run-of-the-mill industrial accident, the prevention of health hazards can be achieved through normal safety programs. The release of poisonous substances as a result of accidents, such as the explosion of the chemical reactor at Seveso, Italy, may be a growing problem, but unless the discharge is the result of a system accident, it can easily be discussed elsewhere, and has been.[3] The contamination of an 18-story New York State office building is a grim harbinger of things to come, but again, it is not our theme.[4] Chemical poisoning is already well covered in Michael Brown's book, *Laying Waste*[5] and the incredible story of PBBs in the cattle feed in Michigan is covered by Egginton.[6]

We will stick rather closely to chemical, refinery, and storage tank accidents. There are plentiful, and serious enough in terms of catastrophic potential. Several problems exist, however. First, the documentation is not extensive. In part this is because fires and explosions—the form of most chemical and refinery accidents—often eliminate the evidence needed for understanding the event. More important, however, is the reluctance of the industry to review accidents in a public, accessible form. We know so much about nuclear power plants because there is a government oversight agency with the authority to inspect utilities. This also holds for airplane accidents investigated by the National Transportation Safety Board, and for marine, railroad, and vehicular accidents. The U.S. Bureau of Mines investigates mine accidents. Nothing equivalent at the federal level exists for chemical plants. Space missions, dams, and a few other systems come under federal review for various reasons, but the chemical industry is in private hands and thus escapes this scrutiny, except in exceptional circumstances. The Occupational Safety and Health Administration (OSHA) collects data on injuries and fatalities, and does some inspection, but undertakes no analyses of the causes of chemical explosions and fires.

The insurance companies that write chemical plant policies have a substantial stake in analyzing accidents, and sponsor research and publications devoted to preventing accidents. But it is largely hortatory in nature and not useful for analyzing accidents. The most important material is produced by the industry itself, and unfortunately, it is circulated privately and is not available to researchers outside the industry. Some material finds its way into safety journals, and I will draw upon that for specific accounts, but there is no extensive and reasonably representative coverage of accidents in any detail as there is for other systems. Many of our accidents will concern European firms simply because they discuss their problems more openly. When I once attended a joint West German-U.S. conference on safety in chemical plants, only one of the invited U.S. representatives came—and he was not sponsored by his organization but came on his vacation time! I even found it extremely difficult to get a simple plant tour of a refinery; after being turned down by firms I finally tagged along with a group of Stanford engineering students. My efforts with companies and trade and technical associations were generally met with the statement that, "We do not want to wash our dirty linen in public." This is somewhat surprising since, as industrial linen goes, the chemical and petroleum industry's linen is very clean.

In terms of injuries, fatalities, and lost work time, the industry's record is one of the best. Working conditions are generally excellent. The average worker appears to be both more skilled and better paid than nuclear plant operators. Plants, refineries, and tank farms have few workers in attendance, so even though fires and explosions are frequent, injuries are few and fatalities very low. Most workers are in protected quarters, such as control rooms. Compared to most industrial activities there is little danger of creating many first- or second-party victims through component failure accidents or system accidents.

The danger is the potential for third-party victims—innocent bystanders or, in this case, people living within a few miles of a facility or driving by one. Even so, there have been only a few major catastrophes. As the industry grows in size, the number and size of facilities will increase, along with, possibly, the toxicity of the materials and the number of people living or traveling nearby. Thus, the potential for third- and fourth-party victims can be expected to rise, if, as I argue, even a highly safety-conscious industry will have a substantial number of system accidents.

The catastrophic potential will become evident in the course of the chapter. But first, some scant data on the frequency of accidents, whether component failure or system accidents, must be reviewed. The industry, variously referred to as the petroleum industry, petrochemical industry, or hydrocarbon processing industry, includes exploration, gas processing, drilling, chemical production, refining, pipelines, and a variety of marketing and other services. Its overall safety record is very good—for example, in 1980, only one in about seven thousand workers was killed outright, and there was one disabling injury for every sixty-nine workers during the year, making it one of the safest industries. The National Safety Council rates industries according to the number of deaths and lost days of work, per million hours worked. The average of all U.S. industries is 2.5, that for the chemical industry, 1.45. Within the industry rates varied. Drilling operations had a rate of 10.75, chemical processing, 1.62, and refineries were 1.41.[7]

Clearly it is not a disastrous industry for employees. The low incidence of work injuries and fatalities does not mean, however, that there are no plant accidents. It means that workers are few in number and well shielded from them. The most serious plant accidents result in, or from, fires and explosions. Statistics are very hard to come by; the American Petroleum Institute, an industry service organization, does not give statistics for all companies, and it is difficult to analyze those it does provide. Yet fires are very common. Bulk plants and tank farms experienced a very sizeable 1,400 fires in 1974 and the same amount in 1975.[8] In 1979 the 166 refinery properties reporting to the survey experienced 205 fires—1.23 fires per year for each plant.[9] Since this overrepresents the large companies, and they have the better safety record, the rate for all properties must be considerably higher.

Fortunately, some useful summary data on one segment of the industry, ammonia plants appeared in 1978, and it is striking. Ammonia plants are an established, venerable part of the chemical industry. Yet the survey indicated the average ammonia plant had fifty downtime days per year, and ten to eleven shutdowns per year. This indicates they were in operation only 86 percent of the time—considerably better than nuclear plants, which average around 60 percent, but hardly reassuring. The industry has a goal of one turnaround (a shutdown for maintenance) every seventeen months, but the average has been one every twelve months because one-third of the maintenance turnarounds were initiated by major equipment failures. In one four-year reporting period there was one fire per plant every eleven months—roughly equivalent to the refinery experience. No indication is given of the seriousness of the fires, but no fire in an ammonia plant (or refinery) is without seriousness.[10]

These data give us more cause for concern. A very mature operation has to be shut down for repairs about 30 percent more frequently than expected, generally from major equipment failures, and its plants have a fire every eleven months. If the problems of the ammonia industry have not been "wrung out" of the system by now, it may be that they are endemic to the system—normal accidents. Our review of accidents in the petrochemical industry will suggest just that. But first, we will review a famous accident that did not even start in a chemical plant, but on a ship. It will tell us, however, what can happen when fires and explosions occur near chemical plants. Just sitting there, they have their catastrophic potential. Twenty-two years later the same location had another serious accident, this time a system accident.

Texas City, Texas: 1947, 1969

In 1947 Texas City, Texas, experienced the worst series of explosions and fires in U.S. history. The fire did not start in a petrochemical facility, but on a ship loaded with fertilizer. The fertilizer contained ammonium nitrate, a highly explosive substance used in manufacturing TNT. Efforts by the ship's crew to put out the fire failed, and the Texas City fire department took over. While a large crowd watched, the ship blew up in an enormous explosion. Fragments of the blast were blown 3 miles into the air, and two airplanes flying overhead were incinerated. The noise of the explosion was heard 160 miles away. Oil storage tanks nearby went up in flames, and so did a large chemical plant. A nearby ship, also loaded with explosive fertilizer, but relatively undamaged, tried to pull away but rammed another ship; the two stuck together and either they could not be moved, or the tugs in the harbor refused to try. The next evening the ships caught fire and another explosion rocked the city and ignited a sulphur warehouse. By now one-third of the city was ablaze or already burned down. The dead numbered 561 and the injured over 3,000; damage was over $100 million.[11]

It is not something one forgets easily, so when another chemical plant in Texas City had an explosion in 1969, the panic in the nearby area was extensive. The explosion was trivial by the 1947 standards, however, and no one, incredibly enough, was seriously injured. Nevertheless it broke windows in houses one and one-quarter miles away, and damaged residential houses that stood as close as 750 feet from the plant. One 800-pound section of the column that blew up landed 3,000 feet away—almost two-thirds of a mile—but fortunately not in a residential area.

There were two striking features of the 1969 explosion. First, it indicates that large explosions need not lead to loss of life. Chemical plants are certainly not labor-intensive; in this case, there were only thirteen people about. Six were in the control room, but the metal sides absorbed much of the blast and the major danger came from the light fixtures that came down with the false ceiling. Three maintenance men outside were shielded by small items of equipment. One operator, 200 feet from the explosion (which, you may recall, sent an 800-pound piece 3,000 feet) was only knocked to the ground. The main fireball was over his head. He crawled out of the fire area and "immediately took steps that limited the amount of flammable material entering the fire zone. He also activated the emergency relief valves on the distillation equipment." His actions substantially limited the extent of the subsequent fires.[12] Operators are still useful to have around despite all the automated safety equipment, and decoupling was still possible.

The other striking feature is that there was extensive experience with this specific process. Indeed, there had not been an incident in any operations with identical column conditions in eighty operating years. Yet after the event the system was substantially changed to prevent another accident in some quite distant future. Three failures that were not anticipated to interact came together. Here is how it happened.

The butadiene refining unit was operating normally when it was shut down for repairs to something called a "stripper make compressor." The repair requires a day-long procedure, and all feed flow to the unit was discontinued after two hours into the process. The refining column is a tall stack, such as those you see when driving by a chemical plant, inside of which there are "trays" stacked up which allow gases to collect and be removed at various stages of refinement. The process is, of course, automatic, with closely regulated temperature and pressure values at the various stages of the refining process. The column is in a state called total reflux when it is shut off from feed flow; it is a closed system circulating the contents, with liquid at the bottom and gases of various weights above it.

The operation appeared normal to the operator, even though it was erratic. The column normally operated erratically under total reflux conditions, so there was no cause for concern. (There is a good bit of erratic behavior in most automated chemical processes, and it is not well understood, nor need it be during routine operation.) Operator attention was required in maintaining a balance between the liquid in the base that generated the vapor and liquid in the accumulator that replaced it. Unknown to the operator, the column was slowly losing material through a leaking valve in an overhead line (failure #1). This is the motor valve, referred to below. But the column pressure and the pressure drop after the feed was shut off were both normal; these conditions would normally disclose a valve leak, but didn't (failure #2), perhaps because of the total reflux condition. "The make flow meter showed a continuous flow; however, the operator assumed that the meter was off calibration since the make motor valve was closed and the tracing on the chart was a continuous straight line near the base of the chart (failure #3: faulty interpretation of information; a forced error). The column base-level indicator showed a low level in the base of the column, but ample kettle vapor was being generated."[13] However, one of the heavy components of the feedstock, vinylacetylene, which concentrates in the lower part of the column as a kettle product, apparently more than doubled. Normally it is 35 percent, and is safe at levels up to 50 percent. In addition, also because of the small leak, the liquid at the base of the column declined in volume and must have uncovered the calandria tubes through which the stock flows. This allowed the tube walls to overheat. (This was not an independent failure but followed directly from the first failure.)

The combination of high tube wall temperatures and the increased concentration of vinylacetylene set the stage for the explosion. It occurred eleven hours after the reduction in feed flow began, and nine hours after total reflux was obtained. There was no warning. There were two explosions, the first being the disintegration of the lower 40 feet of the column, which was scattered around the area, and almost immediately thereafter, the second one resulting from the large quantities of gas released from the ruptures.[14]

In this accident we have a component failure—a leaky valve—along with the absence of information about system malfunctioning, since erratic performance was to be expected, and an unexpected interaction of pressures, temperatures, and vapors within the column; in short, a system accident. The company ordered many changes in the process at all similar units, and these suggest that the process, though in place for a long time, was not fully understood. The changes required lower temperatures for the heating systems, a substitute for sodium nitrite (which is converted to sodium nitrate and makes the butadiene-vinylacetylene mixture less stable), the avoidance of the total reflux, keeping the percentage of vinylacetylene below 40 percent in the vapor phase at all points in the system, and keeping the column concentration of the vinyl below the base concentration.[15] But these changes do not seem to be responsive to the problems with the process. Avoiding total reflux may create other problems, and controlling temperatures had proved to be impossible as it was. One might again have six or more million dollars in damage and a threat to many lives, including nearby residents.

Changes were also made in the fire-fighting and recovery systems, since the company found such problems as tanks without shut-off valves. But by and large, one is impressed by the fire-fighting system and the number of factors that worked in this emergency to prevent an even larger disaster. As is typical, much of the problem was crowd control, since the city is adjacent to the plant. Some, remembering the 1947 explosion, tried to flee; others were drawn to the sight by the fire and thoroughly impeded the efforts of the company to fight the fire and look for survivors. The lesson here, not remarked upon by the investigators, is to isolate these plants.

Flixborough

Chemical plants have huge capital investments. Since they must be operated continuously in view of the investment and the nature of the process itself, operating pressures to stay on-line must be intense. Some indication of this, and of the difficulties of starting and stopping these behemoths, is provided by the official board of inquiry into the famous Flixborough disaster of June 1, 1974, in England. Extensively investigated, it offers a rare glimpse into the world of automation and high technology. It is not reassuring.

The plant produced something called caprolactam, used in making nylon. To make it, cyclohexane was oxidized. Cyclohexane is quite similar to gasoline. It had to be circulated through the plant under pressure and a high temperature of 155° C. (310° Fahrenheit).[16] Oxidation of the cyclohexane was carried out by passing the material through a set of six reactors; air and catalysts acted on the heated cyclohexane, and the desired products were distilled out.

On the 27th of March the Number 5 reactor was found to be leaking cyclohexane. The plant was shut down, depressurized, and cooled, and the reactor inspected. It had a serious crack some 6 feet in length. The plant managers decided to by-pass this reactor. The Board of Inquiry notes that only one of the several people at the meeting where the decision was reached was seriously concerned about restarting the plant without inspecting the other five reactors. Furthermore, no one "appears to have appreciated that the connection of No. 4 Reactor to No. 6 Reactor involved any major technical problems or was anything other than a routine plumbing job." The "emphasis at the meeting was directed to getting the oxidation process on stream again with the minimum possible delay." The design and construction of the by-pass were conducted "as a rush job."[17] In fact, no drawing was ever made, other than one in chalk on the workshop floor; no calculations of strain were made; the designer's guide for the large bellows the by-pass pipe would be connected to was not consulted, and so on. The by-pass was completed after two days, on the evening of March 29. The scaffolding that would hold the 20-inch pipe was rudely constructed in the rush to get the plant back online. The supports were wholly inadequate, the report says, and some were omitted, presumably in haste.

Once in place the by-pass was tested for leaks. One was found, so the plant was depressurized, but they forgot to mark the leak and had to pressurize it again, find it and repair it after depressurization. Pressurizing a large plant is not a negligible task; it can take several hours and require thousands of steps. The plant was then repressurized to test for leaks, none were found, and then the pressure was increased for further tests, then depressurized once again, and finally start-up procedures were carried out. One begins to see, with this account, the pressures the staff may have been under. The report summarizes the condition of the plant at this point as follows:

> An assembly was installed which had been the subject of no design calculations, which did not comply either with the British Standard or with the bellows manufacturer's recommendations; which was subject to a turning moment under pressure; which was wholly unrestrained in an upward direction and wholly inadequately restrained in a downward direction. As a result the bellows were subjected to shear forces for which they were not designed and the 20-inch pipe [the largest they could find on the grounds, incidently; they

> needed a 28-inch one] was under high and unknown stresses resulting from the end loads of 38 tonnes.[18]

However, this solution seemed to work. The plant resumed operation on April 1, was shut down for short periods twice in May, but on May 29 a leak was discovered on one of the vessels and during the course of the day it was shut down. Two days later they were able to start up again. Leaks were again found and so the circulation was stopped and heat taken off; the leaks cured themselves; circulation was started and heat put on. But pressure went up at an abnormal rate, requiring substantial venting. Another leak was found and the process was repeated of cooling and depressurizing. The personnel could not fix the leak because the necessary spark-proof tools were locked up in a shed, it being Saturday! The report felt this was "clearly not a satisfactory situation." Now we are into June 1, four days after the leak that occasioned this series of starts and stops in this highly automated, sophisticated plant.

Full operation was delayed; the plant was short of high-pressure nitrogen. Further supplies would not arrive until midnight. But the plant itself was quite alive, and ventings and rumblings went on because of unusual pressure. The description is quite complex, involving multiple interactions of pressure, steam for heat, nitrogen levels and shortages, and the different conditions in each of the five reactors. The report is very careful to point out that no operator errors were involved in the complex maneuvering.

A number of anomalies were still unexplained at this time, especially the fast rise in pressure and the excessive consumption of nitrogen, but also the need for so much venting (these appear to be independent of the condition of the by-pass pipe). In addition, due to the shortage of nitrogen supplies, the plant was required to "circulate" or operate at operating temperatures and pressure for some hours before oxidation of the cyclohexane could commence. The report notes that the sudden rise of pressure during the final shift might have been due to the accumulation of peroxides in the system, or a nitrogen purge might have occurred, but there was no way of telling.

The explosion occurred at 4:53 P.M. on June 1, shortly after the next shift came on duty. It destroyed all the records, so the proximate cause is unknown. The force of the explosion has been estimated to be equivalent to 15 to 45 tons of TNT. It is estimated that 30 tons of cyclohexane at 300° Farenheit formed a vapor cloud, which ignited. There were twenty-eight employees killed and thirty-six other employees injured; beyond the plant boundary fifty-three people were injured, according to police records, and many more suffered unreported injuries.[19]

The plant was destroyed. Buildings 1,000 feet from the blast were damaged; windows in the surrounding community were broken as far as one and three-quarters of a mile away. At least 3 houses were demolished; at least some degree of damage was sustained by 1,821 houses and 167 shops and factories in this quite rural area. All agree that the temporary 20-inch pipe ruptured and the gas poured out until it was ignited by a nearby hydrogen furnace. But did the pipe rupture because it was inadequate in the first place, or because of the many other anomalies that existed in the process? Prior to the rupture the plant certainly appeared to be in a threatening mood.

The legendary Murphy was wrong. His law, that if anything can go wrong it will, is disproved by almost all post-accident investigations of large disasters. These investigations repeatedly point out that "it was lucky it wasn't worse." In this case the luck was that it was Saturday, since only a light shift was on duty at the plant; the wind was light, or a larger plume could have been created and the explosion could have come later, extending the damage; it was daylight and good weather so the fire brigades could be more effective; the plant was in a rural area, and so on. The rosy face of all accidents is twofold: it could have been worse, and look at the lessons we learned.

Fairly gross negligence and incompetence seem to account for this accident, but I would resist that conclusion. A fair degree of negligence and incompetence is to be expected in human affairs, and under the production pressures generated by a huge facility standing idle waiting for tools or hydrogen or pipes we can expect forced errors. While the temporary by-pass pipe was a poorly calculated risk, the by-pass had worked for a month. There were anomalous conditions prior to the accident, which might have led to overheating or overpressure regardless of the pipe and its supports; these anomalies could not be diagnosed. One suspects it would be quite easy to convene a commission of inquiry to investigate an accident that had not taken place, and in the course of careful inquiry find a dozen or so failures in the DEPOSE components that should have produced an accident long ago. Complex systems must live with these errors of design, equipment, procedures, and so forth. Since the data for the proximate cause were destroyed, we cannot be sure that there was an unexpected interaction of multiple failures, or simply a pipe break, but the account would suggest that there were many independent failures lying fallow for the fatal spark.

Our discussion of nuclear power plants should make the following observations familiar enough. There was organizational ineptitude: they were knowingly short of engineering talent, and the chief engineer had left; there was a hasty decision on the by-pass, a failure to get expert advice, and most probably, strong production pressures. But as was noted in the case of TMI and other nuclear plants, and as will be apparent from other chapters, this is the normal condition for organizations; we should congratulate ourselves when they manage to run close to expectations. Had the pipe held out a short time longer until the reactor was repaired, and a new chief engineer been hired, and had a governmental inquiry board then come around, they could have concluded it was a well-run, safe plant. Once there is an accident, one looks for and easily finds the great causes for the great event. There were unheeded warning signals—the by-pass pipe had been noted to move up and down slightly during operation, surely an irregular bit of behavior; and there were unexplained anomalies in pressure and temperature and hydrogen consumption. But these were warnings only in retrospect. In these complex systems, minor warnings are probably always available for recall once there is an occasion, but if we shut down for every little thing . . .

And finally, there are the great lessons learned. In one account of the accident there are recommendations for less inventory (smaller plants), greater separation between buildings, a different reactor design, and more status for safety advisors ("In highly technical industries like the process industries, it is not sufficient to employ as safety adviser only an elderly foreman who sees his job as taking statements from men who fall off bicycles."[20]) Insurance companies in the United Kingdom, the author complains, do not take into consideration the quality of management at plants and the methods of operation; instead, they establish insurance rates only on the differences in hardware and equipment. [21] It is unlikely that these recommendations will find their way into operating practices.

Vapor Clouds

Flixborough was a vapor cloud explosion. These are the most fearsome of petrochemical plant accidents. In a survey of vapor cloud incidents published in 1977, the author notes:

> Vapor cloud explosions have in recent years been the predominant cause of the largest losses in the chemical and petrochemical industry. Because of

trends toward plants of larger capacity, higher pressures, higher temperatures, and greater inventory holdup, these losses have been increasing both in frequency and severity.*[22]

The designation "vapor cloud explosions" generally means those that occur in the open air. An explosive gas forms a bubble, which may drift for several minutes before igniting. Some of the bubbles noted in the survey were large indeed. Although not all originated in chemical plant accidents, the following details indicate the hazards. In an Illinois accident in 1972, two railroad cars involved in a "humping" operation, where one overtakes another and the collision couples the cars together, collided with too much force and released 118,000 pounds of LPG (liquified propane gas). (Why they were humping together LPG cars is hard to imagine.) Since the cars were rolling, the gas spread as it leaked out. The bubble covered about 5 acres when it exploded. There were no deaths, but 230 people were injured, and the property damage (1972 dollars) was $10.8 million. Two years later in Illinois another railcar accident released isobutane. The cloud was one-half by three-quarters of a mile in size before it exploded—eight to ten minutes after the accident. The "yield" was estimated to be between 200 and 400 tons of TNT, ten times that of the Flixborough accident. There were 7 deaths and 356 injuries; property damage was $21.7 million. A tank valve failure in a New Jersey refinery created a cloud that extended to a highway just 200 feet away (not the twelve-lane New Jersey Turnpike, fortunately), where a car probably provided the ignition source; two were injured. A cross-country pipeline in Texas failed and led to a cloud that ignited after it drifted to a farmhouse, which provided the spark, killing one and injuring four. Sometimes the clouds do not ignite. An accident in a cyclohexane plant in Florida in 1971 produced a cloud that was 2,000 feet long, 1,200 feet wide, and 100 feet high, but it found no ignition source and dissipated. Releases of vapors from such plants are not necessarily due to system accidents, but they do indicate the catastrophic potential as plants and communities encroach upon one another.

*The author of the article is associated with the insurance industry, and thus has access to all the published data on such events. Yet 42 percent of the accidents noted, including some large and catastrophic ones, are referenced only as "private communication." No doubt these were covered at the time in local newspapers to some extent, but it is unfortunate that full technical discussions appear to be available only on a private basis.

Prosaic Synergies

Vapor clouds are fairly exotic. But the petrochemical industry can produce unexpected interactions in a quiet storage tank on a winter's night in bleak Pocatello, Idaho. In 1978 a 7,000-ton ammonia storage tank, at atmospheric pressure, partially collapsed when a vacuum developed in it. The tank was 74 feet high (roughly equivalent to a six-story apartment house) and 79 feet in diameter. Fortunately, the rupture that occurred with the collapse was above the level of ammonia and the tank was only one-third full, so the damage was small. The painstaking investigation revealed the interactions.

The power went out one night when the temperature was 10° F. It is believed that before the power was restored, ice crystals formed in the instrument air system that measured the pressure in the tank and transmitted it to the control room (failure #1). Another pressure indicator existed, a gauge on the tank, but it was also connected to the transmitter (a common-mode failure, #2). A third indicator was a simple manometer on the top of the tank, but with two other indicators this one had been allowed to fall into disrepair, and regular checks on it had not been made (failure of a backup device, #3). Even if the manometer had not fallen into disrepair, it is unlikely that anyone would have been sent out at midnight in 10° weather to climb a 75-foot tank on the remote chance that the instruments might be faulty because of ice crystals forming in the line.

The collapse of the tank was caused by the failure of the vacuum relief valves (failure #4). A tank as wide and tall as an apartment house requires that the proper internal pressure be maintained, but as the contents are drawn out, a vacuum will be created which must be relieved. As the tank and the contents heat up under the sun, on the other hand, the pressure must be relieved. All this is done by a neat device that relieves pressure at a certain point and vacuum at another point. A few months prior to the accident operators had replaced the original valve with a new one. However, it was the wrong valve. It did not have the vacuum relief capacity. Furthermore, maintenance, in checking the performance of the new valve, only checked for pressure relief, not vacuum relief. Normally, this feature is not utilized; as ammonia is drawn out, the compressors are used to pump air in. Even without the severe cold there could have been a failure, for as the ammonia contracts at night, the compressors are run to keep the air pressure up. But they are regulated by the instrumentation system, so its failure might lead to insufficient air pressure.

Next, the heaters failed. Nine of the twenty-two heaters, which are in the ground under the tank (a part of the foundation, I gather) had failed. This is why the ammonia contracted so much; it was not heated sufficiently. It was the combination of these diverse failures (instrumentation, relief valve, heaters) that caused the collapse; any one or even two of them would probably have been corrected by safety devices or have been discovered in the normal course of events.

The recovery operations revealed further complex interactions. Actually, there were two collapses; after the first the compressors had to be shut down immediately for safety reasons. They would have been restarted shortly, but with the instrumentation failure there was no way to ascertain the pressure inside. The tank sagged again, this time rupturing the side. The rupture allowed air in and equalized the pressure. But there was a danger of an explosive mixture, so the tank was blanketed with nitrogen (an inert gas). This blanket reduced the partial pressure of the ammonia in the tank, causing the ammonia to cool below its normal storage temperature of −34° F. Personnel decided to take the risk of having an explosive mixture, and the nitrogen purge was disconnected, because subcooling of the ammonia could bring the contents well below the temperatures for which the steel tank and the foundations were designed.[23]

The accident was a prosaic, trivial event in an industry where there is an average of one fire for every plant every eleven months, and 1,400 tank farm fires per year. But prosaic, trivial failures even in a passive storage facility can interact and come close to causing a large explosion in a large facility. Fortunately, Pocatello is small and remote.

The following concerns a modest fire in an ammonia plant, the analysis of an investigating team, and the revealing comments of industry personnel discussing a paper on the accident presented at a meeting. It nicely illustrates incomprehensible interactions, the tendency to blame the victims, and the production pressures in the industry to get back online no matter what the risks.

A power failure occurred in a Louisiana plant and the system had to start up again. A large start-up heater brings the gases used for the synthesis up to the temperature where the reaction will be self-supporting. The heater is 61 feet high, several feet in diameter, and stands in the open next to many pipes and near some buildings. A compressor sends gas into this heater where it burns and heats coils that carry the gas used for synthesis. Normally, natural gas is burned in the heater, but the compressor for this system was not in service because of a damaged seal, so "process gas" was used instead, as had been done on other occasions.

(This is not quite a "failure," but something equivalent, an off-standard condition that normally can be tolerated.)

The heater was lit at 4:30 in the afternoon. Three hours later the operators were ready to increase its firing, but when they applied more fuel they got a rumbling noise, and the temperature in the heater did not rise to the proper level. They verified the valve alignment of the start-up heater, and then the operator in the control room decided to increase the flow of synthesis gas through the system, hoping this would heat it faster. It did, the synthesis gas temperature began rising, and "all operating conditions seemed normal."[24] Fortunately, the technicians working outside near the heater left the area. Shortly thereafter the coil in the heater ruptured and the five-story heater was engulfed in flames. Two mechanically operated valves were closed and the fire allowed to burn itself out.

Subsequent investigation illustrated that substitutions in complexly interactive systems are not as easy as they are in linear systems. Normally, the substitution of a process gas compressor for a natural gas compressor should create no problems. But no one realized that because the natural gas compressor was not operating, the suction pressure of the synthesis gas compressor (the feedstock that is going to be changed) would be a bit lower than normal. Some unknown connection exists between the three separate compressors—natural gas, process gas, and synthesis gas—such that the inoperative status of the first affects the temperature of the third, even though they are quite independent. Furthermore, another condition must affect this interdependence, since no problem had occurred when the second was substituted for the first on other occasions. The investigation report does not indicate the pathways, but it nevertheless faults the operators for being unaware of them. "The effects of this abnormal condition" (the fact that the natural gas compressor was not operating and process gas was used instead) "were not fully realized by operating personnel." [25]

The report also faults the operators for not closely watching the indicators that monitored the flow of synthesis gas through the heater, which would have disclosed that something was wrong. But the report goes on to acknowledge that "the flow indicators were considered unreliable because there was hardly any indication of flow during both normal operations and in start-up conditions, especially when starting up both converters simultaneously," as they were now doing. (Note this trivial interdependency of the two converters and their effect upon flow.) The operators, then, were blamed for not monitoring a flow that was so faint it could not be reliably measured. It hardly mattered in any case, since both flow indicators had been set incorrectly, and furthermore the alarms for them, indicating low flow, "had been disarmed since they caused nuisance problems during normal operations."[26] Small wonder, after all this, that the flow indicators "were not closely monitored."

Post-accident investigation also revealed that the manufacturer of the start-up heater coils had upgraded the specifications for the coils eleven years before, but the plant had never made the change. Had they done so, the coils would have been stainless steel, as the new specifications required, and there would have been no rupture. During the next maintenance shutdown the change was made, as were a number of other changes. The company relocated the instruments needed during the firing of the heater to a safe location, and added better flow measurements and interlocks to cut off fuel gas when the synthesis gas has a low flow. One foreman was designated to focus solely on the unit when it is under start-up conditions, and "a formalized training manual was prepared." But there was no time for these safety measures when the plant was out of service following the fire. The article notes the production pressures: "As usual, following an unplanned shutdown, downtime has to be kept to a minimum." They patched the coil, and four days later had the plant running.

Considering this haste to get the plant running regardless of unattended deficiencies, and considering the presence of baffling interactions, irrelevant flow indicators, and an engineering failure to upgrade the coils as specified, it might come as a surprise to find that two of the three "causes" of the accident, according to the official report, were attributed to operator error. Operators did not appreciate "the critical parameters of this operation"; and did not realize it was an abnormal start-up condition. The third cause mentioned was the failure to have one foreman assigned only to unit start-up duty[27] —a curious and probably inconsequential assignment of error.

In the discussion that followed the presentation of the paper on the accident it turned out that not all those in attendance agreed with the investigating committee. An engineer from another company, for example, said, "We tend to do too many things too often without telling the operator what he has to do and why we changed our minds." Another said he favored correcting such problems through design—"put in stainless steel, then no matter what temperatures you reach, you should be safe," and "If you depend only on well-trained operators, you may fail." Others noted they had had similar heater failures. An engineer from another company described a similar accident. He said they had a ball of fire of about 100 feet radiating all around the heater, and "this is a most frightening event for personnel. One of our men tried to quit his job to avoid loss of his sanity; and some other people were quite afraid of the whole thing for a long time." Nevertheless, since only the top coil was damaged, "we cut the coil out, pulled the next coil up, and rolled the two together. The next day we were making ammonia again." His remarks provide a nice glimpse into the industry culture of avoiding downtime. The author of the paper analyzing the accident provides a glimpse into the contrasting culture of the workers: "We have an adjacent plant about 100 yards away from the start-up heater, so the crew there had a full view of the fire. When the unit did rupture, they decided pretty quickly that they weren't getting paid enough to stay around, so they bailed out and headed for the gate."[28]

Reassuring workers is a theme that occasionally comes up in the safety journals. It was a distinct problem after an explosion in a liquid carbon dioxide plant in Hungary where five employees were frozen solid instantly, and the liquid carbon dioxide displaced the air and cooled emergency equipment to −118° F. making the rescue of four other operators impossible. Fifteen others were seriously injured by freezing or metal splinters. The failures in this accident were multiple: below-grade steel was found to have been used; improper preheating procedures; a probable ice-plug formation due to extremely cold weather; and faulty safety valves. But note the disdainful comment of the two engineers at the plant who described the accident, referring to post-accident safety requirements: "Certain overrated safety requirements were aimed to calm the panic and stress atmosphere among the personnel under the effect of the explosion."[29]

The final accident we shall consider in this chapter is not noteworthy for its size, but for the frank assessment by the author of the article (an engineer at the plant at the time of the accident) and others on the complex system characteristics.

In one sense it was a fairly simple accident: three power outages in a row apparently caused a fan to operate below the required speed, and in addition, three louvers in ducts were stuck in the closed position. As a result, a steam header in the high-pressure steam system overheated and burst. Fortunately, no one was within 100 feet of the pipe at the time and so the damage, though extensive, was limited to the immediate area. The plant had been operating smoothly. "Nothing in the system appeared overstrained or pushed."[30] But there had been difficulty keeping the 900° F. temperatures in the proper range in the steam superheat system. Each time the plant had a shutdown and start-up the temperatures fluctuated, but no damage was done. High temperature alarms would occur, but the operators learned to ignore them.

One part of the start-up process involved reducing fuel gas pressure. Note the lack of knowledge in the following, and the consequences for operators.

> The procedure assumed that reduction in fuel gas header pressure would directly affect the process exit temperature from the reformer. We have learned since that this is not true over the entire range of fuel gas pressures. What happens is that with a given draft condition, at the time of purge gas extraction, any extra fuel over the air supply simply goes unburned into the convection section and does not impart any change in process exit temperatures. The operator can therefore very easily misjudge the required pressure reduction by only monitoring process temperatures.[31]

In addition, when opening the convection caps to cool the convection section so as to make up for the loss in steam from an exchanger, the temperature would sometimes rise, rather than fall, due to ignition of the unburned primary reformer fuel. This was not understood at the time.

Other problems were present. "In startup, control of superheat is very touchy." This is because zinc oxide desulfurization is used, and even with the safety caps wide open, excess temperatures are generated. The superheater coils, the article continues, are difficult to hold within alarm limits when half or more of the steam supply comes from an auxiliary boiler. Next, the draft fan had insufficient horsepower, and the furnace was operating on the threshold of afterburning. "To make matters worse," the louvers in the ducts were stuck closed (as noted before). "Considering the induced draft fan and auxiliary boiler louver deficiencies, it was fortunate to have not scattered the unit all over the landscape."

Note the number of complex interactions in this part of the system, some of them unexpected, such as the effect of using the auxiliary boiler. Note also the design problems: fan speed, and the heat added by using zinc oxide for desulfurization. Next, we have an equipment failure, the stuck louvers. Consider the amount of instrumentation that would be needed to have proper checks upon all of these components and interactions. (Some instrumentation was added during the rebuilding, but for louvers there is not much you can reasonably do.) Consider, finally, the lack of understanding of the particular dynamics of small parts of the system, such as the afterburning problem that made operator misjudgment easy. The author who, as you recall, was an engineer at the plant, appears to share our concern:

> Now, while this catastrophic failure is fresh in everyone's mind, it is not apt to recur. [But] since the heat exchange relationships are sometimes obscure to the operators because of the many variations, it is feared that some time in the

> future this knowledge will change as people change, and time will erode the memory of the catastrophe.
>
> As the entire event is contemplated, anything above 900° F. is too hot. It is possible to operate within this limit even though all of the recommended changes have not been installed. [He strongly recommended installing a desuperheater.] If the present state of knowledge had been achieved prior to December 11, the failure would not have occurred; it can be chalked up to lack of operating experience and incomplete mechanical diagnosis.
>
> On the other hand, *if a system is so complex and integrally meshed as to require superhuman operators** to constrain the process within safe limits, then it needs some modification. As the new generation of reformers with air-preheat is about to be born, steam temperature sensitivity is likely to continue unless some help is given the plant operators.[32]

After the paper was presented, a discussant noted it is not the newness of the plant that is the problem. Even in the older plants, he said, "We struggle to control it. . . . Runaways will take place and control by these caps is not the answer. . . . The way it is now we are in difficulties and I don't think anybody is sophisticated enough to operate the plant safely."[33]

The problems in this mature, but increasingly sophisticated and high-volume industry, appear to lie in the nature of the highly interactive, very tightly coupled system itself, not in any design or equipment deficiencies that humans might overcome.

Conclusions

Petrochemical plants are familiar to all of us, much more so than nuclear power plants. They have been around for a century and have done very little outright damage. The biggest killer, an explosion in an I. G. Farben chemical plant in Germany in 1921, took 550 lives; the biggest series of explosions, in Texas City, Texas, killed 56 and injured over 3,000. As disasters go, petrochemical plants have contributed very little to human suffering, though we are excluding air pollution and other forms of contamination here. Yet even some industry personnel are concerned that their scale, complexity, and proximity to human communities has been increasing steadily. The DuPont powder plants along the Brandywine River in Delaware in the nineteenth century had walls facing the river built of insubstantial material so that the frequent explosions would be vented over the river, and not damage adjacent powder sheds. That pre-

*Emphasis added.

caution is no longer possible; every major metropolitan area is encircled by or adjacent to vastly more lethal concentrations.

Along with the growth has come increasing safety. At least it would appear that as production has greatly increased, the annual loss from accidents has declined. Despite the many instances of neglect and haste and incaution in the accounts I have given, the record of the industry appears to be considerably better than that of the nuclear power industry, though the record is suspiciously difficult to establish. It is an industry with thousands of years of operating experience for most major configurations, and even the new installations can draw upon, and for the most part do not significantly deviate from, several hundred years of operating experience for the major pieces of equipment.

But it may be that we are beginning to reach the point in the experience and learning curve where the potential for nonlinear interactions is increasing geometrically. The new processes appear to be more complex than the old (though it may have always seemed thus), the throughputs significantly higher in volume, and all but the basic feedstocks more complex and unpredictable. Ammonia and gasoline plants are not going to change dramatically though the processing will undergo certain changes, but new products with unpronounceable names (and exciting properties) are spilling out of the industry. Given that the very nature of the transformation system requires nonlinear interactions and ever tighter coupling, the chances for system accidents no doubt will increase. The danger to the nearby metropolitan areas appears to be on the increase also. We could have a vapor cloud explosion that could destroy a square mile of populated earth, rather than merely wipe out an isolated farmhouse and its family.

It is unlikely that, reaching the point where an adequate comprehension of the possible interactions flags, the designers and the corporate officers will draw back. The culture of the industry runs counter to such restraint. Texas City, Texas, rebuilt its plants and tank farms quickly enough, sure that it would not happen again. "Cancer alley" in New Jersey continues its sprawling growth. Land is expensive in San Francisco Bay; the plants there are unlikely to put more space between their tanks, or between themselves and the crowding communities. Parts of Texas and Louisiana look like a vast Go board from 12,000 feet. New processes abound. The swollen control room of the large facility is being decentralized in the face of the complexity, with "supervisory controls" or "distributed controls" as the new buzz words. These are computers (microprocessors) that take over most of the control problems at the point of production, with only high-level functions being fed back to the central control room. This computerization has the effect of limiting the options of the operator, however, and does not encourage broader comprehension of the system—a key requirement for intervening in unexpected interactions.[34]

Yet my purpose in this chapter is not to call a halt to the expansion of the petrochemical industry. It is not even primarily to indicate the significant catastrophic potential that goes largely unremarked by those who worry about these things (admittedly because nuclear power and nuclear weapons accidents are vastly more dangerous)—though that is quite important. The purpose of the chapter is more prosaic: to demonstrate in something other than nuclear plants the existence of system accidents in systems that have many nonlinear interactions and tight coupling. The analysis of accidents in the petrochemical industry reinforces the utility of the argument of Chapter 3, where the theory of system accidents is laid out. If the accounts in the present chapter resemble those of Chapter 2 on nuclear power, that is precisely the point.

A petrochemical plant, like a nuclear power plant, remains a fairly self-contained system, however, even though an excursion will reach out into the environment. The environment may impinge—an airplane might hit the plant, or extreme weather create problems—but by and large that is not an issue. In the next chapter the environment will play a larger role, since aircraft have to use the environment to a greater extent. In addition, while we have to note the proximity of a tank farm to a heater explosion in the chemical industry, the two do not move about. The environment of an airplane, however, includes other aircraft, as well as air pockets, windshears, thunderclouds, and the disorientation of high-speed flight. The system, then, becomes more complicated. In succeeding chapters, these distinctions will be both more important and more troublesome. Thus, moving on to aircraft, we will move further into the environment and into the problems of defining systems.

CHAPTER 5

Aircraft and Airways

This chapter will take us closer than any other to our personal experiences, for almost all of us have flown on commercial airline flights. Having survived, we know it is at least fairly safe. Yet aircraft and the airways have system accidents, as we shall see, and there are many examples of production pressures, malfeasance, incompetent designs, and regulatory inaction. Why did you fly without giving much thought to any danger, then? There are some unique structural conditions in this industry that promote safety, and despite complexity and coupling, technological fixes can work in some areas. Yet we continue to have accidents because aircraft and the airways still remain somewhat complex and tightly coupled, but also because those in charge continue to push the system to its limits. Fortunately, the technology and the skilled pilots and air traffic controllers remain a bit ahead of the pressures, and the result has been that safety has continued to increase, though not as markedly as in earlier decades. But while crashes have declined, and mid-air collisions almost disappeared, little has been done about cabin fires and cabin missiles after a crash, which have killed hundreds needlessly. With a bit of inconvenience, the systems could be even safer with respect to crashes, and much safer with respect to post-accident recovery.

This introduction will review some of the risk data in historical perspective, suggest some system and industry characteristics that promote safety, briefly note some production pressures, and mention the role of the environment. That will clear the ground for an examination of some technical terms we must use, and the role of automation in flying. Then we shall turn to accidents, first in airplanes, and then in the airways. It is necessary to separate the two since flying (the aircraft and crew) is more hazardous than navigating the airways (the aircraft, crew, other aircraft, and ground control). The boundary between the two is fuzzy, but accidents rarely occur at the boundary.

In the course of this discussion of accidents we shall frequently come up against the possibility of production pressures as a contributing cause. As fuel becomes more expensive, aircraft more versatile, and deregulation of the airlines introduces new commuter airlines, such pressures may be mounting. Some near-accidents have occurred simply because, to save fuel, pilots are required to start up their third or third and fourth engines at the last minute before turning onto the runway for takeoff. Pilots complain of pressures to fly with lower reserves of fuel than they think they might need, because the extra weight requires more fuel to fly the aircraft. There are reportedly pressures to avoid declaring that there is no braking action on icy runways because the airport would have to be shut down. Pilots can be made to work fourteen hours out of twenty-four, for days on end, leading to extreme fatigue; made to fly low over a hospital in order to avoid official noise monitoring equipment; forced to fly with faulty equipment or be fired.[1]

As Safe as Driving

Going about in the air has always prompted a concern with safety.[2] Hang gliding was popular in the 1890s, and one of the foremost manufacturers of hang gliders installed a wooden hoop to break the fall of the pilot should he hit the ground too hard. Alas, the designer himself was killed in a glider that lacked this ingenious device. The Wright brothers made about a thousand glider flights to test designs and safety equipment before their first powered flight in 1903. In fact, they even built a simulator to train pilots in rudder movements. Their first flight carried a flight data recorder (engine revolutions, distance traveled, and flight duration). An upgraded version added heading, and altitude and acceleration became mandatory for airliners some fifty-five years later.

Aircraft and Airways

It was only five years, and a few thousand powered low-altitude flights, before the first fatal air crash occurred in 1908. By 1910 flying had become popular; there were almost 2,000 pilots in the world, most of them in Europe, and there were 32 fatalities that year. The odds of one chance in eighty of being killed did not slow the growth of aviation. The military value was quickly seen, as Europe prepared for war. By 1913, just ten years from the first powered flight, France had 1,400 military airplanes, Germany 1,000, Great Britain 400, and even Russia had 800. The United States, which was not preparing for a war, continued to lag, with only 23.

Accidents that are familiar to us today existed from the beginning. The first windshear accident (Orville Wright, pilot) occurred in 1904. The first mid-air collision was in 1910, just seven years after the first flight. The first pilot to be killed by a bird strike came in 1912. Exploding fuel was the biggest killer, and crash tests were conducted. After World War I, which of course gave great impetus to the industry, the U. S. Air Mail Service was founded. The service lasted for nine years, before it was taken over by private interests. Whatever it was that was so vital to transport quickly, it was delivered at a high cost. Life expectancy for a Mail Service pilot was four years. Thirty-one of the first forty pilots were killed in action, trying to meet the schedules for business and government mail. It may be the first nonmilitary example of a phenomenon that will concern us in this chapter: production pressures in this high-risk system. Though nothing comparable exists today in commercial aviation, such pressures, are, as we shall see, far from negligible.

Founded in 1918, the Air Mail Service had its first strike in 1919 over safety. One resulting procedure: "if the local field manager of the Post Office Department ordered the pilots to take off in spite of their better judgment, he was first to ride in the mail pit [a kind of second cockpit] for a circuit of the field." [3] The year 1922 had no fatalities. Yet there was a forced landing every twenty hours of flight. The demand for fast mail delivery (or the challenge to the Mail Service) led to the Night Transcontinental Air Mail Service shortly thereafter, using bonfires lit across the country and flare pots at the many landing fields. One in every six air mail pilots was killed in the nine-year history of the service, mostly from trying to fly through bad weather, such were the production pressures. It is hard to believe that the mail was all that urgent, but the air mail pilots were heroes of their day. In contrast, today the risk to a commercial airline pilot is comparable to that of the average citizen, and they pay normal life insurance rates.

Since those fiery early days of aviation, the number of flights has increased astronomically, and the fatalities per flight, or per passenger, or passenger mile, has decreased almost as dramatically. Safety comparisons with other systems are hard to make. In many respects, commercial air travel appears to be much safer than automobile or rail travel. Many fewer people are killed in the first than the other two. But an equally useful statistic would be the number of fatalities per hour of exposure, or per million miles traveled. Unfortunately, we do not have these statistics for automobile accidents. Jerome Lederer, often called the father of modern flight safety, suggests that if we used the statistic of fatalities per 100,000 hours of exposure, highway travel would be the safest mode of transportation.[4] At the least, it seems, the airline pilot has about as much of a chance of being killed in a year of flying as does the average driver in a year of driving. Nor is this a significant chance, despite the 50,000 people killed each year on the highways. One's chance of being killed while driving a car is only one percent in fifty years of driving.[5] We simply do a lot of driving and very little flying, giving us the impression that the risk of the latter is much smaller. In terms of exposure (fatalities per hour) they may be very similar. Both have dropped dramatically since they were introduced.

Comparing types of flying there is an obvious hint as to why it has become safer: the more commercial the activity, the safer it is. Jetliners are the safest mode, followed by corporate jets, then commuter airlines, then general aviation, and at some distance, military flights. Again, the figures have to be examined carefully. Considering the rate of fatal accidents per million miles of aircraft flight, commercial flights (major airlines and commuters) are sixty-five times as safe as general aviation flying. But if we consider fatalities per 100,000 hours flown, thus "correcting" for the fact that a large airliner has 350 people at risk, and the small sport craft only the pilot at risk, there is not much difference between the two. Comparison of either with military flights must be done cautiously; it is not necessarily true that safety is less carefully considered in the military. Military flights take place in weather no commercial flight would fly in, and in simulated combat conditions; much of the effort is training (they do the training, in effect, for the commercial airlines); the aircraft are advanced, high-performance craft, and they must make difficult landings, as on ship decks. The risk will be there no matter how much attention is paid to safety. Navy statistics indicate that a fighter pilot who spends a twenty-year career flying high-performance jets has, incredibly enough, a 23 percent chance of dying in an aircraft accident.[6] This does not include the possibility of dying in combat. Though this is about twenty times as risky as driving, military pilots do a lot more driving than flying, and more of them die in auto accidents than in aircraft accidents.

There is, of course, an enormous incentive to make commercial aviation safe. Airline travel drops after large accidents; airframe companies suffer if one of their models appears to have more than its share of accidents. Public reaction appears to be stronger when identifiable, rather than random, victims are produced in an accident. There are passenger lists in airline accidents, but in a chemical accident only a small, random proportion of those living nearby are affected. Lawsuits follow immediately if the investigation of the National Transportation Safety Board (NTSB) hints at vendor or airline culpability. In addition, there is a strong union at work to protest unsafe conditions—ALPA, the Airline Pilots Association. It even conducts its own studies and makes its own safety recommendations. The Federal Aviation Administration is charged with both safety and facilitating air travel and air transport, and spends significant amounts of tax dollars pursuing safety studies and regulations. Finally, an independent board, the NTSB, conducts investigations and prods the FAA to set new safety requirements. No other high-risk system is so well positioned to effectively pursue safety as a goal. The nuclear power industry, for example, lacks a strong union, has random public victims with delayed effects, has no safety board that is independent of licensing and regulatory functions, and does not see an immediate effect on its profits if safety flags (though a far more severe incentive exists to avoid a catastrophic accident which could shut down the industry).

In addition to these incentives, there are "structural" conditions that foster safety. A trivial one is that industry elites and regulatory elites and politicians all fly themselves, and thus have a personal stake in safety. More important, experience accumulates fast: there are thousands of flights every day, with all the dangers of takeoff and landing and mid-air collisions; new aircraft are introduced every few years, permitting the introduction of safer designs. The performance of operators is closely monitored, even recorded, and the environmental conditions that exist are a matter of record (in contrast as we shall see, to the marine system). Gradated training is possible, from small aircraft to large, from military to commercial. The salaries of operators are very high in the large airlines ($70,000 with a short workweek), attracting exceptional personnel—although a pilot can work long hours for a mere $20,000 a year in the commuter airlines. Finally, the military services and to a small degree, the space program, in effect provide much of the capital for development, testing, and production.

Despite these advantages over most other high-risk systems, there are still system accidents to contend with, as we shall see. But even here there are differences. I have located flying on the Interaction/Coupling chart (Figure 3.1) as being quite complex and tightly coupled, and I think it will always be thus. But it is not as complex and tightly coupled as the nuclear power system because, with one exception, it is not basically a transformation system. The exception is the high-speed, high-altitude flight where the craft "transforms" the "envelope" of air around it, and thus may exceed it or fall behind it, with very serious consequences. This part of the system has most of the characteristics of transformation systems: poorly understood dynamics, unobservable processes and intrinsically poorly monitored and instrumented processes, and critically narrow limits of safety. But aside from that, the complexity and coupling stem from more prosaic sources, such as proximity and common-mode connections (there is no way to make the system linear in this respect), very limited time for recovery from failures, almost no buffers or redundancies other than those designed-in and subject to failure, limited slack, and unifinality, only one way to achieve the goal.

With millions of operating years of experience, repeated trials, tests without catastrophic consequences, and considerable government support, the industry has been able to maximize the potential for technological fixes, including buffers and redundancies. Two engines are better than one; four better than two; the jet engine less complex than the piston engine; and of course the industry makes use of exotic new materials and instrumentation. System accidents in flying will remain, but they have been reduced substantially. Unfortunately, the technological fixes have frequently only enabled those who run the commercial airlines, the general aviation community, and the military to run greater risks in search of increased performance. As the technology improves, the increased safety potential is not fully realized because the demand for speed, altitude, maneuverability, and all-weather operations increases. The corporate Learjets, we shall see, keep rocketing their executives to the limit, when cutting the maximum speed back by 10 percent would allow much greater margins for safety. The goal of the airlines appears to be to land with 100 feet visibility—too little for the pilot to see the tail of his own airplane. I am surprised they do not have studded snow tires so that airports never need be closed because of ice!

The airways system is more linear, and loosely coupled. The task is much simpler than flying itself; it is just to keep aircraft from bumping into each other in the airways and from bumping into the ground. No transformation processes are involved and complexity and coupling can be reduced through technological changes. For example, bumping into the ground is known as controlled flight into terrain, or CFIT. This was the most frequent form of serious accident until the introduction in 1975–1976 of two significant innovations, the minimum safe altitude warning (MSAW) system of the third version of the automated radar terminal system (ARTS-3), and the ground proximity warning system (GPWS) carried in the large jets which sounds a horn and shouts "pull up, pull up." Put succinctly, if maddeningly, CFIT accidents succumbed to ARTS-3's MSAW and to the GPWS in large turbine-powered aircraft. Of course, the aircraft and airways systems interact. Distracted by company paperwork, flight attendant inquiries, or by minor equipment failures, a pilot misunderstands controller orders, and the runway is missed or another aircraft struck. The density of traffic results in many flight numbers sounding alike, and if a pilot is preoccupied with a faulty landing gear warning horn, the two failures can produce a catastrophe. However, I have found such interactions to be quite rare. Generally, system accidents stem from either the aircraft or the airways system alone, not in the interaction of the two. We will consider the airways system and air traffic control at length in the second part of this chapter.

In sum, then, commercial flights are very safe, and they are safer than other forms of flight because they cover longer distances and at higher speeds carrying more people, thus reducing the risk of departures and landings per person, and hours of exposure per mile traveled. In terms of miles traveled and hours of exposure, travel on major scheduled flights is about as risky as automobile travel. The safety of both automobile travel and airline travel (and military and general aviation as well) has increased dramatically in this century, but since the 1960s and 1970s the safety curve has flattened out; we appear to be in the area where further increases are very hard to achieve.

Aircraft

That Wonderful Flying Machine

Everyone is familiar with the picture of a cockpit of a modern jet airliner, with hundreds of knobs, switches, and buttons, and perhaps a hundred dials and visual aids. The newer cockpits have CRTs and audible aids in addition. There are also several levers to be manipulated, and a kind of steering wheel studded with buttons; the wheel, or "yoke" or "stick" will shake or vibrate as an additional alarm. A malfunction may produce, simultaneously, a few flashing lights, a squawking horn, a recorded male voice intoning "pull up; pull up" or "slow down; slow down," and a shaking of the yoke.

Within arms' reach of the crew there are an enormous number of devices to alter the aircraft's behavior.* They can check or flip or set or increase or decrease perhaps a thousand parts, scores of units, and a dozen or so subsystems in the aircraft system. Yet only rarely do they touch more than a few of these access points; the aircraft is a model of automation, as well as of complexity. Much of the crew's intervention is prior to takeoff—setting and checking—and after landing. Once the airplane is aligned on the runway, ready for takeoff, the officers usually get by with only a large handful of devices. Computers control computers in the sophisticated corporate jets, military aircraft, and jet transports. Despite the automation, the complexity of the system keeps the crew extremely busy at peak times.

We will refer to the pilot as he; there are only about 100 female airline pilots out of 30,000, probably because there are none at all in the major supplier, the military. Once the captain of a jetliner is ready for takeoff, he can engage an automatic flight control system (AFCS), and then function largely as a system monitor until the airplane lands at its destination. The AFCS will perform such tasks as maintaining the rate of climb or descent, the direction, the wing level, and the altitude. It will carry out predetermined maneuvers, such as turning left 30 degrees after passing a beacon near Denver and going from 30,000 to 35,000 feet. Theoretically, the whole flight could be programmed into the computers and executed automatically, without any attention from the crew.

In practice this does not happen. The crew may not know, until they are clear of the terminal control facility of Air Traffic Control (ATC), what altitude has been assigned to the plane, and they will probably have to change it at various points in the flight. The approach path to the airport will be set at the last minute by ATC. The heading may be changed, which will affect the radio navigational aids that will be used, and the means of determining the location over the earth will be changed. Thus, the pilot typically does not leave all the functions in the hands of the AFCS, but will reset them or take over manual control.

However, the computers do handle the very delicate matters of maintaining the proper altitude under varying atmospheric conditions such as temperature and air density, of "flaring" out the airplane when it reaches the proper altitude (leveling it off), of increasing or decreasing power the

*The crew includes the captain, co-pilot or first officer, and in large airliners, an engineer. They operate from a "flight deck"—the cockpit.

proper amount with the automatic throttle system, and of running the inertial navigation system (INS). This latter, which only the newest airliners have, is based upon a system of gyroscopes and accelerometers which are sensitive to all motion from a predefined starting point on earth. The pilot can intervene, but does not have to.

I have drawn upon a brief article by Edwyn Edwards, a human factors engineer, for the above information.[7] Edwards continues by arguing that all this automation has not reduced the workload of the pilot a great deal; instead, it has increased the operational effectiveness of the system. Airplanes are able to fly faster (the cruising speed has doubled in a twenty-year period) leaving less time for navigation, communications with ground control, and system management. Traffic density has increased greatly. Airplanes can fly in worse weather and with very low ceilings. The spacing between aircraft has been reduced substantially. Fuel management has become a high priority for the pilot, requiring precise navigation, flap control at landing, and engine thrust management. Cabin comfort has increased in priority, affecting climb rates, rate of turn, and avoidance of turbulent patches of space. On a cross-country flight I watched the captain of a wide-bodied jet observe on a CRT a thunderstorm developing 100 miles ahead, ask the first officer to radio the ATC en-route controller for permission to deviate around it, receive the permission, type the proper values into the autopilot and reset it, all in a matter of moments. (This happened while the engineer was shifting radios about in the flight deck, because, on this "thoroughly routine" flight, as they characterized it, two of the three radios that communicated with ATC were malfunctioning. Altogether, there were about six malfunctioning or nonoperational pieces of equipment on the flight deck, all dutifully written up by the captain without surprise or comment at the end of the trip.)

All of these automatic systems make the craft more "efficient," in terms of commercial or military criteria. Indeed, in the newest jets, such as the Boeing 767, the flight crew has been reduced to two, and a large number of knobs, switches, and dials have been replaced by CRTs and buttons linked to computers controlling computers. But with each bit of automation, more difficult performance in worse weather or traffic conditions is demanded. Thus, argues Edwards, there is no net reduction in workload on the crew. It appears, indeed, that workload has become more "bunched," with long periods of inactivity and short bursts of intense activity. Both of these are error-inducing modes of operation.

With each of the automatic devices (cabin pressurization, temperature control, transponders that emit signals to ground control, fuel monitoring and adjustment systems, warning devices such as horns or stick shakers, and the whole vast complex of units in the autopilot subsystem) follows the inevitable residual potential for error. While each new device reduces some chances for error, it also introduces its own bundle of error possibilities, hopefully smaller than the bundle it eliminates. Accidents have been associated with forgetting that the autopilot is on, or is off; with confusion over which localizer or beacon is in use; with forgetting which one of four ways to measure altitude is in service; with computer failures in the INS and other systems. In one awful accident, the headings entered into the autopilot were changed at the last moment before a flight without informing the captain, and the airliner flew into a mountain in Antarctica.[8] It has also been suggested that the high degree of automation has meant that the skills that pilots need when they have to intervene in the automatic system have become rusty through lack of application and use. Even staying awake on a long flight, especially over the Atlantic, becomes a serious problem. Engineers speak of a "control loop," in which the "man in the loop" is the problematical element. This is the human component in a series of sequentially interacting pieces of equipment that control or adjust a function. But when the pilot is suddenly and unexpectedly brought into the control loop (in other words, participates in decision making) as a result of (inevitable) equipment failure, he is disoriented. Long periods of passive monitoring make one unprepared to act in emergencies. The sudden appearance of several alarms, all there for safety reasons, leads to disorientation. As Earl Weiner puts it in a perceptive article about CFIT, "The burning question of the near future will not be how much work a man can do safely, but how little."[9]

Some studies have argued that automation has gone too far, or gone as far as it can. (Although that conclusion is probably reached every decade, in every system!) A government study even recommended that one key device be disabled for all but a few long-distance flights. This resulted from a study of thousands of near accidents reported voluntarily by aircraft crews and ground support personnel. The study concluded that the altitude alert system (an aural signal) had resulted in decreased altitude awareness by the flight crew. This resulted in more frequent "altitude busts"—instead of leveling off at 10,000 feet the craft keeps climbing, or keeps descending. A study of altitude busts noted that they rarely occur in bad weather when the crew is most attentive.[10]

Nevertheless, these systems must depend upon automation if they are to be reasonably safe at the level of aircraft density, speed, inclement weather, crew size, fuel savings, and cabin comfort that the airlines now demand. And there is no doubt these automated systems are highly effective, in terms of both efficiency and safety. In marked contrast to marine transport, as we shall see, the technological fix has increased both performance and safety. Equipment failures appear to be few compared to nuclear plants or chemical refineries, and they very rarely lead to accidents, at least in large commercial jets. But accidents will happen, and still do.

What, then, is left to cause accidents? The equipment is exemplary, safety and redundancy are built-in to a high degree, the designs are reasonably sensitive to human factors problems. One obvious possibility is human error. There are more occasions for it in flying than in running chemical or nuclear plants or navigating ships. The crew may be inactive on long legs of a flight, especially the very long legs of transoceanic flight, but other than that they are very busy with hundreds of small duties and a few large tasks. There is plenty of room for at least small errors, then.

Fortunately, there is one careful study of errors on the flight deck (but unfortunately nothing comparable for other systems). It suggests rampant errors without any catastrophic consequences; that is, the errors primarily concern small adjustments, and recommended but not required procedures and sequences. The study was made for a European airline flying short hauls (such as London to Glasgow or Frankfurt to London). They found an incredible rate of more than one error every four minutes.[11] The vast majority of these errors are caught very quickly, or are insignificant. But I doubt that even very close observation of industrial plant operators when they are starting up, shutting down, or changing the production system, would reveal an error rate of one every four minutes for a crew of three.

Thus errors are common. Where accidents are involved, the studies indicate 50 to 70 percent of the cases stem from human error (the rate is over 90 percent for ground controllers). One Air Force study, by Major Santilli, came up with a similar figure, but noted two circumstances. First, the "mishap" rate for the Air Force is exceedingly stubborn; after a sharp drop from 1950 through 1968, it leveled off through 1977. This is frustrating, he notes, since human error should in principle be reducible, and human error is held to be the source of so many accidents. (I would disagree with the major here; in principle human error in operating is not as reducible as in more unpressured activities such as design, fabrication, and maintenance.) However, he himself is skeptical of the classification, arguing that the designation of human error, or pilot error, is a convenient catch-all for "mishaps whose real cause is uncertain, complex, or embarrassing to the system."[12] His analysis seems apt; uncertainty and complexity are accident causes we have identified, and "embarrassment" is another way of saying "blame the victim" rather than the masters of the system.

Here's one example of the industry blaming the victim for its own failures. According to the official report of a New Zealand inquiry board, New Zealand Airways Limited tried to hide its own ineptitude by deliberately falsifying or destroying evidence concerning that 1979 crash, mentioned earlier, of a DC-10 sightseeing airplane into a mountain in Antarctica. New Zealand Airways blamed one of the 257 victims, the captain. The initial inquiry, which found pilot error as the cause, was contested by the widow of the pilot, but particularly by the pilots' union, and a subsequent, more thorough investigation was conducted.[13] In the military, there is no pilots' union, and even widows probably find it hard to protest the official sentence, so embarrassment may more easily be disguised. The controversy over this event still continues. A new employee of the airline recently reasserted the conclusion of pilot error in a flight safety bulletin, citing the fact that the airplane was flying too low. But the Mahon report noted that the airline virtually advertised the fact that low flights were made in order to improve sightseeing; the captain was expected to fly low.[14]

Thus, we can agree with the major that the attribution of pilot error is a convenient catch-all. Pilot or crew error does exist; it is bound to exist. Pilots are no more infallible than designers or contractors. But the complexity and the coupling of the system appear to account for a significant number of accidents. Let us turn to some accident stories.

Kitchen Trivia

This book opened with the example of a trivial failure in the kitchen, indicating unexpected interactions and tight coupling. We will return to the kitchen for two examples of trivial failures in our exploration of flying. They are useful illustrations of complexity and coupling, the rigors of the environment, and of how nothing is trivial.

A commercial airplane (an Israel Aircraft Industries Model 1124) was flying at 35,000 feet over Iowa at night when a cabin fire broke out. It was caused by chafing on a bundle of wire. Normally this would cause nothing worse than a short between two wires whose insulations rubbed off, and there are fuses to take care of that. But it just so happened that the chafing took place where the wire bundle passed behind a coffee maker, in the service area in which the attendants have meals and drinks stored. One of the wires shorted to the coffee maker, introducing a much larger current into the system, enough to burn the material that wrapped the whole bundle of wires, burning the insulation off several of the wires.

Multiple shorts occurred in the wires. This should have triggered a remote-control circuit breaker in the aft luggage compartment, where some of these wires terminated. However, the circuit breaker inexplicably did not operate, even though in subsequent tests it was found to be functional. Though the report does not speculate on this, it is possible that multiple shorting created circuits that defeated the circuit breaker. The wiring contained communication wiring and "accessory distribution wiring" that went to the cockpit. The consequences of these complex interactions and failures, namely the proximity of the bundle and the coffee maker and the simultaneous failure of the emergency safety device (the circuit breaker), are graphically and economically covered in the NTSB report.

> Warning lights did not come on, and no circuit breaker opened. The fire was extinguished but reignited twice during the descent and landing. Because fuel could not be dumped, an overweight (21,000 pounds), night, emergency landing was accomplished. Landing flaps and thrust reversing were unavailable, the antiskid was inoperative, and because heavy breaking was used, the brakes caught fire and subsequently failed. As a result, the aircraft overran the runway and stopped beyond the end where the passengers and crew disembarked. . . . There were no injuries.[15]

There is nothing inherently complicated about putting a coffee maker aboard a large airplane, but a simple part in a complexly interactive system can have such extensive consequences that Murphy is almost vindicated. A kitchen problem nearly caused another accident in a Capitol Airways DC-8 night flight from New York to San Francisco, during the awesome winter of 1981–82. The temperature at Kennedy Airport in New York was 2° F., and the plane was delayed. Mechanics were changing a fuel pump, and received frostbite, causing further delay. After the plane got into the air, the passengers were told that there would be no coffee because the drinking water was frozen. They settled down for a long flight; it was now 2 A.M. New York time. The flight engineer then found he could not control cabin pressure. Later investigation disclosed that the frozen drinking water cracked the water tank, heat from ducts to the tail section then melted the ice in the tank, and because of the crack in the tank, and the pressure in it, the newly melted water near the heat source sprayed out. It landed on an outflow valve in the pressurizing system, which allows excess cabin pressure to vent to the outside. The water, which had gone from water to ice and then back to water, now turned to ice because the outside of the air valve is in contact with −50°F. air outside the plane at 31,000 feet. Ice on the valve built up pressure in the valve and caused it to leak, and the leak made it difficult for the compressors to maintain the proper cabin pressure (and oxygen content). This is not an interaction that a design engineer would normally think of when placing a drinking water tank next to the fuselage of a jetliner.

The captain decided that there was, in the words of the NTSB report, a "remote possibility" that the loss of cabin pressure (and thus oxygen) would be accelerated (and having no idea of the cause of the pressure problem, it *was* a possibility and not very remote at that). So he ordered an emergency descent, reversing the inboard engines, ejecting oxygen masks for the passengers, and diving 14,000 feet in three minutes. Perhaps preoccupied with the anomalies, or fearing loss of cabin pressure, he didn't warn the flight attendants or the passengers; he just took it down. (It was discovered later that even if he had tried to warn them, the public address system in the front half of the cabin was inoperative—a not unusual failure in airliners.)

The flight attendants and passengers naturally panicked. Those lovely speeches about how one should put on oxygen masks in the unlikely event that there were a loss of cabin pressure were useless. The masks did not all drop. A stewardess edging up the aisle told passengers to use a sharp object to remove the masks ("as if," one passenger later commented, "everyone carried a sharp object"). Other masks stuck together. Some were yanked loose in the panic. But it didn't matter; there was no oxygen, and by the time the masks dropped, none was needed. For safety reasons (oxygen is highly flammable), oxygen is on an automatic system and will only come on when the cabin altitude reaches the Mount Rainier equivalent of 14,750 feet. The plane was below that altitude in just under three minutes.

Some passengers are suing Capitol Airlines because of the handling they received when the plane made an emergency landing at Denver; they did not receive food or drink or lodging, no special attention was given to children traveling alone, and it was some hours before they learned how they were to get out of Denver (Capitol has no facilities at Denver). Some refused to fly out, rented a car on their own, and drove the 1,235 miles to San Francisco!

Thus, even delay in replacing a fuel pump at the terminal can have unexpected consequences in a system so tightly packed that unexpected interactions are likely. One other point: had the airplane crashed and burned, and the minuscule evidence regarding the shorted wires been destroyed, it is conceivable that "pilot error" or crew error would have been given as the "most probable cause" by the NTSB.[16]

A Few DC-10s

DC-10s have been involved in some dramatic, catastrophic, and highly publicized accidents. Those described below will serve two purposes: to indicate the complexity and coupling of jet airliners and to indicate—to put it politely—the exercise of calculated risk by private business. Jet airliners are very safe, as we have detailed earlier. But they are still subject to system accidents as well as component failure accidents. Any actions by private companies that knowingly compromise safety increase the risk of both types of accidents, while prudent management, on the other hand, can reduce the frequency of such accidents.

The crash of the American Airlines DC-10 at Chicago's O'Hare Airport on May 25, 1979, with 273 victims, after the engine tore off as a result of an engine pylon failure was not, strictly speaking, a system accident. There were multiple failures but they had a common cause, even though they were in independent systems. However, it is worth a brief review of the accident, since it points out the tight coupling of airplanes, and some important liability and calculated risk problems. The ultimate "cause" of the accident was determined to be poor maintenance practices by American Airlines wherein the engine was removed for servicing in a one-step procedure, which could possibly (not inevitably) cause damage to the pylon that holds it onto the wing. As deplorable as this shortcut was, the pylon separation itself is not what concerns us; it merely exposed a design failure that had been exposed before.

Losing an engine, even one that comes completely off the wing, should not disable the DC-10 completely, for the plane is designed to fly with two of its three engines. The unfortunate aspect of the accident was that when the engine and pylon ripped off, they severed cables that controlled the leading edge slats which are extended on takeoff to provide more lift to the wings. The slats on one wing then retracted.

But that in itself is not fatal, since by applying full power and other maneuvers the pilot can fly the plane with two engines and retracted slats on one wing. However, the crew must know that the slats are retracted. But four hydraulic lines were also severed in the process, and this eliminated two warning signals of the slat position: a flap mismatch signal, and a stall warning signal. If either of these had been functional, the pilot might have had time to apply full power and pull out of the stall/roll. (Though I have not been able to determine this, it is also possible that if the two warning systems themselves gave a warning that the warning system was inoperative, the crew would have averted a disaster. Unfortunately, most warning systems do not warn us that they can no longer warn us.)

As the NTSB report put it:

> The loss of control of the aircraft was caused by the combination of three events: the retraction of the left wing's outboard leading edge slats; the loss of the slat disagreement warning system; and the loss of the stall warning system—all resulting from the separation of the engine pylon assembly. Each by itself would not have caused a qualified flight crew to lose control of the aircraft, but together during a critical portion of the flight, they created a situation which afforded the flightcrew an inadequate opportunity to recognize and prevent the ensuing stall of the aircraft.[17]

This quote deals only with the loss of control of the airplane, which is not the same as the "probable cause" of the accident. The fine distinction between loss of control and probable cause can determine hundreds of thousands of dollars in retrofitting and vastly more in the assignment of blame in legal proceedings. The "probable cause" of the accident was carefully worded by the Safety Board in another part of the report. It blamed the accident on a "maintenance-induced crack," but not also on a design failure that allowed the slats to retract if the wing were punctured. Because of this careful distinction by the NTSB, McDonnell Douglas, the manufacturer, was not required to change the design, nor could the company be charged with a design deficiency. The company insisted that since the plane can fly with asymmetrical slats, it was not required to introduce changes to prevent slats from becoming asymmetrical.[18] Perhaps so. But flying it under those conditions is so difficult and dangerous that they saw fit to put an asymmetrical slat warning light in the cockpit, which the pilot must respond to with an immediate change in his controls.

But then it turned out there had been a similar accident in 1977 with a DC-10 in Pakistan. And on January 31, 1981, a DC-10 that was leaving Dulles Airport had one fan blade fracture. The engine cowl and the fan case came off and struck three leading edge wing slats. "The slats, flaps, and (landing) gear were retracted at the time the aircraft was climbing. There is the possibility that if the aircraft has been configured differently or had been at a different speed or altitude, the aircraft structure may have been substantially damaged by the separated components."[19]

On September 22, 1981, it happened again. A DC-10 engine blew up on taking off from Miami, with the slats still forward, and the debris severed the cables holding the leading edge slats in a forward position.

Fortunately, the crew knew of the engine blowing and aborted the takeoff.

McDonnell Douglas was aware of the danger; for aircraft designers it is not a recondite matter at all. But in various studies performed by McDonnell Douglas, the chance of (1) loss of engine power, (2) with resulting slat damage, (3) during takeoff was estimated to be less than one in a billion.[20] Yet this highly improbable event had now occurred *four* times in DC-10s. After even one occurrence they might have thought that their probabilities had now run out, and made changes. McDonnell Douglas finally came around, and was, as of the spring of 1982, installing a device (essentially, a ratchet) that costs only a few thousand dollars and can be installed in a few hours, which will prevent slat retraction in such emergencies. It took three years since the Chicago accident, five years and 273 deaths since the Pakistani accident for the company to make the modification. The locking mechanism is a part of the standard design of other jets, such as the Lockheed L-1011 and the older Boeing 747 jumbo jet.[21]

One of the worst accidents in aviation history was the crash of a Turkish airline DC-10 near Paris on March 3, 1974, with 346 deaths. For two days after the accident the authorities held to the theory that a bomb had exploded in the airplane; there were serious problems with bombs in those years. Yet a day after the accident John Godson, an English journalist, who just happened to be writing a book to be called *The Rise and Fall of the DC-10,* noticed a news item that a cargo door had been discovered far from the other wreckage. He sent a letter to a British newspaper proposing a theory of the crash, which was eventually accepted. The cargo door blew open, resulting in rapid decompression of the cabin. This caused the cabin floor to collapse. The collapse wrecked all the major control cables and hydraulic lines, which has been placed under the cabin floor.

The reason he suspected this was that McDonnell Douglas had been warned of this by a Dutch engineer in 1969, when the first prototype of the DC-10 was being built; by a McDonnell Douglas subcontractor in the spring of 1970 who predicted one of the cargo doors would come off during flight during the twenty-year life of the airplane because of the latching and locking design; by a static ground test of the airplane in May 29, 1970, when the rear cargo door blew out in a pressure test and the cabin floor collapsed (an inadequate modification of the design was made as a result); by eleven entries in maintenance logs up to June 11, 1972, concerning difficulties with locking the door; and by a narrowly averted disaster near Chicago on June 12, 1972, when the cargo door blew out. John Godson simply put five blazing warnings together with cargo door evidence in the Paris crash and made a reasonable hypothesis. The FAA might have done the same, but it didn't. McDonnell Douglas had plenty of reasons to act after the first, second, third, fourth, or fifth warning. Only the NTSB seems to have been concerned.

As a result of the earlier June 12, 1972, Chicago accident the NTSB recommended that the FAA issue a mandatory directive regarding redesign of the door. (One solution, since it was a cargo door and not a passenger door, would be to have it open in, or slide up, rather than open out.) But the head of the FAA and the head of Douglas Aircraft entered into a nonmandatory gentlemen's agreement that a simple plate would be added and this would be sufficient. Godson suggests that this money-saving solution may have been aided by an unusually large donation by McDonnell Douglas to the reelection campaign of President Richard Nixon just three days after the accident. However, this could have been a mere coincidence since the FAA, as we shall see, has long been criticized for siding with the aircraft industry when the industry opposes the recommendations of the NTSB.

Presumably, the metal plate might be sufficient. But first the simple installation had to be made. It took United Airlines 90 days to install them; American Airlines, 268; National Airlines, 285; and Continental, 287. Through an oversight, apparently, some new planes waiting to be delivered never had the modification made. Laker Airlines, after the Paris crash, found that the modification had not been made on two planes it had received some time after the modification was decided upon in the gentlemen's agreement. The president of Douglas Aircraft claimed that the manufacturing records of the planes indicated the modification had been made. But an examination of the records showed that it had not been done. The modification, of course, also had not been made on the Turkish airline plane.

But this was not a simple component failure accident; it took some independent failures and a confluence of several events to bring the plane down near Paris. First, McDonnell Douglas had to decide not to thoroughly redesign the door; next, they had to fail to put a patch on all the still undelivered planes. The Turkish airline was not at fault since they would assume the plate had been added, if they even knew of the troubles. This still would not bring about an accident. The door had to be improperly closed. Unfortunately, the unmodified plane just happened to be at Orly Airport in Paris during a crisis. An air traffic controllers' job action in London was holding up all flights from Europe. Many were cancelled, but the Turkish plane had a chance of getting in if they hurried. The plane filled with people anxious to get back, and the departure, at night, was rushed. (Production pressures might be added to design, equipment, procedures, operators, supplies, and environment in our DEPOSE list of possible component failures.)

The baggage handler responsible for closing the cargo door was presumably under production pressures. He was experienced, and spoke two languages. Unfortunately, however, he was unfamiliar with this version of the DC-10 cargo door, and the instructions for locking, while clearly printed in two languages next to the door, were in two languages he did not understand. (The procedure was complicated and difficult; a flashlight would have to be focused upon a small window to be reached by moving the platform the baggage handler stood on, and even then it would be hard to see the marker.) He closed the door, but the lock did not properly seat. So he pressed hard on the handle, forcing it. It appeared to seat at this point. Still, even with this faulty human engineering design, there was a safety device—it should not have been possible to pressurize the cabin with the door not properly locked. Unfortunately, this ESD, introduced by McDonnell Douglas after the failure of the static tests, apparently failed (as it had in the June 12, 1972, accident), and the cabin was pressurized as the plane rose, until the door blew out.[22] It is even possible that if the plane had not been so full, putting so much weight on the floor, enough of the controls would have survived the floor collapse to save the airplane.

One obvious conclusion of this review of some DC-10 accidents (Godson gives other frightful case studies) is that with complex, tightly coupled systems with catastrophic potential, there is precious little room for management error, let alone other errors in the DEPOSE system.*

Buffet Boundaries and Small Jets

Pilots must contend not only with management errors and equipment malfunctions, but also with the constant and unpredictable forces of nature. Indeed, they must use these forces to be able to fly at all. The best evidence for the complexity of the airplane-environment interaction comes not from the obvious storms and windshear effects and icy runways, but from high-altitude flights in clear weather conditions. Small jets are more vulnerable to these problems than large carriers. Problems

*Lest McDonnell Douglas feel singled out in this section, let me note an NTSB study of four instances of tires blowing out in a twenty-month period on the supersonic Concorde. Each was considered a close call (in one instance the aircraft was severely damaged though there were no injuries). New precautionary directives were announced after the first instance but ignored. The technological marvel of a plane even had an inoperative cockpit voice recorder for several flights. See NASA Air Safety Reporting System Staff, July 1978.

with one high-performance corporate jet have been well studied by the NTSB. This involves a popular two-engine jet airplane used for air taxi and corporate flying—the Gates Learjet Series 20, in its many versions, particularly the Learjet 25.[23] One FAA study examined fifteen accidents involving stall warning and accidental overspeed at high altitudes, and lack of maneuverability when the airplane goes beyond or below its "operating envelope" or its "buffet boundaries."[24] We will go into this in detail because these problems of the Gates Learjet Series 20 are more extensively studied yet similar to problems found in the more familiar jets used by the airlines.

An airplane creates air turbulence around it as it flies. What typically happens in these accidents is that the plane is flying between 16,000 and 41,000 feet, and momentarily exceeds its "operating envelope," the area of turbulent air around the plane that gives it lift and provides stability. This can increase the plane's speed to around 80 percent of the speed of sound, and the pilot loses control. The causes are speculative and varied: a mysterious fault in the graphite gear box freezing the stick; clear air turbulence encountered at high altitude and speed that disturbs the envelope; unauthorized modifications (such as an overspeed warning cut-out switch, found on eight airplanes the FAA examined); and other possibilities. The basic problem is that the craft is unstable when flying at high altitudes, close to the speed of sound (though it is purchased because it can fly so fast and high). While various safety devices were required which push, pull, and shake the stick, sound alarms, and disengaged automatic systems, these devices themselves create problems because they mislead, fail, or require reactions of extraordinary strength and speed. Here is a dense pack of examples.

Under certain conditions overspeed warning can come on after the design speed has been exceeded rather than before, so the pilot is misled by the safety device. Under some circumstances, the required Machmeter can show 0.80 (80 percent of Mach, the design limit) when the true speed is 0.86. The Machmeter is off more than one percent at high speeds because it was calibrated on unpainted aircraft, and the paint reduces the drag (a one percent error is large near the speed of sound, where the margin for error is very small). Because of various "nose down" problems with the design, at very high speeds the pilot has to pull the shaking stick back, but the total force needed under these high g conditions can be as high as 150 to 200 pounds, and since pilot reaction time has been found to exceed three seconds the pilot cannot pull the stick back fast enough. The overspeed warning cannot be tested because of passive failures that will not show up in ground tests. This untestable warning signal does not tell the pilot the causes of the problem and there are at least six different causes: gusts, "unannunciated autopilot softover," "pitot" static system error, inattention, fuel burnoff, and flying into a colder airmass. Just for a pitch problem, which can also rip the envelope, there could be five different sources of equipment malfunctions alone causing it, and five separate actions must be taken to find the correct source, within a second or two. At these speeds in thin air the problem of flying becomes similar to the transformation problems in nuclear and chemical reactors, in that little information is available and knowledge is limited; compounding the problem, the time available for action is far less than in the other systems, and there may also be, as we shall see, the problem of disorientation.

The next accident, an account of problems with a four-engine corporate jet, the Lockheed Jet Star Model 1329, is more prosaic, but it gives some idea of the world of corporate jets and involves a system accident, unusual risks, and a safety change that was responsible for killing eight people.

Here are the final few weeks of the airplane, owned by Texasgulf Aviation, Inc. The safety improvement involved new, solid state units in the generator control units and new wiring. The airplane was flight-tested after installation and one generator failed. Repairs were made. In the next test flight, all four generators failed at one time or another, and were manually reset during flight. A problem in the new wiring was "found" and repaired. In the next test, one generator tripped (shut off), and was reset, but no maintenance was performed as a consequence of the malfunction. Having completed the tests, the plane went into service. A generator failed twice on the first flight. On the return trip, three of them tripped, were reset, but tripped again after ten minutes. The plane landed with only two of the four generators operating. The system was "repaired," and in the subsequent test flight, one generator dropped off the line again, was reset, and continued to work. No maintenance was performed after this flight.

Ten days later the plane was dispatched to Toronto, and first one and then another generator tripped, was reset, tripped again, and was off for nine minutes before coming back on. Eventually, all four experienced intermittent failures during the flight. The aircraft landed safely in Toronto, but it is not clear what, if any, maintenancc was performed there. It then took off for the Westchester County Airport, near New York City.

The crew had trouble with the landing gear after takeoff. At Westchester, in very poor weather, with wind gusts of 60 knots, the plane could not keep on course and had to be frequently redirected by the controller.

During this time the crew lost one of their navigational radios, and the transponder, which signals ATC as to location and heading, was silent for ninety seconds, probably because of the generator failure. At least one generator was working just before the crash, because a motorist saw the landing lights on the wing (the approach to the Westchester Airport goes right over an interstate highway with one of those curious signs with the fruitless warning: "watch out for low flying aircraft"). The plane crashed a mile short of the runway, and a half mile to the right of it, not far from the highway. All eight people perished.

The NTSB is not certain of the proximate cause of the crash. The four generators serve as redundant elements for lighting and navigation and control purposes; even if only one is working, that is enough to fly and land the plane. Beyond that, there is a backup battery emergency system for some controls. The NTSB suspects that repeated failures of the generators (and lighting and controls, for a moment at least) distracted the crew so much that they missed the approach. But that explanation is difficult to accept, they admit, and they add the following, which particularly interests us: "It is possible that an unknown fault occurred in the generator control circuitry so that an electrical malfunction, which invalidated the design logic of the normal or emergency electrical system, persisted and could not be corrected by the pilots at the time of the approach." That is, the redundancies of the new, safer solid-state system did not work, were defeated, even possibly defeated themselves, as a result of unanticipated interactions.[25] This may even have meant misleading information was being generated, which would account for the difficulty the crew had in following the controller's orders at various points. The NTSB learned of at least three other Jetstars that experienced similar problems after the safety modification. The example strongly suggests a system accident, and reveals intense production pressures in the face of repeated warnings concerning a key piece of equipment.

Disorientation

Flying into bad weather, where one cannot tell up from down, and right from left, is so disorienting that some pilots ask to be routed around heavy clouds or thunderstorms. Of course, often there is no choice. If the pilot is inexperienced, he or she may tear the airplane apart. This apparently happened to the pilot of a small two-engine propeller Beech plane in July 1981. He ran into bad weather, which had been predicted, and was flying by his instruments. According to the accident investigation, the most likely scenario is as follows: The plane ran out of fuel in the main tank, and the pilot had failed to switch to the auxiliary tanks ahead of time. The first thing he probably did, which would be correct when experiencing abnormal engine operation, was to increase the richness of the fuel and advance the throttle. When this failed, he bent down to one side to turn on the fuel boost pumps and switch on the auxiliary fuel. If, at the same time, because of the loss of airspeed plus turbulent weather, the plane rolled or turned, and he suddenly returned his head to the normal position, "a disorientation would most likely have occurred." The report continues: "A false sensation of diving or rolling beyond the vertical plane would have been produced. As a result, there may have been a strong, instinctive tendency to pitch or roll the aircraft in the opposite direction." Since the pilot had received no instructions in operating a multi-engine aircraft, "it is easy to visualize the pilot's reflex action as being abrupt and excessive." The wings and the tail section broke off the plane, probably as a result of a violent maneuver, and the plane fell to the ground, killing the three occupants.[26]

In a safety recommendation regarding pressure/vacuum pumps that operate directional gyroscopes and attitude indicators, the NTSB reviewed five accidents, four of them with Cessna 210N model planes. In all five cases, the pilots lost their attitude and directional instruments, were in or had to enter clouds, became spatially disoriented, and had their planes break up as a result of losing control or making sharp maneuvers, or went into steep dives from which they could not recover. The only instruments left to them were turn indicators, inclinometer, airspeed indicator, altimeter, and vertical speed indicator. It is very difficult to "fly blind" using these quite crude indicators, which respond slowly. The pressure/vacuum pumps have a long history of failure. At least 325 of them produced by the only two companies that make them failed over a four-year period, and the actual number is probably much larger since only a small percentage of the failures are reported.[27]

Disorientation played a key, fatal role in the Antarctica crash of the New Zealand Airways sightseeing trip. Recall that the initial board of inquiry found the captain at fault. On a clear day with 40-mile visibility, he flew into the side of a huge mountain. Subsequent investigation disclosed an anomaly. Up to the moment of the impact, the passengers were taking pictures of the Antarctica scenery. Even they might have seen the mountain ahead. The judge appointed to conduct the second inquiry learned of a peculiar phenomenon called "white out," familiar to pilots who fly near either of the earth's poles. It is not the white out that skiers and mountain climbers experience in a snowstorm, but is far more insidious since it occurs on clear days. Because the air in these polar regions is so extremely dry, the light reflected from very dry snow crystals is altered in quality and diffused in such a way that the contours of the hills, and even the presence of black objects such as rock faces, are wiped out. The passengers (and the crew) could see and photograph to the side, where there were large bodies of water, but ahead all they saw was a flat horizon line—which was what the crew would have expected, since they did not know that the headings entered into the automatic pilot two days before the flight had been altered by the airline. It took the combination of these two independent "failures" to produce the frightful accident of flying into a mountain on clear day.*

Summary: The Aircraft System

Enough evidence has been presented to suggest considerable complexity and coupling in the aircraft system—the equipment, crew, and the immediate environment. Some of it stems from trivial sources, such as kitchen equipment; some of it is so exotic as to be highly speculative, such as exceeding the buffet boundaries. Much of the evidence in between deals with unexpected electronic interactions or proximity problems. What prevents the aircraft system from being more risky than it is at present is probably the extensive operating experience gained in several decades of flying. Unlike nuclear plants, or even chemical plants, there are repeated "trials" of the equipment under the peak loads of start up and shut down—several a day for commercial flights—under realistic and often extreme conditions. With seventy years of experience, each new model builds upon the lessons of the previous ones. While we have seen that warnings are sometimes not heeded, with DC-10s and with Learjets alike, by and large warnings are plentiful and are taken seriously.

In the aircraft system, more than in any other industry we shall consider, there has been the time, incentive, resources, and talent to design-in buffers and safety devices, and provide comparatively exemplary training for unusually expert operators. It is also apparent that for the commercial success of air transport, accidents must be reduced. But the hard core of system accidents, while small, will probably not get smaller. This is because with each new advance in equipment or training, the pressures are to push the system to its limits. The new Boeing 767 airliner is designed to land with only 100-foot visibility—the crew can bring the craft down without even being able to see from their forward cockpit to the tail of their own craft! At the least, they are likely to get lost on the runway.

As we now turn to the larger system—the airways—the complexity and coupling in the aircraft system itself is compounded by that of other

*See note 8, this chapter.

aircraft and the ground controllers. Even here, the technological fix has been remarkable. This larger system is not a transformation one, but an additive one, or a fabricating one, so to speak. It should be simple to keep airplanes from running into each other, and to orient them for the proper landing position; it is a quite mechanical task. But, as we shall see, even with small objects in a big three-dimensional sky, the potential for system accidents is present.

The Airways System

The Orange Berets

The Los Angeles area is covered with freeways and dotted with airports of all sizes run by a variety of groups and governments. Driving about one gets the impression that all the world is moving, on land and in the air. Freeways run hard by airports, and airports have approach patterns that shake the water in backyard swimming pools. Like the automobiles that bunch up in the morning and evening rush, the airplanes, commercial and private, circle and peel off in close order when landing, and stand in patient lines on the taxiways, waiting for the light to change so they can join the others in the sky.

As the city mindlessly expanded, so did the freeways and the airports. A recent expansion was the Orange County Airport, renamed the John Wayne Orange County Airport in honor of its prominent citizen. As airports go, it isn't much in terms of size and belies thereby the illustrious images of both John Wayne and Orange County. There are only two runways, parallel to each other and cheek by jowl at that. The terminal doesn't soar, and one thinks of Madison, Wisconsin, for a comparable building size. The area is, of course, flat, but with a major freeway at the head of the two runways, and another on the side, the airport gives the sense of being crunched, as if it should buckle in the short space between the two parallel runways. To crawl along the freeway at the head of the runways during what is called the "jet rush" by the local controllers is a thrilling experience; huge jets roar overhead interspaced with little Cessnas and Bonanzas of the county rich; they can't be much more than 150 feet overhead, the height of a medium office building in any normal city, On the other freeway, if there is a break in traffic, you can race down your concrete strip as they race down theirs; they always win. In the control tower at the end of one of the runways the language of the controllers and pilots is laconic and brief, the twang of John Wayne himself untangling some simple dilemma with the certainty of movie morality. Despite its small midwest town size, the John Wayne Orange County Airport was, in 1980 at least, the *fourth* busiest airport in the country, with over a half a million takeoffs and landings in that year, or 1,500 a day.[28]

During the jet rush on a clear afternoon in February 1981, the controller momentarily had only six planes to deal with, three Boeing 737s, a Beech Baron, a Bonanza, and a Cessna. But it was tricky to have the small, slow general aviation planes interspersed with the large fast jetliners. One of the 737s, operated by Air California, was coming in for a landing, and another one, also operated by Air California, was about to take off. Let us call the first "X" and the second "Y." X was cleared for landing, and Y was cleared for takeoff. The controller then saw that the distance between the two was not sufficient, the "separation," it is called. So he told X to abort the landing and go around again, that is, pull up and make a full circle; and told Y to abort the takeoff and pull off the runway. Y was slow getting off the runway to the taxi strip, and X was slow in aborting its landing (according to the majority of the NTSB). X had lowered its landing gear, then retracted it as it tried to pull up, failed in the maneuver and decided it had to land, but apparently did not lock the gear down as it did so. The gear tore off, the two engines separated from the wings, and the plane skidded about and came to a stop, conveniently just 600 feet from the fire station, where it burst into flames. The evacuation was efficient and timely, given the fact that seats were ripped out and in the aisles, and overhead baggage compartments were strewn about. There were four serious injuries and no deaths. The plane subsequently exploded twice.

Going behind these bare facts will give us an insight into the life of a controller and the pilot he or she deals with. It is a story of production pressures, juggling errant aircraft, pilot dilemmas, and perhaps unsafe airports. It is a good, though complicated, glimpse into a minute in a day in the life of a controller and captain, which should make us grateful for every safe landing in Orange County.

In a busy airport with only two runways, one may assume that production pressures must be quite heavy. The controller exceeded the required separation in directing the activities, but was exonerated by three members of the NTSB. They blamed the pilot of X for delay in aborting a landing and improper handling of the aircraft, and left some blame for the delay on the part of Y in aborting its takeoff. But another member of the Safety Board dissented sharply. The separation was clearly inadequate, and the controller was at fault in trying to bring one plane in while another was still taking off on a short runway. The controller, however, said that the expected separation was adequate and was quite normal for the John Wayne Orange County Airport. (Required separation is 6,000 feet; in this case it was only 3,600 feet.) We can leave the NTSB to its dispute; what interests us is the problem the operators faced, one we shall again see in the chapter on marine systems. It exists in systems of moderate or high complexity and coupling.

Let us look closely at the pilot of X, who is certainly not blameless, but who was confronted with an unexpected problem. First he was cleared to land after two other aircraft, another 737 and a small Beech Bonanza. But he did not see the Beechcraft plane, and asked the controller (X's number is 336): "And Orange County this is three thirty six, we don't have our secondary traffic; could you tell us where that is." The tower replied; "He's probably going to end up behind you, I've got him on a three sixty on the downwind [we'll] see how it works out." The Beechcraft was supposed to make a 360-degree turn and come in before X, but turned too wide, so the controller decided to bring him in after X. In the next minute X went through its landing checklist (landing gears, speed brakes, flaps set at 15 degrees, notifying flight attendants, flaps to 25 degrees, and so on), while the controller talked with a new plane on the scene, got another 737 in position to land and cleared, changed the sequencing of the Beech Bonanza to come after X, gave clearance for a Cessna to take off on the second runway, and gave clearance (3 seconds later) to Y to taxi into takeoff position on the main runway. It sounds like a lot, but this is actually a light load for a controller; in one minute many things can happen at an airport. In the next twenty seconds he warned the departing Cessna about wake turbulence from Y, which was about to takeoff, acknowledged the conclusion of the landing of the previous 737 that X was now following in, and prodded Y to get going because "Boeing seven thirty seven a mile and a half final" (in its final approach).

At this point Y acknowledged he saw X coming in as he was making a running turn from the taxiway on to the runway: "In sight we're rolling." But aboard X, a few seconds before, the captain said to the first officer, "He'll never make [the turn]."* Let us start counting seconds from this point. Seven seconds after this warning comment from the captain, who had the controls, Y radioed to the tower, "In sight we're rolling"; two seconds later the captain on X said over the radio, "Go." Two seconds

*Phrases in brackets indicate questionable text—the transcribers were not certain these were the words.

later a Cessna called the tower saying it was making a wide right turn. The controller was momentarily distracted from the X and Y problem by this, because he had told the Cessna to go downwind, not to turn. He said to the Cessna: "I want you on the downwind, sir. Okay, I see you way out there. Why don't you make a three sixty report again on the base. You get to follow additional traffic Air Cal" (that is, come in behind X).

While this was going on, the captain of X reduced power to "flight idle" and lowered the flaps to 40 degrees to slow the plane up. Y was in sight on the runway about three-quarters of a mile ahead, about to take off; slowing up seemed prudent. Three seconds after that, the captain of X said (in cockpit, not on the radio), "Come on." And the first officer said, "Ah #" (in accident reports, # is defined as a "nonpertinent word," though it certainly seems pertinent at this point in the event). This was eleven seconds after the "He'll never make [the turn]" comment. Two seconds later, the first officer said, "They shouldn't have cleared him out there." Then the controller, having straightened out the Cessna, returned to X and Y and saw the danger. He told X to "go around, three thirty six, go around." The captain said he then began advancing the engine power towards takeoff thrust and motioned to the first officer to raise the landing gear and raise the flaps from 40 degrees to 15. He said the throttle was full forward, but the engines were not coming up (it would take six to eight seconds for them to reach the maximum thrust). Three seconds later the captain said to the officer, "Can we hold, ask him if we can . . . hold"—that is, can we go ahead with the landing. The first officer asked the tower, "Can we land tower?" The tower was talking to Y, telling them to abort their takeoff. When that conversation was finished the tower then said to Y, "Air Cal three thirty six, please go around sir. Traffic is going to abort on the departure." (That is, Y is going to get off the runway.)

The captain said he believed he would have to land, so he reached over and put the landing gear handle down. There wasn't enough power, he said, to pull up for a go-around. He reduced the power as they touched, and then the landing gear collapsed. The cockpit recorder, however, suggests that rather than respond to the first "go around" order (and they are orders, no matter how polite the post-strike controllers are), he only trimmed the flaps, and it was not until eight seconds later that he applied full power. "The Safety Board believes that the captain was still committed to land at this time and did not add power for the go-around" until he was told to the second time. The Board majority apparently believed that the captain actually retracted the landing gear as the plane was skidding, but their acount is ambiguous at this point. They also felt he was coming in too fast initially, and then slowed up so much that it would have made a go-around very difficult should one have been required at the last possible moment (as it was). In my view, these are retrospective hecklings. The majority of the board was ready to demand perfect performance of the captain, but ignore the violation of the separation rule by the controller. The one dissenting Board member did just the reverse, placing all blame on the controller, and none on the captain.[29]

There is blame enough for everybody, suggesting the system is at fault. (1) A controller violated separation rules that would turn the John Wayne Orange County Airport from the fourth busiest in the country to the fiftieth if they were followed to the letter. (2) Two general aviation airplanes using the same airport failed to follow instructions, with the consequence that the first aircraft changed the sequencing at the last minute (a minor event, but we know how minor events can disturb a mind set) and the second distracted the controller trying to execute a very tight sequencing during "jet rush." (3) The captain of X spotted the trouble, but received no confirming analysis from the controller, who was distracted, so the captain hoped that Y would get going fast. (4) Y delayed its takeoff for unknown reasons (though this delay is disputed by the dissenting member), making the situation worse. (5) The captain of X was truly committed to landing, having slowed up so much in order to give Y a chance to take off. (Had he been going faster, his abort of the landing, which would have been possible, might have brought him into collision with Y, had it taken off, which it was still supposed to do at that time.) (6) X's last-minute abort maneuver failed, probably because, as the safety board reports, the flap change from 40 to 15 degrees was not accompanied by an appropriate compensation in the pitch attitude and angle of attack. I suspect the captain was, in effect, both aborting and landing at the same time, and did neither correctly. Such mistakes should not be made by airline captains—or ship captains or nuclear plant operators—and they almost never are. But such mistakes still can and will occur.

(Failure to lower landing gear or lock it into position is not uncommon. A story is told in the Navy that there were 14 instances, over a few years, of pilots who tried to take their planes off of aircraft carriers with the outside quarter of their wings still in the vertical position—the parking position which saves space on the carrier. Three of the pilots were skilled enough or lucky enough to manage to maintain control after takeoff and fly around the carrier and return for a landing, with their wing tips still up. Two of these actually managed to land safely in this way. The third forgot to put down his landing gear and crashed!)

One can think of four or five other failures or out-of-place events in addition to pilot error that would have caused this accident at the John Wayne International Airport. We can fault the controller on at least two points, the Cessna on one, perhaps crowded conditions on the radio, and perhaps the crew of plane Y. We could also fault the Orange County supervisors (but not, alas, John Wayne). They allowed an airport of this small size to be used to handle all the private planes of their wealthy county residents and also to provide commercial flights to other California and Southwest cities. Most importantly, we can fault a complexly interactive and quite tightly coupled system that attempts to work at the maximum limits of safety. The NTSB has to work within those limits, and their assignment of probable cause makes sense only within those parameters; once we step back and look at the larger system of an intermix of traffic (in terms of plane size, pilot experience, and density), and location, the NTSB assignment of blame appears irrelevant and quarrelsome.[30]

Familiarity

One of the problems with managing interactive systems is the complacency that comes with familiarity. Familiarity is what allows systems to function smoothly; things that we are familiar with we do well. Doing well at a familiar job, however, means that we are not endlessly alert, ever searching for that extremely rare event. Our systems would break down quickly if the operators were required to be ever vigilant—for no other reason than vigilance must be time-shared, so to speak. The pilot of an airplane must share his attention with a vast array of instrument panels, his radio, and that part of the sky that can be seen from those small windows. When two planes collided and fifteen people died near Loveland, Colorado, in April 1981, each plane had about forty-five seconds to view the other aircraft. The surviving pilot testified his attention was on the spot on the ground where the parachutists he carried would land. The Board suspects that the pilot and co-pilot of the other plane could have been plotting their course, observing their instruments, or attending to other cockpit duties. Though they do not speculate on it, the ground controller could have been preoccupied with other planes in his airspace, and not have noticed the intermittent dot, without identification or altitude numbers, that indicated the position of one of the planes.

Numerous rules and regulations were violated in this case: the Cessna that carried the sky divers did not have authorization to fly above 12,500 feet; a notice of the flights planned for the day was not properly distributed by ATC; the Cessna pilot did not establish the required radio communications and did not have the required transponder (a radio signal sent to ground control continuously) for flights above 12,500 feet (a Mode C transponder that sends out altitude information); the FAA management at the Denver Center knew of altitude violations but did nothing about them, and so on. Any one of these violations, if corrected, might have prevented the accident, and several of them certainly would have. But these violations were not as important as the numbing effect of familiarity, as the following account indicates.

The sky-diving service had been in operation for a year and a half, and had conducted 10,000 individual parachute jumps in the last year. (Assuming they flew everyday of the year—extremely unlikely—that would be almost thirty jumps a day.) Their airplanes routinely went above 12,500 feet without the proper notification and without the proper equipment; the controllers routinely acknowledged messages from the sky-diving office that they would be jumping one mile from the airport from as high as 18,000 feet, from 1:30 until one hour after sunset. The routine might have gone on for years without any problem; after all, there is a lot of airspace up there, and the chances of two planes colliding on a sunny day in view of each other is extremely remote.

In addition to familiar routines, such as the above, the very safety devices contribute to complacency and inattention. The charter airplane involved in the above example, a two-engine Handley Page Jetstream, was under positive control from ATC, and knew that its altitude-encoded "blip" from its transponder was under the watchful eye of the controller down below. The Cessna also thought that its transponder was "squawking" appropriately, and warning the controller. What the Cessna crew didn't know was that the controller was not using the setting on his screen that would pick up a transponder that was not a Mode C transponder. Instead, because of his setting, he only picked up an intermittent signal, which did not show a track on the screen but only an intermittent dot, without any identification tag. It could have been a small private plane at 5,000 feet, instead of a five-passenger Cessna at 13,000.

The controller could have had the screen set to pick up and track continuously planes at below 12,500 feet, and two and one-half years before, a Denver controller had recommended that this be required. His or her superiors all concurred in the recommendation, all the way up to the chief of the Denver Air Route Traffic Control Center, who vetoed it because it would increase the clutter on the screen. (Four days after the accident he reversed himself and required that screens be set to track planes below 12,500 feet even if the controller is working only with those above it.) Even this change, however, would not have helped. The controller would have expected the Cessna to be below 12,500 feet, because above that, a Mode C transponder is required (but wasn't, tacitly, for the sky-diving firm's operations).[31]

The complacency that comes with high-tech solutions to problems is noted in a NASA study of seventy-eight near midair collisions in the airspace controlled by the terminal. The study found that half of the collisions involved an aircraft that ATC did not know about—no transponder, no flight plan, only a spot, if that, on the radar. The study comments:

> If [these] reports are representative, many pilots under radar control believe that they will be advised of traffic that represents a potential conflict and behave accordingly. They tend to relax their visual scan for other aircraft until warned of its presence; when warned of a conflicting aircraft, they tend to look for it to the exclusion of within-cockpit tasks and scanning for unreported traffic. . . . The air traffic controller cannot inform the pilot of traffic that is not visible on his radar scope, nor can he provide separation from such traffic. It is plain that at least some pilots receiving [complete ATC] services believe that they will be told about all traffic that represents a threat, yet controllers can handle traffic only with regard to threats they can see. . . .[32]

The saying is that it is not the one you can see that will hit you, but the one you don't see and were not looking for because of the one you were alerted to. This is a feature of all automatic systems, from cruise control in an automobile to autopilot in a jumbo jet. We should not outlaw or condemn the systems; without them we would have to both slow up and risk more accidents. We should just note the residue they leave for incomprehensible or unexpected interactions.

Our final example concerns an extremely thorough study of a series of near midair collisions at the Atlanta, Georgia Airport in October 1980. The NTSB report is seventy-six pages long, and reconstructs in dramatic and graphic terms a twelve minute sequence, with charts and ATC-pilot dialogues.[33] The events were very complicated, but essentially the controller did not take control of a plane handed off to him by another controller as it entered the terminal airspace he controlled, and as a result of changes in the landing pattern of the other planes (a routine event), collision avoidance alarms were sounded four different times. One of the near-collisions involved four planes that were occupying the same 2-square-mile area. In two of the events, a pilot had to make severe emergency avoidance maneuvers; one of the pilots exceeded the limits of his three engines. Planes passed each other within a few hundred feet. The workload was not high for the controller (and the weather was clear and bright), but within the twelve-minutes, fifteen aircraft were in the space he was controlling. Five of these were involved in collision avoidance alarms, some twice. In the ATC room, in addition to four collision avoidance alarms, low-altitude alerts sounded three times. The sound is the same, and controllers tend to ignore them since they are preoccupied with their duties. No accident occurred, but this example gives some circumstantial evidence for the interactive complexity of the ATC system.

Getting Cooperation

The vast sky is surprisingly populated with small, unconventional, and sometimes uncooperative objects, making the controller's life difficult. We have generally considered only airlines and some private craft, but the airspace includes military activities too, along with model gliders, sail planes, parachutists, and once even a lawn chair with occupant attached to several weather balloons that sailed uncontrolled to several thousand feet in California in 1982, and was reported by an airliner. The following incident involves just a part of the traffic handled by the controller: two corporate jets that almost collided, another jet, a jet military tanker, and two military fighter aircraft being refueled.[34] The second incident deals with private planes invading commercial airspace. These incidents offer further evidence, if any is needed, of the interactive nature of the airways system, as well as glimpses into the "cowboys" that fly there.

The controller returned from a relief break. While he was away, the other controller at his station had handled both the communication function and the radar. He was behind, and preoccupied with a midair refueling mission, so not all the tags and other information were in order, and the controller in an adjoining sector was also behind. The radar at one station (Keller, Texas) was out of service for scheduled maintenance, further increasing the workload. In the next few minutes, the following steps were taken and events took place: Two additional fighters requested permission to descend, through traffic, because of low fuel; they were asked not to, but insisted that they had to. The tanker requested and received a route change. One corporate jet requested a route change. There was confusion as to which of the two corporate jets was climbing to a new altitude, because the data tags (little blocks of light with numbers on them adjacent to the symbol for an aircraft on the CRT monitor) were not complete.

The controller called one of the corporate jets to clarify the situation, but the reply was blocked by transmissions from the fighters getting refueled. The fighter pilots were asked to "standby" (shut up). He then lost the radar targets for the two jets. One of the two corporate jets tried to transmit to the controller, but the transmission was blocked by the fighters, who did not "standby." The controller again politely asked the fighter pilots to shut up and called one of the jets. It reported that a corporate jet had just passed in front of him; it did not say whether it had to take evasive action or not, but it was clear that it was close. The conflict alert feature of the radar had not come on because one of the planes was technically in a "coast" track, and thus apparently (the report is not clear about this) not subject to the alert.

Here is a second glimpse into the life of a controller: One of the big problems they complain of concerns what controllers around the New York City sectors call FLIBS (Fucking Little Itinerant Bastards)—general aviation aircraft. The stories told about FLIBS are endless and probably exaggerated. They land at the wrong airports, and are surprised not to find their cars in the parking lots. They land, or take off, on the wrong runways (so do commercial aircraft, but less frequently). They leave their transmission button on the radio depressed after talking to the tower, which means they cannot be called, nor can anyone else use that frequency. Reportedly, another pilot has to be sent up occasionally to flag them, or scare them sufficiently so that they will try to call the tower to complain, and thus discover they have disabled the communication system.[35] They fly "under the influence"; run out of fuel frequently, and so on. Here is a small account of one incident.[36]

A single-engine light transport called Kennedy Airport traffic control and requested Air Traffic Control clearance through the area from Islip Airport (out on Long Island) to one of the radio beacon turning points near Kennedy. This terminal control area (TCA) requires an altitude-reporting transponder, and the little plane did not have one. But the controllers were not too busy, so they issued the requested clearance, telling him to maintain an altitude of 2,500 feet. The pilot acknowledged the clearance, but the controller was watching him and saw that he did not travel in the right direction. (The controller can shift his screen to pick up aircraft without altitude-reporting transponders.) The controller asked what was going on. The pilot stated that he had now lost both of the directional beacons that he was navigating on.

The controller's report goes on:

> As aircraft was now down to having only one required item for a TCA clearance (an airplane), TCA clearance was revoked with a 180 degree turn [or-

> dered] and reason broadcast. Pilot exited TCA and kicks VOR equipment [the device that picks up the beacon's signal] and called controller again, saying he had got one VOR operating again and could he have that TCA clearance again.[37]

The clearance was issued to proceed to the desired beacon, which would mean a heading of about 250 degrees. The pilot thanked the controller, but again selected the wrong heading. The controller's report concludes: The "aircraft took off on about a 330 heading terrorizing four Instrument Flight Rule departures off LaGuardia airport," which is slightly to the northwest of the Kennedy beacon.

These cases give us sufficient feel for the airways system. It is now time to look more closely at air traffic control, and the reasons for its remarkable safety record in recent years—virtually no mid-air collisions where the system had both aircraft under control.

Air Traffic Control

The air traffic control service has two primary functions: safety, and expediting the production of commercial passenger service. The two are in some conflict, even though each needs the other. The increase in safety brings more aircraft into the airways system, and increases the density, thus the danger. (Density involves the number of aircraft, the number in a corridor, the number and intersection rate of corridors, the number of takeoffs and landings, the separation allowed between aircraft, the communication activity among airplanes, among ground personnel, and between airplanes and the ground facilities. Adding one aircraft multiplies the other indices of density.) An increase in numbers, and thus density, interferes with the economics of commercial travel and freight, because it lengthens and delays flights. With the many-fold increase in fuel costs, it is important that the jet transports fly the most direct route possible, at the most economical altitudes, and with the least delay of departure or delay of landing when they reach the airport. A recent study by a major U.S. airline, for example, indicated that a 3 percent reduction in fuel used could result in a 23 percent increase in profits.[38]

ATC enables the commercial transport system to meet these goals, and has done a remarkable job of doing so. Each morning a central office of ATC surveys the weather across the United States (and foreign countries) and advises the airlines on possible delays in departure and landing because of the combination of weather and density of traffic. It does not tell the airlines when to take off, but they are able to adjust their schedules and the number of flights on a route to minimize departure and landing delays. The problem of "holding" over an airport, circling until there is room to land, has been greatly reduced. Instead, the "holding" is done prior to takeoff—which is cheaper, less inconvenient for passengers, and certainly safer. The problem of flying nearly empty planes has been reduced somewhat, since the airlines can cancel flights more expeditiously with the ATC information.

Meanwhile, in the sky, ATC has laid out more skyways, set up more beacons, and divided the sky into more efficient packages, so to speak, through which the airplanes can fly and be handed off from one control facility to another as they move along and finally land. This has enabled the density of aircraft to increase substantially, and their traveling speed to increase, and the separation of airplanes to be reduced from 20 miles to 5 (until the separation was increased as a result of the reduced number of controllers following the 1981 air traffic controller strike).[39]

ATC also has an overriding safety goal. Part of this is accomplished through advice on landing (helping the airplane to position itself correctly, warning it about weather or errors in altitude or position) and clearing it for takeoff. But the most important part of the safety function is to prevent airplanes from bumping into each other. On the airstrip this is fairly straightforward, though we have just seen how it can get rather complicated in our account of the John Wayne accident. The tower controllers can see the aircraft on the ground and know who is coming in and taking off. In the air it is more difficult. Mid-air collisions are quite rare, but near misses are not.*

Since mid-air collisions (including those on landing approach and shortly after takeoff) are likely to represent the most complicated interactions that ATC has to deal with, I will focus upon these. The goal of preventing mid-air collisions conflicts with the production demands placed upon the airways system. Fewer aircraft, at greater separations, flying in a larger number of jetways, spread out more evenly over daylight hours, and flying at slower speeds would greatly reduce the collision threat, but greatly increase the production costs. The problem, then, for ATC has been to keep collision risks low while increasing the occasions for collisions. This they have done with remarkable success. The density increases steadily, but the number of mid-air collisions has been reduced to near zero (especially those where both planes are controlled by ATC).

*Figures are very unsatisfactory here; I could not even learn, for sure, how many mid-air collisions there were between aircraft that were both under the control of the ATC system in the 1950s and 1960s. Near misses are estimated from a variety of sources and are subject to numerous errors, including the official FAA computer-generated count of "system errors," where aircraft come closer than the rules allow. In the mid 1970s these averaged 400 per year. See Spahn, 1977.

The pressure to increase density and decrease collisions is great. In fact, the FAA is legally liable for all the damages if two planes that its ATC facilities control collide in mid-air. This kind of an "incentive" is likely to be effective, even if there are no others.

The Reduction of Complexity and Coupling

In the Coupling/Interaction chart airways are placed near the center—neither very tightly coupled nor complexly interactive. I suspect that they once were much more interactive and coupled, and that the change is a result of organizational and technological changes since the early 1960s, constituting a striking example of the possibility of reducing complexity and coupling in nontransformation systems. I will review the changes in terms of the criteria for complexity and coupling that were laid out in Chapter 3. The airways system is potentially quite complex in the following respects. Unexpected interactions can occur when aircraft enter the airspace that are not under the control of ATC or even not seen by them. This is a proximity problem, similar to when a short in a cable disables a nearby cable that holds the safety device meant to correct any fault that might occur in the first cable. As the ATC system expanded, with the potential for more unexpected interactions in an airspace, it met this problem by restricting access to various spaces of air. It told corporate and recreational aircraft that they could not fly there unless they had the proper instruments and training. This reduced the number of planes in some airspaces than would otherwise be there, given the increase in traffic in general. It also gave the controllers more information about those planes that continued to fly there—since, for example, they had to have altitude-reading transponders.

Before the widespread use of radar, all information about position, speed, altitude, and flight plans had to come via radio contact. A failure here affected numerous sources of information. The danger was reduced with radar because it could operate independently of voice communication. It was further reduced when transponders could supplement radar and voice communication. In addition, the FAA required backup radios in the aircraft (as well as having backup systems for the controllers). Nothing is perfect, and we have seen problems with both the data tags and the radio channels in the examples given above. But the potential for common-mode failures has been minimized.

Dependence upon multiple function units or subsystems was reduced by the segregation of the traffic (as well as by the use of transponders). More corridors were set up and restricted to certain kinds of flights. Small aircraft with low speeds (and without instrument flying equipment) were excluded from the altitudes where the fast jets fly (though controllers relent on this point, as we have just seen). Military flights were restricted to certain areas; parachutists were controlled. In this way, the system was made more linear. Of course, the density in any one corridor probably also has increased, offsetting the gain to some or even to a great degree. But if we could control for density, we would expect to find a decrease in unexpected interactions.

The technological fix of transponders that encoded altitude and the shift from direct radar information for the controller to representations on the screen created by computer analysis greatly increased the direct, on-line information sources. The radar sweep gives only intermittent information on the position of an airplane. Though the screens introduced in the 1970s were more "indirect" in one sense, since they were a representation of information from the radar or transponder, the screens gave continuous read-outs of position, altitude, and direction. Most important of all, they did not require communication with the aircraft to determine altitude (and in earlier versions, communication to get heading and speed).

Many characteristics of nonlinear interactions still remain; the system may never be linear. There are many control parameters with potential interactions; limited substitutability of roles, or isolation of failed components (though it should be noted that if a collision does occur, both airplanes promptly exit from the system!); there are still unfamiliar and unintended feedback loops created (wrong identification of aircraft, insertion of airplane into the wrong sequence, et cetera); tight spacing of units (allowing unexpected interactions) at crowded airports; and limited substitution of supplies and materials on an ad hoc basis.

Reading accounts of near mid-air collisions hardly supports the idea that the system is anything but tightly coupled, but I will argue that it is only moderately so. Tight coupling reduces the ability to recover from small failures before they expand into large ones. Loose coupling allows recovery. In ATC processing delays are possible; aircraft are highly maneuverable and in three-dimensional space, so an airplane can be told to hold a pattern, to change course, slow down, speed up, or whatever. The sequence of landing or takeoff or insertion into a long-distance corridor is not invariant, though flexibility here certainly has its limits. The creation of more corridors reduced the coupling as well as the complexity of this system. Time constraints are still tight; the system is not loosely coupled, only moderately tightly coupled. But aircraft are maneuverable. They are also quite small. Near misses generally concern spaces of 200 feet to one mile. Those near misses reported to be under 100 feet are exceedingly rare and the proximity may be exaggerated. Even with 100 feet, though, there are 99 spare feet. If one tried, it would be hard to make two aircraft collide. In other high-risk systems it is comparatively easy to produce an explosion, or to defeat key safety systems and produce a core melt.

There are also important alternative methods of meeting the goal of preventing collisions and directing the aircraft to where it should be. Rerouting because of crowding is common. We have seen how accurate weather and traffic information leads to delays of departure (or flight cancellation) rather than delays in holding patterns. If crowding does occur (perhaps because of sudden bad weather) the system can be "expanded"—more space used; speeds reduced; aircraft held on the ground, or routed to other sectors. The extensive information available to ATC allows this to happen, along with their authority to change speed, routes, and altitudes.

Some other aspects of tight coupling probably cannot be reduced with either technological fixes or organizational changes. There are few opportunities for nondeliberate, fortuitous buffers that foster recovery from a dangerous situation, nor significant substitutions of supplies, equipment, or personnel. The substitution of personnel will probably be further limited in the future. The airlines, for financial reasons, are pressing for a reduction of flight crews from three (pilot, co-pilot, and engineer) to two (pilot and co-pilot) on the grounds that computers and other devices reduce the engineering load. (However, studies of accidents and near accidents disclose that crew overload is a pressing problem in departure and landing.) The FAA is pressing for more automation in its system, thereby reducing the number of controllers extensively. Both of these, I would suggest, will lead to much tighter coupling—that is, less resources for recovery from incidents.

Another move on the part of the airlines and the FAA will undoubtedly increase unexpected interactions and decrease recovery potential. At present, aircraft about to land will line up in a descent corridor that is many miles long, which is entered at the far end. Substantial separation is needed to prevent fast airplanes from overrunning slow ones. Under a proposed MLS system (microwave landing system), aircraft could turn into the glide slope at any point, with the slowest traveling the shortest glide path. The capacity of a runway would be considerably increased, and fewer miles would be flown by some aircraft. The workload of the controller would go up (and the chances of recovery from a slip-up would go down), but another innovation is being considered that might handle that.

The pilot could have a cockpit display of traffic information (a CDTI, it is called), and several pilots could then work out their own landing sequence, with the controller offering assistance and oversight. A form of "distributive management" or decentralization, it was endorsed by pilots in a simulation study, but not by controllers. The simulation indicated increased efficiency (more landings) and more even (safer) separation for the five aircraft.[40] The Orange County supervisors would probably love this innovation. The implications of CDTI extend beyond microwave landing systems. By providing the pilot with a radar screen that locates all the airplanes around him or her, perhaps giving their speed, direction, and altitude, it would redistribute authority between the controller and pilot, increase pilot workload and decrease the need for controllers, introduce more equipment that could fail, perhaps lead to overconfidence and greater risk-taking, and to the "non-collision course collisions" (see Chapter 6) that plague the maritime industry. It may be one of those dubious technological fixes that designers of technological systems seem unable to resist.

FAA, The Carriers, and Safety

Our final topic in this chapter concerns the technological forces that drive this particular high-risk system. I offer the following hypothesis most tentatively, because I am not certain that the many clear exceptions to it do not overwhelm the supporting cases, but it is a hypothesis worth consideration. The hypothesis is that the air transport industry (aircraft manufacturers and the airlines) supports safety regulations and requirements primarily when the increase in safety permits an increase in production efficiencies, and that the FAA concurs in this strategy. The industry is not against safety, and does a lot to increase it on its own; it is, after all, a prerequisite of the system that it be reasonably safe. But it will voluntarily undertake safety modifications primarily under two conditions: (1) when the modifications make increases in production efficiency possible (building more economical aircraft and engines, for the equipment side of the industry, and increasing density and decreasing operating costs, for the service side) and (2) when they can be added to new aircraft without significant cost, especially if there is fear that a retrofit of the equipment might be required by public pressure (largely through Congress) or (more remotely, as we shall see) by FAA requirements. This means that voluntary safety modifications or additions will not be made simply because there is evidence they are needed. The industry will concur in and not protest and delay *mandatory* safety efforts primarily when these increase efficiency of the system (including higher utilization by the public).

All that this careful wording really says is that no one in the industry is going very far out of their way to protect the lives of employees and customers and innocent bystanders (first-, second-, and third-party victims). Perhaps it will always be thus, and we should not be surprised to find this attitude in an activity that is primarily for-profit in nature, and furthermore must be organized through large, formal organizations (which inescapably will be indifferent to some degree to the fate of these victims). Yet the rhetoric of the industry and the FAA sharply contrasts with this view, and thus, the hypothesis needs exploring and airing. This section will do this, though after perusing it the reader may board her next commercial flight with less than her customary ease.

Safety involves two factors—accident prevention, and damage mitigation after an accident. The industry and the FAA have been preoccupied with the former, because each improvement there has meant greater density, higher speeds, and more customers. The latter, damage mitigation, has little or no effect upon these economic variables. It merely reduces the injury and death rate from accidents. The largest injury and death rate sources in damage mitigation come from evacuation delays, and more important still, from cabin missiles, obstructions, and toxic fumes and explosions.

First, let's examine the matter of timely evacuation of aircraft after an accident. A key to this task is a functioning public address system and means of communicating with the flight attendants. American passengers are remarkably compliant when faced with uncertainty in these awesome technological marvels. They will sit still until told to get out. If the electrical system is damaged, or all power shuts off to reduce explosion dangers, or the craft runs out of fuel, there *is* no way to tell the attendants or the passengers to get out. In 1971 a Boeing 747 caught fire after an aborted take-off and landed again. The first officer made an evacuation announcement, but inadvertently made it over the radio rather than the public address system. When nothing happened in the passenger cabin, he tried the proper system, but it was inoperative because all power had been turned off to reduce the risk of fire and explosion. The flight crew then entered the passenger cabin and shouted the order, but only those passengers in the front part of the cabin heard. All eventually evacuated safely, but since this had happened before, the NTSB recommended to the FAA that self-powered audio and visual alarm systems be installed. The FAA agreed it was needed, but believed further study should be made of the best system to require.[41] This report was presented in 1972; and the FAA continued its "study."

After some more accidents where the public address system failed and passengers were needlessly injured or killed, the NTSB tried again to get the attention of the FAA through a special study of their own, with recommendations, in 1974. Nothing happened. The Safety Board reiterated its recommendations after the 1975 crash of a DC-8 in Portland, Oregon, where the system was inoperative. Six years later, on January 19, 1981, the FAA finally issued a proposal for requiring self-powered warning systems, but still had it under review at the end of 1981. Air carriers would have a full two years in which to comply, thus providing a minimum of thirteen years for the studies, recommendations, and implementation for a system as simple as a battery-powered speaker system. The cost of the equipment is estimated to run from $500 to $5,000 per aircraft. Some aircraft already have the system in operation—United had it on four of its five aircraft types, but other airlines did not. Even the regular public address system does not have to be repaired before twenty-five hours of flight are up for some aircraft, and for others, such as McDonnell Douglas aircraft, there is no limit to the time that the airline can take to repair a malfunctioning address system. The NTSB deplored these lax regulations, to no effect.

More serious is the matter of cabin safety.[42] The predecessor agency of the NTSB, the Civil Aeronautics Board, recommended to the FAA in 1962 that the testing that was going on in the FAA regarding seat failures during a crash be expedited. This followed a 1962 crash where it was believed that twenty-eight persons could have been saved if their seats had not ripped out. The FAA replied that it recognized the need for further studies, and it was pursuing them "consistent with available manpower and funds." (Stronger bolts could have made a large difference, and extensive studies to determine that were hardly needed.) The regulations regarding the crash force that seats would have to withstand were then ten years old, and aircraft had become larger, faster, and were crashing with more force. By the end of 1981, when the NTSB undertook another Special Study of the problem, the old 1952 standards were still in effect.

The Special Study found that since 1970, only examining those crashes where all or at least some passengers could be expected to have survived the force of the impact itself, 60 percent of the crashes exhibited failures of cabin furnishings. Of the more than 4,800 passengers involved in these crashes, over 1,850 were injured or killed. Many of these deaths and injuries could have been prevented, the study concluded, had cabin furnishings not failed, particularly in the 46 percent of these accidents where there was fire.

Of the forty-six accidents where cabin furnishings failed, seats or the seat belts failed in 84 percent of the accidents; overhead panels and racks failed in 77 percent; galley equipment in 62 percent. Most of these failures occurred when the g (gravity) forces were well below the figure that the FAA sets as the maximum survivable force, and under which the equipment should survive. (In the John Wayne Orange County aircraft accident we described earlier, there were four serious and twenty-nine minor injuries caused by seat failures or other cabin furnishings, though the g forces were well below the standard of the 1952 regulations.) However, the study conclusively showed that the FAA maximum was far too low; people survived much higher g forces than the forces the FAA set as the maximum for survivability, and thus the seats and other furnishings should also be required to survive these higher forces. While this fact had been well established for some years, even by the FAA's own studies, it was still disputed by the FAA in congressional testimony in 1980. The FAA currently has a huge study underway, started in 1980 and not expected to be concluded until 1985. (It might recommend stronger bolts.) The NTSB comments: "Although it should be possible to conduct many worthwhile experiments and to gather new data in this test, the Safety Board questions whether the FAA will be any more willing to accept such crash data as being representative of modern aircraft." It also argues that "the major emphasis of FAA's ongoing crashworthiness programs should be on applying available technology. . . ."[43]

The problem is not only flying missiles, flying seats with occupants, jumbled debris preventing evacuation, and inoperative exits. Toxic fumes are probably the major killer. When a Saudi Arabian flight exploded and burned on the runway at Riyadh in 1980, killing the crew and 301 passengers, it was the smoke and toxic fumes created by cabin furnishings that proved lethal. Cabin materials, when heated or burned, produce deadly hydrogen cyanide and hydrogen chloride, which produces hydrochloric acid, phosgene (a nerve gas when ingested), and an explosive high-temperature mixture which consumes all oxygen and leaves only carbon monoxide.[44] At least 371 persons in recent years are known to have survived crashes only to die as a result of fires involving cabin material. The first such fire occurred in 1961, but the FAA has been reluctant to require the use of flame-resistant materials. The chairman of the NTSB, James King, said in 1980, "Ever since the 1961 crash

. . . the FAA has promised action. No action has been forthcoming."[45] The National Research Council of the National Academy of Sciences reported in 1977 that safer resins and foams were available for use. Even such painless improvements as the elimination of carpet as a wall decoration would help, it said.

The FAA has overlooked short-term improvements in the search of an elusive, perfect solution, notes Jeffrey Smith in an article in *Science*.[46] More important, the General Accounting Office of Congress notes, the FAA has issued proposals twice, but withdrawn them because the industry was opposed. The FAA convened a panel of 150 of the world's top experts in aircraft fire safety, but about 100 of them were from the industry itself, and the FAA. After two years it concluded that the FAA was on the right track in this area. With this backing the FAA proceeded cautiously, contracting with one of the aircraft manufacturers to develop a highly sophisticated fire chamber at the cost of about a half a million dollars. Even this use of public funds was not enough; the FAA decided it was not sufficiently sophisticated and that more money and at least another year were needed. Meanwhile, it continues its own testing, which involves holding a Bunsen burner (1952 model, no doubt) to cabin material to see if it burns. The problem is extreme heat, which decomposes the material, not the presence of a cigarette lighter. A radiant heat panel test has long been advocated by the National Academy of Sciences and other groups.

What is going on here, in an agency that developed and installed the sophisticated air traffic control system and is launching an even more automated and advanced one? The conflict between the NTSB and the FAA is, perhaps, to be expected, since the NTSB is the independent agency set up to review accident reports, conduct background studies, and recommend to the appropriate federal agency (the FAA in the case of air transport) changes in regulations, more intensive research, and so on. The forerunner of the NTSB was the Safety Bureau of the Civil Aeronautics Board, but it was made independent of the regulatory agency when the National Transportation Act was passed in the mid 1960s, so that the same person (in this case, Jerome Lederer) was not both promulgating civil air regulations and investigating the accidents they might cause. The FAA replaced the CAB, but over the years it has been criticized as being too close to industry. The General Accounting Office of Congress, a House government operations committee, the Ralph Nader-affiliated Aviation Consumer Action Project, and other groups have recently charged that not only is the FAA too industry oriented, but the Reagan administration has cut back on the funding of its primary critic and watchdog, the NTSB, and the FAA has relaxed many rules and restrictions (for example, commuter airline pilots can now work a 70 hour week; pilots for the large airlines are restricted to 30). The air transport industry, through its various trade associations, vigorously supports the policies of the FAA. But why would the industry, and the FAA, drag their feet on cabin safety, if the critics are correct in their charges? It appears that the air transport industry welcomes and supports efforts to allow more efficient, economical, and reliable flights, and these efforts improve safety. But bolting seats down better, or using inflammable material for decorations will not increase efficiency, nor is it likely to increase ticket sales. The airline cannot even be sued for negligence in these respects. (A jet airliner may fly fifteen to twenty flights with an inoperative public address system without liability or penalty.) It is not that the proposed improvements would cost much. Industry is sometimes willing to put them into new planes, and even retrofitting costs are not high. It just seems as if they are either a nuisance, or that tough regulations would establish a precedent for the FAA that the industry fears. How else can we explain a two-year study by a group dominated by industry and FAA representatives that concluded the FAA was doing a fine job in not upgrading thirty-year-old standards that resulted in perhaps hundreds of needless deaths?

Conclusions

The aircraft and airline industries are uniquely favored to support safety efforts. Profits are tied to safety; the victims are neither hidden, random nor delayed, and can include influential members of the industry and Congress; a vigorous union fights the industry's temptation to call "operator error" and instead looks for vendor and management errors; a remarkable voluntary reporting system exists (section on ASRS, p. 168), experience is extensive and the repetitive cycle of takeoffs, cruising, and landing promotes rapid training, precise experience with failures, and trials with errors for new designs and conditions. For getting from one point to a distant other, there is nothing safer.

This achievement has occurred despite persistent, uneradicable system accidents. But in contrast to nuclear power plants and chemical plants (and recombinant DNA research), the system is not a transformation system, with hidden and poorly understood interactions that respond to indirect controls with indirect indicators. (An exception occurs in encountering buffet boundaries.) The airways system is high on interactive complexity and on tight coupling, but these will respond to a considerable extent, although not completely, to management and technological innovations, which have been forthcoming. There are exceptions. Ugly industrial norms surface in the case of McDonnell Douglas and the DC-10s, with failure to heed warnings, sloppy work, and perhaps worse. Production pressures may be quite excessive in commuter airlines. With some slight inconveniences and expense the system could be made even safer—though not a great deal.

The air traffic control system interested us considerably because we could trace the changes that reduced complexity and coupling, resulting in about as error-free a large system as we are likely to see in our society. The contrast with the marine waterways we will consider later is astounding even though the problems are not all that different. Still, we wondered aloud about some of the automation steps that are planned; the marine chapter will serve as a warning about the limits of technological fixes, even as this one has served largely as a celebration of them.

A Note on the Air Safety Reporting System

The Air Safety Reporting System, ASRS, was established in 1975, and receives over 4,000 reports a year on safety-related incidents and near-accidents. Similar systems had been established in Europe, tried in the United States, and used by at least one U.S. airline, United Airlines. In fact, in 1974, a TWA flight crashed on a Virginia mountaintop as a result of a confusing map and misinterpretation of ATC reports. In the subsequent NTSB investigation it turned out that United pilots had been warned of the hazard by their program but TWA had no such program. The FAA had sponsored a program in the late 1960s, ostensibly nonpunitive in nature, but pilots and controllers did not support it. After the Virginia crash, they sponsored another, but this time allowed the respected National Aeronautics and Space Administration to supervise it. NASA selected the Battelle Memorial Institute as the contractor. This insured considerable independence from the FAA, and with guarantees of immunity except in extreme cases, the program succeeded.

Controllers, pilots, or others can write in or call in an account of a dangerous situation; often they are reporting errors they have made themselves. The FAA cannot then penalize them if they broke Federal rules or laws (unless criminal activity is involved), though pilots are still subject to discipline by their airlines. Reports are de-identified almost immediately (usually in less than four days) after verification contacts with the person submitting the report. If a pilot breaks a rule but submits a report regarding it, he cannot be penalized. This, of course, opens the system to abuse, and it was apparently difficult for the FAA to agree to this. However, extensive experience with the system indicates that in less than 10 percent of the enforcement actions based upon known violations is that action compromised or hindered by the limited waiver of the disciplinary action.[47]

The fruits of the program seem to be substantial. Reports pour in about unsafe airport conditions which are then quickly corrected. Changes in ATC and other types of procedures have been made on the basis of analysis of the ASRS reports. A staff of veteran flyers ("old eagles" they call themselves) who have mastered social research techniques (a proper sequence; it would be harder to teach a researcher about flying) write informative reports on topics such as controlled flight into terrain, distractions, inflight emergencies, communication problems, and so on. We learn that, relying on the ASRS reports alone, there are two potential collisions every day involving air carriers.[48]

Hall and Hecht note that 48 percent of the reports are submitted by pilots, 44 percent by controllers.[49] (The number of controller reports rose sharply just before the controllers' strike in 1981, and dropped even more sharply afterwards. Some suggested that the rise indicated they were using the system to build a case for changes, others suggested it was for disruptive purposes. The drop-off could be due simply to the excessive workload during and after the strike.) As is true of all accident reporting systems, this is clearly a "political" data source in some respects, but neither I nor others involved find any reason to doubt its overall accuracy. Indeed, the extent of mea culpa in the reports is striking, as is the objectivity of the analysis. Once de-identified, a report is part of the public record. I have used these reports to a limited extent myself to investigate incidents where airline management was somehow involved; the cooperation of the ASRS was exceptional.

It would be extremely beneficial if such a virtually anonymous system were in operation for the nuclear power industry and the marine transport industry.

CHAPTER 6

Marine Accidents

Introduction

Marine accidents bring us to a wider system than we have had to encounter so far. It is a fascinating country, and in our tour, we will encounter a frying pan that destroys a luxury liner in hours, captains playing "chicken" in sea lanes with forty ships about, "radar assisted collisions," monumental storms, tugboats blocking radio channels by playing Johnny Cash music, and tankers over a city-block-long negotiating channels only two feet deeper than they are. In the midst of these calamities are owners egging their captains on and insurance companies that fail to inspect the ships but shout "stop the carnage." On the side sits the Coast Guard, charged with safety in U.S. inland and coastal waters but understaffed and underbudgeted. The Coast Guard operates vessels too, but these appear to run aground or collide for no good reason just as often as the Exxon tankers or the cargo ships of the Greek magnates. Also on the sidelines is the National Transportation Safety Board, this time its Marine Board, investigating and hectoring. The Coast Guard and the NTSB write many of the reports we will draw upon. In the face of the carnage—a ship a day lost—they carry on ineffectually about excessive speed, failure to use safety equipment, and endlessly cite violations of rules and regulations so complex that even the lawyers for the shipping companies cannot figure them out.

That collisions, groundings, and tanker explosions occur for no good reason is the central paradox we will deal with. There would seem to be every reason for the accident rate to decline, instead of rising, as it has, since ships are equipped with technological marvels from collision avoidance devices to satellite navigation systems, carry larger and more expensive cargo, cost more to build, and are increasingly subject to national and international regulation. There would seem to be adequate economic incentives, adequate technology, and perhaps the basis for effective regulation. None of these turn out to be true. A loss rate of a ship a day proves to be about as risky, for the owners, as smoking. Those among us who still smoke will note that we do not count that risk as excessive. The owner has only a 17 percent chance that his or her ship will drop dead before its alloted time, and in a boom or bust industry that is not much incentive for safety. The technology has simply raised production pressures, increasing efficiency, as narrowly measured, but not reducing the social costs. Regulation succumbs to economic and nationalistic pressures, and is highly ineffective.

There are substantial costs to this state of affairs, from the consumer who pays the cost of accidents in higher prices, to the seamen who pay it with their lives, and most of all to the people of the earth who suffer the risk of toxic spills, gigantic explosions, and massive pollution. The death of the ocean by oil pollution, first prophesied at the end of World War I, will certainly be speeded up when 400,000 tons of oil are let loose from one wrecked tanker. It is no consolation that the greatest pollution threat is said to be not from the accidents, but from the daily slop of careless loading and unloading, washing out tanks, whether legally or illegally, and small leaks from machinery. Only 10 percent of the pollution caused by tankers is estimated to come from accidents, but that amount, as well as the other 90 percent, is well worth worrying about.[1]

Tankers carrying liquefied natural gas (LNG) have the capacity to blow up a part of a city, just as the explosion of two ammunition ships after a collision in Halifax Harbor during World War I destroyed two-thirds of the town and killed 1,600 people. (The Great San Francisco Earthquake, a decade earlier, killed only 452.) Some tankers resemble floating chemical storage plants with dangerous chemicals being maintained at delicate temperatures and pressures. A fair bit of the world's sulfuric acid, vinylidene chloride, acetaldehyde, and trichloroethylene, etc. moves by tankers, virtually unregulated, and often old, in poor condition, poorly designed for their cargo, traveling through winter storms.

One wonders if these toxic chemicals, even if essential to the industrialized countries, must be moved so cheaply and dangerously. If a freighter with hazardous cargo compounds sinks, and a storm breaks the containers, a disaster with 25 tons of mercury compounds will follow. This happened off the coast of Uruguay; the *Tagnari* sank in 1971. The wreck was not raised, despite the cargo, because the cost of recovery would have been more than the salvage value, so it was left to break up. It broke up in 1978. Villages had to be moved inland; thousands of dead marine animals were swept up on the beaches, and the red tide that formed killed more animals and apparently some of the people.[2] Even with the best of care we have cause for worry. In one frightening study by the Coast Guard, a computer simulation of the effects of a tank rupture 2 meters in diameter for fifteen hazardous chemicals stored at dockside produced alarming casualties. A relatively small tank of chlorine, a highly volatile and toxic chemical that boils at −29° F., ruptured at Coney Island docks, was predicted to kill 75,050 people outright; in Los Angeles, where the population density is much lower, there would be 18,740 deaths.[3] Such a catastrophe is not a remote possibility in the next thirty years. Thus, irrespective of the financial costs to the industry, marine accidents have enormous public costs.

Although it is obvious that there is a great problem, it is not clear that any of the usual solutions such as better inspection, training, equipment, personnel, or international policing agencies will make much difference. The problem, it seems to me, lies in the type of system that exists. I will call it an "error-inducing" system; the configuration of its many components induces errors and defeats attempts at error reduction. Discrete attempts to correct this or that will be defeated by something else; only a wholesale reconfiguration could make the parts fit together in an error-neutral or error-avoiding manner. Despite problems with specific parts or units of the airline and airways system, the components of that system reinforced a safety perspective. In fact, even if one wanted to, it would be hard to restructure that system to make it error-inducing. The pilots' union, the flying congressmen, the easy identification of victims and perpetrators and the easy access to courts, the "elasticity of demand" for the service (enough people can avoid DC-10s or avoid air travel as a whole for at least a short time to have an economic impact), the federal presence and experience with international controls, and even the voluntary reporting system—all these conspire, so to speak, to promote safety. Even deregulation is likely to cause only a slight decline in safety.

Much of the marine system is perversely inverted. The identifiable victims are primarily low status, unorganized or poorly organized seamen; the third-party victims of pollution and toxic spills are anonymous, random, and the effects delayed. Elites do not sail on Liberian tankers. The marine courts exist to establish legal liability and settle material claims, not to investigate the cause of accidents and compensate seamen. Shippers do not avoid risky "bottoms" but pick the cheapest and most convenient, and cannot choose to stop shipping for a time because the last cargo was lost. The federal presence is minor and appears inept in the United States; its major impact is to subsidize the shipbuilding and the shipping industry of the U.S. It sets standards for those ships that want to use our ports, but the United States ranks fourteenth among nations in ship safety, so the standards cannot be very high. And finally, the only international association concerned with safety is advisory and concerned primarily with nationalist economic goals.

I do not see any single failure as responsible for an error-inducing system such as this. Socialist countries are a part of this system, so private profits are not the primary cause of the rise in accidents and the increase risk of creating third-party victims. Our domestic chemical industry is run on private profits, but while hardly error-free, is not error-inducing. Production pressures are very high in the marine system, but they exist in other systems where they are moderated by units and subsystems that can establish responsibility and enforce accountability. Nationalism might explain a great deal, but not the behavior of U.S. ships in U.S. waters that appear to go out of their way to collide. The insurance industry is a passive contributor, passing on the costs to the final consumer, but insurance plays the same role in many other systems that are error-neutral or error-avoiding (construction, petrochemicals, precious gems). The federal presence, in terms of research, licensing, inspection, regulation, and sanctioning is just as weak or spotty in many other areas where there is some risk, at least to first-party participants, such as sports, recreational activities, and housing developments where natural hazards are probable (earthquake zones, lands subject to flooding, hurricanes, or tornadoes). But these "systems," if we may stretch the concept, are not error-inducing.

Rather, it seems to be the combination of system components that promotes error inducement, such that improving or changing any one component will either be impossible because some others will not cooperate, or inconsequential because some others will be allowed more vigorous expression. The safe, well-designed, Shell tanker can still be rammed by an itinerant cargo ship with a long list of violations; better radio communication can mean less communication because of the chatter; collision avoidance systems are swamped by higher speeds; larger tankers, which would reduce occasions for arrivals and departures where the biggest dangers lie, mean more and bigger explosions because of mysterious processes inside the huge tanks—as Shell found.

In an error-inducing system the tendency to attribute blame to operator error is particularly prominent. Such studies as there are all report operator error as the cause of 80 percent or more of the marine accidents. We have learned to be suspicious of this in other systems, and the marine system is no exception. Yet it is abundantly evident to me, from reading some 200 detailed accounts of marine accidents, that operator error does abound. Captains simply zig when they should have zagged, and take risks that even high production pressures would not countenance. I think this may be associated with the special nature of error-inducing systems. Risky behavior is often attributed to the "traditions of the sea," including that of risk acceptance. Granted that fearsome environment is no place for an officer averse to risks, but does "tradition" really explain much?

Consider instead that this is a system where incidents (failures that do not lead to subsystem or system failure) are higher than in any other that we shall encounter, with the exception of underground mining. Yet accidents are really quite rare for any one officer or ship. This means that recovery is almost always possible. If this is the crew's perception then it does not matter that much whether their decision is very carefully considered, their attentiveness is always high, whether they prudently assume that the other ship might do the wrong or unexpected thing, or whether every piece of equipment is up to standard and well maintained. We shall encounter frequent examples of all these failings, but I find it hard to believe that ship officers are somehow or other more deficient than the operators of other systems in intelligence, attentiveness, skill, and concern for their own safety. Since they come up against many more problems than operators of other systems and almost always come through them with no damages or only minor damages, when they do have a major accident observers will easily see them as having taken too many risks or as behaving stupidly. The plethora of problems they encounter stem from an error-inducing system, where, for example, some ships will not spend a few thousand dollars on a simple navigational aid (Loran), and others will spend hundreds of thousands on equipment that enables them to take ever bigger risks—and creates the expectation they will take those risks.

Considered in this light, we still have no choice but to call many errors unforced in the immediate sense; the officers or crew should have known better. But rather than 80 percent operator errors, I would make the wild guess that about 40 percent of the accidents have unforced operator errors as their source. (These would be component failure accidents in our scheme, with the operator being the component that failed.) Perhaps 5 to 10 percent of accidents are system accidents. Forced operator errors and other sources of simple component failure accidents would account for the rest. The occasions for forced errors are worth mentioning here: a captain can be on duty for forty-eight continuous hours; a fourteen-hour day for a mate on a coastal voyage is not uncommon, and the risks as well as the demands are higher on coastal than oceanic voyages; the communication problems are immense, on the bridge and between ships, since the native language of so many officers and crew is not shared by other officers and crews—personnel often come from Pakistan, India, China, Greece, Turkey, the Philippines and Indochina, and the official language, English, is garbled indeed; the ships sail with faulty and dangerous equipment; captains are fined for missing schedules regardless of the weather or traffic; crews rotate every voyage in many cases, and thus have little incentive to maintain equipment or even to learn how to use it well, leaving the captain with vulnerable equipment and dwindling resources for recovery from failure. None of these problems can be easily corrected; each runs up against some other part of the error-inducing system.

Our complexity and coupling concepts and that of system accidents will still play a role in the analysis. Some technological fixes on individual ships had the unanticipated consequence of changing a loosely coupled set of ship interactions to a tightly coupled one, making recovery more difficult when failures occurred. Interactive complexity will explain some explosions and collisions. The notion of an error-inducing system itself is derived from the complexity and coupling concepts. It sees some aspects as too loosely coupled (the insurance subsystem and shippers), others as too tightly coupled (shipboard organization); some aspects as too linear (shipboard organizations again, which are highly centralized and routinized), others as too complexly interactive (supertankers, and also the intricate interactions among marine investigations, courts, insurance agencies, and shippers).

But in general the system is only moderately coupled; failures appear to be continuous, but recovery is possible because time constraints are not all that tight, resources can be redeployed in an ad hoc fashion, damaged ships can continue their voyage. While there are unanticipated interactions of failures (complexity), much more of the system is linear than, say, flying, even though our examples will emphasize both complexity and coupling.

Though notions of complexity and coupling will help us, the principle of an error-inducing system is perhaps more important. Ships operate where most of nature and most of man conspire to ravage them. The navigation rules have developed to aid the courts in finding fault rather than aiding the ships in avoiding accidents; production pressures are often extreme; the working conditions are debilitating (forty lonely years of boredom and intermittent strain for the captain; worse for the officers, and still worse for the crew, as we shall see). The equipment is complex and barely maintained; captains refuse to establish radio contract with a foreign vessel that is about to hit them because they are foreign (and won't understand anyway), while the Japanese shipbuilders constructed ever-larger tankers but forgot to weld several of the ribs to the sides; and perhaps above all, though all are important, there is the authoritarian organizational structure that belies the interdependency and complexity of the operators and the system.

And there is nature: wild storms, 70-foot waves, ice-covered decks and equipment, shifting and narrow channels, suction effects in channels, atmospheres that can lead you to believe a sound is coming from the wrong direction or that shut it off altogether, fog so thick you cannot see the main deck below. Homer was correct; humans tempt the gods when they plow these green and undulating fields. The very notion of system accident loses some of its distinctiveness here. There are so many sources of failure, they are encountered continuously, often as a matter of course, and become indistinguishable from normal operations much of the time. It is almost too picky for me to point out that for this grounding there was deceptive weather, poorly calibrated radar, a wreck that shifted, a possibly inebriated mate, and a tight schedule, and then exclaim, "System accident, multiple independent failures!" That's the sea.

I will argue that the error-inducing character of the system lies in the social organization of the personnel aboard ship (and thus our voyage starts with the captain and the traditions of the sea), the economic pressures operating (raising some questions about risk avoidance and risk assessment by captains), the structure of the industry and insurance (leading us to examine system statistics more carefully), and the difficulties of national and international regulation. Then we will examine the technological developments and attempted fixes, receiving little encouragement that the system will be altered by these for the better. One major example, that of what I will call "noncollision course collisions," will raise a problem we have encountered before, the social construction of reality, or building cognitive models of ambiguous situations. Why do two ships that would have passed in the night suddenly turn on one another and collide? Finally, we will expand our discussion to the larger system, in this case nations gathering around a shipwreck, dramatically illustrating not so much the interaction of failures, though that is apparent, but the resistance to solutions in an error-inducing system.

The Captain

Until very recently, ships have always been the preeminently centralized human system of any size or complexity. By comparison, even military operations are moderately decentralized, because field commanders must act at a distance from headquarters and adjust tactics to changing and unforeseen circumstances. But aboard ship, the captain is in supreme command. The captain has a small crew, except in large naval vessels, permitting personal control and surveillance; he is in charge of a single system that cannot easily be "decoupled" and placed under split control. It is a manageable system, for a single authority, even if the ship is four football fields long. Strong traditions support the centralized control, and our language makes liberal use of the term "captain" in a variety of contexts. The only break in this tradition, as discussed below, is in large tankers where reportedly a management team is replacing the traditional hierarchy as engineering and navigation become more complex and important, giving equal power to engineers and electronic specialists.

With so much riding on one person, it is not surprising that many of the worst marine disasters stem from incompetent captains. History has preserved a singular number of examples of unforced operator error. A sampling follows.

The captain of the *Medusa,* a French frigate that foundered in 1816 with the loss of 152 persons (most of whom were needlessly lost), was drunk most of the voyage and ignored the warnings of his officers about dangerous waters. A famous but wretched painting, *The Raft of the Medusa,* hanging in the Louvre, memorializes the tragedy.

A captain of a luxury liner approaching the treacherous Halifax Harbor, Nova Scotia, in 1873 refused to look at the charts, misread a shore light, and went to bed; 560 people perished, and almost none of the women and children were saved, only male passengers and crew. In 1893 the ironclad monster gunboat H.M.S. *Victoria,* commanded by a brilliant and daring tactician, Sir George Tyron, led a squadron of thirteen ships into Tripoli harbor. By prearrangement the ships were to execute a manuever which his subordinate officers had unsuccessfully questioned, since it would bring the two flagships, the *Victoria* and the *Camperdown,* both prides of the English fleet, into a collision. The maneuver was carried out and the *Victoria* went to the bottom with 358 men. The captain appeared to grasp his mistake only at the last moment, and even after the collision ordered away rescue boats, not believing his ship would sink.

In 1904 an excursion vessel (the *General Slocum*), with 1,500 passengers, caught fire in New York harbor. The captain, already notorious for earlier navigational blunders, did not beach the boat on the nearby land, but headed into the windy sound for a distant rocky island. The headwind not only fanned the fire, but sent the flames aft to where the passengers had fled; there they found rotted life preservers and the crew found the fire hoses rotted. Rescue was difficult on the rocky island coast, and over 1,000 perished, almost all of them women and children. Over 1,500 perished in the *Titanic* disaster in 1912, partly as a result of an overconfident captain imperturbably sailing into a field of icebergs at night, thinking he had an unsinkable ship. An iceberg sliced open five watertight compartments; the designers had assumed that no more than three could ever be damaged at once.[4] As we shall see, captains have not changed much in modern times.

These systems, then, are somewhat unique in our collection since one person, the captain or his delegate, can wreak so much havoc. Perhaps for this reason the attribution of "operator error" or "human error" is higher in this than other systems. If one person has unquestioned, absolute authority over a system, a human error by that person will not be checked by others. Airline pilots are also "captains" and have ultimate control, but the second-in-command is called a "co-pilot," not a "mate," and has more responsibility than the first mate on a ship. In an air crash in Japan in 1982 a co-pilot tried to jerk the controls away from a malfunctioning captain, but this is unusual. Generally, co-pilots and pilots seem to be in agreement, even when both are wrong. They check each other's interpretations of anomalous circumstances. But it is not unusual for a deck officer to remain aghast and silent while his captain grounds the ship or collides with another.

Centralized control over a small number of personnel in one work location is not all that unusual as organizations go—think of professional sports teams, orchestra conductors, monasteries, and so on. But these cases do not encounter one unique problem that faces the captain: If the system suddenly expands to include another ship, who is to be in charge of the new system? One of the more baffling and stubborn problems in marine systems is the refusal of ships to cooperate with one another when in imminent danger of collision. It is as if the power and authority of the captain of one vessel is challenged by the power of the captain of another when these systems are suddenly joined into one. Neither captain is now in charge, and the subsystems are tightly coupled in an unplanned interaction. The complicated and ambiguous rules of the road do not solve the dilemma. In the airways, the ground controller is in charge of an imminent collision; in the error-inducing marine system, the authoritarian role, thought functional for ship emergencies, is dysfunctional for inter-ship emergencies.

Production Pressures

Ship captains may exhibit more clearly than most occupational roles the problem discussed by economists in the area of "risk homeostasis."[5] The theory is that people have a taste for risks, so if you make the activity safer, they will just make it riskier, by doing it faster, or in the dark, or without a safety device. The theory is extremely simplistic and the data hardly support it.[6] It appears to work only for some exotic and specialized activities such as auto-racing or mountain climbing, and even here, other variables are possibly more important. However, if we remove the disabling assumption that risky behavior is a function of the preferences of the individual at risk—the automobile driver, or mountain climber—and replace it with an analysis of the system in which the behavior occurs, it becomes more interesting. The ruling preferences may belong to those who control the system but are not personally at risk.

For example, one story that makes the rounds, though I have never seen it documented, is that when better braking devices were put on large trucks to reduce the dangers of brake failures on long hills, the number of such accidents did not decline. The devices certainly worked, but "working" meant that drivers were then able to go even faster on the long downgrades, because they had the extra margin of safety. They either exceeded the new limits on brake failures, or exceeded the limits on vehicle stability. I believe that story, but not because I suspect that truck drivers have been itching to hurtle down the great downgrades of the country's interstates at exciting speeds, but because going faster means more money to them if they own their rigs, or less hassle from the boss if they do not. The safety devices allows them to increase their income, or to keep their jobs when production pressures rise as safety devices are added.

Given these very real and reasonable pressures, the problem is not the psychology of the driver, but the failure of designers and engineers to produce a reasonably priced fail-safe device for braking, or more stable rigs. The truckers do not want to use low gears and crawl down the other side of the Donner Summit, they want to go as fast as the turns and the highway patrol will allow them. Granted there are irresponsible drivers (irresponsible to themselves and their families as well as to third-party victims), just as all of us are irresponsible at times. But the work truckers do puts even irresponsible drivers in a situation where irresponsibility will have graver consequences than it does for most of us. Again, it is the system that must be analyzed, not the individuals. How should this system be designed to reduce the probability or limit the consequences of situations where irresponsibility can have an effect?

The role of engineered safety devices, production pressures, and risk is similar for merchant ship captains. There has been an extensive increase in the safety devices aboard ships over the decades, especially with the widespread use of radar and other electronic navigation devices since the 1950s. But a director of Shell International Marine Limited, and a captain, is not so impressed by the results. He writes:

> Instruments for course keeping, position finding, depth recording, have all improved very considerably over the last several years and the twin radar sets now commonly fitted in tankers mean that there is data readily available on the position of all other vessels in contact, regardless of visibility; yet ships continue to collide, to strand and occasionally to founder. It appears that one must conclude that improve instrumentation is being used to enable navigators to prosecute their voyage with greater economical efficiency, and certainly with greater ease, but the risk per ship would seem to remain about constant.[7]

By carefully choosing his words, greater economic efficiency and greater ease, the writer appears to distribute the blame between the owner and the captain—efficiency for the owner means more profit, ease for the captain means less work. I will assume the latter motive does exist; it is rather widespread in the world. But there is also no question, from the accounts of accidents and the general literature, that the motive of profit certainly exists. A captain can save his owner's money, and perhaps get financial rewards as a consequence, by not using a pilot in waters where it is optional, or not calling for tug assistance. More important, if a tanker misses a high tide, it may lay about outside the harbor for four days, waiting for a high enough one, at a cost of seventy thousand dollars a day when the demand for oil is high. (In the mid-1970s a tanker's two-month run from the Persian Gulf to Rotterdam and back could bring a profit of $4 million.) Captains are judged on their ability to keep to schedules; the pressure of tight schedules is great. A ship is capital afloat, and profits are not regulated as they are in the utility industry. As with all modern industrial activities, the money is to be made by keeping it working.

The evidence for economic pressures is not, by its very nature, easy to come by. In the accident reports of the Coast Guard and the NTSB such considerations are treated very gingerly; they could hardly come up with recommendations to be written into our maritime laws that owners shall not be greedy, so it is hardly worth conjecturing about it. Owners can easily dispute any such charges, and would be expected to do so: "We never told him to run a risk like that; our regulations are clear that safety comes first." Yet we may infer these pressures from some of the extended accounts of accidents to be given later in the chapter.

Meanwhile, however, there is evidence of a sort from a survey of mariners conducted for a National Research Council panel studying "Human Error in Merchant Marine Safety." The National Research Council (NRC) is the research arm of the National Academy of Sciences, a professional organization of scientists that works quite closely with various branches of the federal government, especially the military, but also in other areas. In 1974 the panel commissioned interviews with 153 seagoing personnel, and then commissioned a questionnaire that was mailed to over a thousand seamen; the response rate was quite low, 25 percent, but there were 359 questionnaires returned in addition to the 153 interviews. It is limited, but still the best body of survey information I know of on the marine environment.

The results concerning economic pressures that result in risk taking are quite unambiguous. "The ability to make schedules is viewed by the largest group of respondents as the single most important factor in a company's evaluation of a captain's performance."[8] When asked how often a captain could refuse to take a ship out or delay sailing without trouble from the front office, 38 percent said seldom, or one to three times, without getting into trouble; 26 percent said it depended on the situation; and 23 percent said it was up to the captain. The question was an open-ended one—they did not check alternatives, but wrote in their answer—which makes interpretation difficult, but over a third appeared to indicate that refusal to sail in bad weather or with a faulty ship would bring strong censure. In addition, those who said "it depends" could have meant, "It depends upon how much pressure might be put on the captain."

When asked how the company feels about meeting schedules in poor conditions, one half said that there was strong pressure to meet schedules, despite conditions. Eight-seven percent agreed that a captain must do all in his power to meet an expected time of arrival (ETA, a watchword in the industry); 52 percent agreed that calculated risks are part of the game and should be treated as an operational expense; and 75 percent agreed that scheduling ships to ports with minimum tolerance for maneuverability is in the nature of such a calculated risk.[9] One example of minimum tolerances for maneuverability occurs when ships a city block long go into port with only two feet under their keel, with highly unpredictable suction effects and a virtual complete loss of maneuverability. Using the concepts from Chapter 3, this increases the time-dependent nature of the system and reduces the slack available (tighter coupling), and through increased proximity, brings into play poorly understood processes (the suction and bank effects), which rely upon indirect or inferential information sources (thus, more complex interactions are fostered).

Fully 99.6 percent of the questionnaire respondents who had sea experience said they had sailed on a ship that they personally knew was unseaworthy. Granted there may be some exaggeration here, but 99.6 percent is an impressive figure, suggesting strong commercial pressures that overcome safety considerations.

The interviews supported the above questionnaire data. Some quotes:

> When the *X* suffered severe cargo damage, the captain slowed down in heavy seas. He was fined because he did not make the schedule. If there is a guaranteed cargo delivery, there is bound to be hull damage.
>
> Sometimes a shore person [such as the representative of an owner or charterer] will suggest we sail with no tugs or sail in limited visibility. This reduces his port operating expenses. The young captains are more subject to this pressure because they don't know how much water the company official draws. If we come into an anchorage in fog, his budget gets an expense of a launch and reliefs.
>
> A company dropped a safety program in 1969 which offered a good bonus to tugs and crews with the least accident claims. It was observed that the result was decreased productivity, slowdown in task completion, the desire to opt for less hazardous jobs, to tow upriver rather than carry a big floating crane, etc.[10]

Consider the following accident. Like all the examples I will use it involves a variety of failures, but economic pressure is clearly important in this case. An experienced, meticulous captain of a large tanker chose to take a less safe, but more direct route to Angle Bay, British Petroleum's deep-water terminal on the western tip of Wales. He would save about six hours by passing through the Scilly Islands; the normal route avoids them. He had been informed by British Petroleum's agent that if he did not reach Milford Haven, at the entrance to Angle Bay, in time to catch high water, he would have to wait five days because of the considerable fluctuation of the tides. As it was, he figured he needed to arrive four hours early to shift cargo in the calm waters off Milford Haven. At sea, to reduce drag, more oil was in the midship tanks than the fore and aft tanks; thus the tanker drew 52 feet 4 inches at the deepest part of the hull. This was too deep a draft to make it into the harbor even at high tide, so some oil had to be pumped from the midship tanks to the fore and aft ones. This would save two inches! (One wonders what happens if they miss the precise center of the channel, or if there is a swell that might raise and lower the monster four inches.)

Why did the captain insist on transferring the cargo in calm waters rather than on the voyage in, when the latter would save him four hours? There is a risk of spilling the oil at sea, the captain apparently said. This explanation was treated with open amusement by the chairman of the board of inquiry. "He didn't want to dirty his deck, to come into port looking sloppy," he said after the hearing to reporters.[11] Whatever the explanation, it seems that the four hours saved would not be sufficient to meet the tide.

The captain decided to pass through the Scilly Islands, a rash of sandspits and rocks comprising forty-eight tiny islands. Four are inhabited, mostly by fishermen, and there have been 257 wrecks there between 1679 and 1933. Tales of false lights and plundered ships abound. Edward Cowan, in his lively tale, *Oil and Water,* notes that the following petition has been attributed to the Reverend John Troutbeck, a chaplain in the Scillies in the later eighteenth century:

> We pray thee, O Lord, not that wrecks should happen, but if wrecks do happen Thou wilt guide them into the Scilly Isles for the benefit of the poor inhabitants.[12]

Navigation in the passage the captain took, in good weather, even at night, is "perfectly simple" as long as one's position is frequently checked, says the navigator's bible, the *Channel Pilot.* But in the "perfectly simple" passage, he came across fishing boats (which one would expect to meet on occasion) and was unable to make his final turn to avoid some underwater rocks just when he wanted to; unfortunately, in his rush he was making full speed in the channel. Six minutes later, after another bearing was taken, he realized he had overshot the channel. When the helmsman received the order to come hard left on the wheel, nothing happened. The captain had forgotten to take it off automatic the last time he turned it himself. He then threw the switch to manual so it could be turned and helped the helmsman turn the wheel, but it was too late. The *Torrey Canyon* dumped its cargo of 100,000 tons of oil over the coastlines bordering the English Channel.

The accident involves the usual number of "if only" statements. If the captain had not forgotten to put the helm on manual, they might have turned in time; if the fishing boats had not been out that day, he could have made his turn earlier; if he had prudently slowed down once he saw the boats, he could have turned more sharply; once deciding to risk going through the Scilly Islands, he used a peculiar passage through them, and another might have been safer (even faster), and so on. We simply don't know why he did various things, and we do not, of course, know whether we should believe his explanations even if we had them. Production pressures are clearly present, however. They contributed to a decision that increased the proximity of subsystems and reduced the amount of slack available, moving it towards the complex, tightly coupled cell of our Interactive/Coupling chart.

Accident Statistics and Insurance

There were 71,129 ships in service worldwide in 1979, and 400 of these were lost, giving a probability that any one would be lost of 560×10^{-5} (560 in 100,000, the equivalent of 5.6 in 1,000).[13] In 1979, the rate at which ships were lost was about the same as the rate at which individual smokers could expect to die from smoking (500×10^{-5}). (Lloyds does not give the number of lives lost; some countries may not have reported it. If all were lost on a ship there might be about thirty-five deaths; many total losses do not involve any loss of lives.) I roughly calculate that each ship can expect to go about six lifetimes before it is a loss, assuming a ship is good for thirty years. A firm with six ships, that will last an average of thirty years, will expect to have one complete loss every thirty years. That may not be much.

The United States has a small fleet, but it lost twenty-one vessels of over 100-gross tons (about a 60-foot vessel) in 1973; its rate of loss in 1974 was 14th in the world. We did better than Liberia, Greece, Italy, and Panama, but far worse than the USSR, England, Japan, France, Germany, Netherlands, Norway, and Sweden. The statistics are getting worse. The average number of ship accidents per year has been rising for decades. From 1970 to 1979 those involving only commercial vessels in U.S. waters rose 7 percent annually—from 2,582 to 4,665. The 1979 figures are 81 percent higher than the 1970 figures. The ton-miles also rose, but only by 6 percent a year; the 1979 figure was only 33 percent higher than the 1970 figure, going from 306 billion ton-miles to 409 billion. The best measure, the accident rate per ton-mile, increased 74 percent over the decade.[14]

During this time a number of technological innovations were introduced to reduce accidents—more widespread use of radar, some limited Vessel Traffic Services (VTS), which is something like an Air Traffic Control system, and stiffer requirements regarding equipment and navigation rules. Cargo, vessel, and property losses for 1974 were estimated at $155 billion, worldwide. Surely one would think it would either be in the interests of the owners, or in the interests of the insurance companies, to reduce this staggering loss as much as possible. The NTSB, or Safety Board, investigated eighty-two marine accidents in the eleven years from 1970 through 1980. From this small sample of accidents (but these were the most important ones), it tried to stem the tide by sending 640 marine safety recommendations to federal agencies, pilot associations, and maritime organizations.* Issuing eighty recommendations a year to various groups was certainly a heroic effort but, judging from the rise in accidents, was probably ineffectual.

One important element in this error-inducing system is the presence of production pressures (though by itself this is an insufficient explanation, since comparatively error-free systems also have such pressures). If one-half, or even only one-quarter of the accidents were associated with production pressures, there would seem to be an incentive to reduce them. Why don't owners and insurance firms at least stabilize the loss rate, if not turn it around, by insisting on caution even at the expense of productivity? It would seem to be in their economic interests to do so. Efforts to answer this question have been discouraging, and what follows is speculation on my part.

There are a large number of ship owners and ship charterers; this makes experienced-based insurance fees hard to determine. If you have

*NTSB. *Special Study: Major Marine Collisions and Effects of Preventive Recommendations.* MSS–81–1, 9 September 1981.

only a few ships, and none has had an accident in ten years, should your rate go down because you have done so well? No. There is not enough experience upon which to base a rate reduction. Since a ship has only a small chance of experiencing a devastating accident in its lifetime of thirty years, say a 17 percent chance, we would not expect a firm with six ships to have an accident for perhaps twenty to thirty years. (The percentage of tonnage lost each year in the United States has run between 0.2 and 0.3 percent. It would take over 100 years of operation with the same number of ships to lose between 20 and 30 percent of the ships.) Well, if you can't tell with the small firms, how about the large firms, particularly the tanker fleets. Shell is likely to have hundreds of tankers at any one time. We can construct an experience rating for them, but we can neither apply it to the small firm nor know whether it is high or low for the large ones. We could wait, over the years, and see whether some large firms drop their accident rate and others do not, but for comparatively rare events that are likely to fluctuate with numerous variables, it must be hard to set experience ratings. For example, with the worldwide recession, the number of tankers at sea must have dropped precipitously. So the accident rate per sailing must be reinterpreted. The accident rate per ton mile is a good measure for the overall problem, but not for a company that ships short distances, has many more landings and sailings, and thus more exposure to risk.

Marine insurers do insure each ship, or type of ship in many cases, individually, and do give different rates to different companies, presumably based upon some kind of performance rating. But due to the low probabilities involved, these differential rates must be difficult to establish in a way that would reward careful owners and penalize incautious ones. An English organization, the Nautical Institute, in a memorandum published by Lloyds of London, expresses concern over present survey (inspection) procedures, classification of vessels for insurance purposes, and the practice of limiting liability for damage to others to a figure based solely upon the size of the ship. The following plaintive quote suggests the disorder of the insurance practices: "Marine insurers and charterers should be required to inspect tonnage [ships] they underwrite and hire."[15] In another context, the Nautical Institute finds it necessary to admonish that "insurers should . . . conduct conditions surveys before underwriting." If insurance practices do not require inspection of ships, they quite possibly do not reflect in any reasonable way the safety experience of the owners (which would be difficult in any case, due to the fragmentation of the industry and the low probability of total losses). Thus the industry itself might take a substantial share of blame for the following summary, from the Nautical Institute's important and widely discussed memorandum:

> This study indicates an unnecessary and excessive loss of life at sea, of ships, and cargoes, and an increasing potential danger to the marine environment. In addition the effect of the rising level of gross tonnage lost, resulting in higher cost of goods and services, places an unacceptably high financial burden on the public.[16]

As a consequence, I suspect that a ship's performance has little to do with its insurance rates. If this is the case, there is little incentive to reduce production pressures or increase expenditures on safety. The international consortium that tries to deal with these matters, the United Nations' Intergovernment Maritime Consultative Organization (IMCO), "has been regarded by many people . . . as a forum dominated by shipowners who want to minimize their capital outlay and operating costs despite the greater risk of chronic pollution and accidents."[17]

The direct economic costs of accidents, then, are borne by the ultimate clients of the system—those who buy the shipped goods. If insurance rates go up, the charges go up with them. A large company may expect to save some money by reducing the number of accidents, since insurance probably does not cover all the associated costs, but the savings is not likely to be great, and the frequency of accidents for even a large company will be small. Marine insurance rates are not rapidly rising since the industry is competitive and fragmented, so that the increment to the final consumer of higher insurance rates is trivial in each purchase and certainly hidden and consumers are not outraged. Considering that lives are lost and the seas are polluted in the process, one could wish for a better final accounting, but we are not likely to have it.

Furthermore, as an example of its error-inducing character, marine transport is a system in which safety behavior is hard to enforce. Captains are not under surveillance; ships' logs can have false entries, and it seems everyone has resisted the installation of the equivalent of aircraft "flight recorders." Storm intensity must be judged by the captain. The shipowners, then, are likely to mistrust the captain's report of dangers and hazards that required extra trip time; indeed, they fine them over their protests. This is in marked contrast to air transport, where numerous independent measures of weather difficulties, crowded airlanes and airports, and mechanical difficulties exist. The airlines can "trust" their pilots, or rather, do not have to trust them; they can have independent verification of causes of delay. Even trucking has a better information and surveillance system than marine transport. Shipowners probably feel that a missed expected time of arrival is due to the captain maximizing his comfort rather than their interests.

It is possible that shipowners, then, are operating in a system (which includes the presumably ineffective insurers) that encourages them to force captains to increase complexity and tight coupling beyond the extent to which it would otherwise be necessary. The matter is complex, and I will not go into a full discussion of it here. But one outcome could be that owners, whose expenditures for both insurance and accident losses only imperfectly reflect safety efforts, seek to maximize efficiency by using as a substitute for surveillance and supervision of ship officers a very rough measure of level of accidents. That is, as long as a captain meets the production level expected, no action is taken even if it is known that he takes large risks to do so. If the captain falls below this production level, pressure is increased. If the result is an accident, the captain is blamed, and penalized through fines or dismissal. The rest of the system cooperates by attributing 80 percent of the accidents to human error.

If this is reasonable, it supports the argument that the marine system is error-inducing. The surveillance of captains, and rewards and punishments for captains, and the financial incentives to owners, charterers, and insurers, along with other factors we shall discuss such as the weather and the international character of the system, do not encourage safety and indeed encourage risk. The encouragement of risk induces the owners and operators of the system to discard the elements of linearity and loose coupling that do exist, and to increase complexity and tight coupling. To some degree this discarding of intrinsic safety features occurs in all systems, but it appears to me to be far more prevalent in the marine system than in, say, nuclear power production and chemical production. I think the difference lies in the technological and environmental aspects of the system (fewer fixes available and a more severe environment), in its social organization (authority structure), and its catastrophic potential (which is less in the marine case, thus inviting less public intervention).

Incidently, the captain might have good reason to maximize comfort rather than the owner's interests. Captains complain about long hours involved with entering ports, unloading, loading and leaving, since the whole is one continuous operation. Ships rarely lay over long in port anymore. "I have been up for forty-eight hours continuously piloting, docking, and undocking," said one master in an interview. A record was kept for a twelve-day coastal voyage of a small tanker. The chief mate worked an average of fourteen hours a day; several of the other officers worked twelve or more hours a day (the captain was not observed).[18] Large tankers in 1973 cost between $6,000 and $8,000 a day to run, and $30,000 to $50,000 a day if depreciation and other book costs are added. Thus they are expected to be at sea 340 days of the year, with only 25 in port for loading and unloading. Crews are often changed by helicopter, and supplies dropped by helicopter as they round the South African Cape. Shell Oil once calculated that cutting one hour in port on the 13,000 or so port calls its tankers make in a year could save $2.5 million a year.[19] One can imagine that the captain and the crew are hard driven under such circumstances during the turnaround. The turnaround is also the most dangerous part of the voyage. The use of a tug or a pilot might well be a desirable luxury for the captain. The captain, then, might have reason to avoid the hard work and strain of the production pressures, and the owners reason to suspect he will and therefore are likely to use sanctions.

Safety requirements are also hard to enforce because of the international system. The marine world does not take kindly to international regulation, though passing agreements and rules of the road go back for centuries. It is not a system that breeds cooperation. Captains are masters of their fate and do not want their freedom impinged by another captain; nations are similar. Nations have little experience with national regulatory systems in the marine world; most have only a few ports, and these are visited by foreigners as well as nationals. In contrast, the air transport system was regulated from the beginning by each nation because the planes fly primarily within the nation. Pilots and owners grew up with this regulatory system and made it work. Then it was comparatively easy to extend the practices to international contacts. The marine world has not had this experience; national regulations were minor and slow to come; international regulations are even less significant. The United States managed to persuade the IMCO to pass weak rules regarding segregated ballast systems to reduce the pollution from oil tanker accidents only by threatening to require this system for all tankers entering U.S. ports. A weak compromise was worked out.[20]

The Nautical Institute notes that despite the efforts of this organization, "There has been little sign of worldwide improvement to regulate safety." It summarizes another study in this way:

> As the Rochdale Report points out, the results of such regimes not infrequently lead to "a ship" beneficially owned in one country, directly owned by a company resident in another country, registered under a flag of a third country,

managed by a company in a fourth country, but on long-term charter to interests in a fifth country and even sub-chartered to interests in another country.[21]

This complexity reflects strenuous efforts of owners and shippers to avoid the efforts of some countries to impose safety constraints, as well as to evade tax and other fiscal constraints. The very economics of the system conspire to induce errors, it would seem.

Meanwhile, in the face of international indifference, the Coast Guard and the National Transportation Safety Board (NTSB) do what they can. But they can do little in this error-inducing system. The accident reports of the NTSB are detailed and fascinating. (Some of the titles alone read like a country-and-western hit tune: "Collision of M/V Stud with the Southern Pacific Railroad Bridge over the Atchafalaya River, Berwick Bay, Louisiana.")[22] But their tone is defensive and hectoring: We made all these recommendations, and it still goes on; the appropriate agencies should increase their efforts to have the international maritime community consider this or that. Their recommendations hardly seem to cope with the economic and other realities of the marine system; rarely do they inquire into what we have called production pressures, they appear to avoid the topic of alcoholism doggedly, and they fail to note that no one can be expected to be alert, wise, foreseeing, and cautious all the time.

For example, in 1978 a container ship, overtaking a tanker, collided with it in the Galveston-Houston Ship Channel. There were no injuries, though the damage was estimated to be $1.4 million. The recommendations were: Don't allow deep-draft vessels to pass on bends (but the channel is almost nothing but bends, and vessels have been passing on them for decades); require the helmsman to inform the navigator that he has executed the rudder change order, in addition to repeating it as it is given (this would have been totally inconsequential in the accident); don't allow deep-draft vessels to ever meet on bends (this would tie up traffic forever); require the pilots to tell the captains what maneuvering agreements they have agreed upon (nice, but inconsequential). What happened was that the two pilots, with the combined experience of fifty years, operating in good weather, agreed on a passing. One of them pulled too close to the other and was going too fast, in retrospect, and misjudged the closing distance and the responsiveness of the ship.[23] It can happen. It is bound to. The recommendations are futile.

These are huge ships in very narrow channels, sometimes with only a few feet of water under their keel, subject to "bottom" effects and "bank" effects and "hull" effects (suctions created by water having to pass through very narrow spaces). They are at risk even if the weather is clear. Unfortunately, they often carry hazardous cargo, and the shore is lined with loaded chemical barges, tank farms, chemical plants, and human habitations. The deeper the channels are dredged, the larger the ships that use them; the more they are widened, the heavier the traffic becomes; the better the navigational aids, the faster the ships go.

The NTSB is particularly good at citing the violations of the "rules of the road," a compendium of rules that govern international sailing.* But a survey of mariners tells us just how useful that is. Of those in the sample responding to the question, fully 29 percent said they had been in a situation where strict obedience to the rules of the road contributed to a marine casualty or a near-casualty. Almost half felt justified in violating the rules of the road to meet *normal* expectations and operations. A risk analysis study cited in the report put it this way, contradicting the ritual findings of the NTSB:

> According to court decisions, 99 percent of all collisions are caused by failure to obey the rules of the road, and no one, not even an admiralty lawyer, fully understands the rules and their various legal interpretations. The legal interpretations could not possibly be understood by a master or watch officer who may have only seconds to decide which rule should be applied to a given set of circumstances.[24]

This makes the citing of violation of "rules of the road" somewhat less than reassuring in the safety reports.

Captain Dickson, of Shell, hints at this when discussing the inadequacy of international steering and sailing rules. "It is probably fair to say," he writes, that these rules "are regarded by people at sea as very clear in their application to determine responsibility after a collision but of dubious value in relation to collision prevention."[25]

I should make it clear that the NTSB and the Coast Guard are working against very heavy odds—indifferent owners, an international community of quarreling states, "cowboy" captains who take ridiculous risks, and a lack of power even, it seems, to extract testimony. In addition, the budgets of the two agencies have been cut recently, and the Reagan administration has argued that the services the Coast Guard provides should be paid for by the shipping companies. (Since a major service they provide seamen and the rest of us is forcing the shipping companies to pay attention to safety, this policy is not likely to lower accident rates.) But while looking at the vastness of the problems in the

*The official title of this United States Coast Guard Publication is Commandant Instruction M 16672. 2.

marine environment, at the interaction of so many interests and worlds, to cite excessive speed as a problem which captains might be accountable for or to say they should have been more alert, or used "better judgment," or to cite the rules of the road, simply does not seem to be appropriate. A little investigative reporting into the accidents by looking behind the scenes might do more for marine safety.

For example, in this system, since groups are suing one another and large damage settlements are involved, the Safety Board can be bold in its general recommendations ("be alert," "use the compass,") but avoid altogether considering hearsay reports (such as of drunkenness) or even inferential data, such as the past record of the company or crew. Such considerations would disturb the parties to the suits associated with accidents, and distract from the matter of who will pay by asking what really happened. Take the pilot of the *Summit Venture* that brought down the Sunshine Skyway Bridge in Tampa Bay on May 9, 1980, and thirty-five people going across it, all of whom died. He hit a sudden rainsquall which wiped out his radar (it was "filled with rain return" as they say), and didn't see the bridge until it was too late. The NTSB blamed him for not anchoring, the National Weather Service for not predicting brief and sudden thunderstorms more accurately in this Gulf city, and faulted the pier, which should have had a crash wall to handle a bulk carrier two football fields long.* That is politically safe. But a *New York Times* story noted on May 14, 1980, before the Safety Board even met, that the pilot was involved in seven other reportable incidents in his four years on the job at Tampa, and one of these was bumping into the same bridge three months before. The Board's report never mentioned this information. Rather than calling for better weather predictions or reinforcing all the hundreds of bridges in the Gulf area, the Board might have considered the licensing, review, and sanctioning of pilots. But in an error-inducing system it is safer to cry operator error or poor weather forecasts; neither of these targets, nor the bridge pier, will disturb the system. This one accident cost $31 million and thirty-five lives, third-party victims all.

Kitchen Trivia and Garden Hoses

It is time for some stories. The first of these two accidents will indicate the simple and human things that can go wrong, and the second, the complex human and environmental things that can go wrong. From the

*NTSB, MAR–81–3, 10 April 1981.

trivial to the typhoon, they give us a sense of the range of the problem that the mariner confronts.

Consider an electric skillet, cooking oil, and the thermostat left on high. How often has that happened to you, the smell hanging around the kitchen for hours? It happened on the Italian cruise ship *Angelina Lauro* in Charlotte Amalie Harbor, St. Thomas, in the Virgin Islands. Only here the fire, which started in a large tilting skillet used for deep fries in the crew's galley, got into the greasy ducts and thus managed to get through a fire division bulkhead (safety device) and spread to other concealed places that did not have sprinklers in them. The responsible personnel rather botched the firefighting job from the beginning, and the fire detection and sprinkler system did not work well. Most of the passengers were ashore, fortunately, and after a few hours the dense smoke forced everyone from the ship. It burned at the dock for four days, and the ship was almost totally destroyed.[26] So much for domestic errors aboard reasonably tightly coupled systems.

Far more complicated is what happened to the U.S. *Steel Vendor,* a cargo ship plying the rich wartime trade in 1971 in the Pacific. On the way from its homeport of Houston, Texas, to the Philippine Islands, the ship had repeated boiler difficulties. The boilers were repaired, but the trouble kept reappearing. One assistant engineer was discharged in Manila, apparently following a controversy over equipment repair. Then the ship left Manila for Vietnam. A typhoon was building up, but the captain thought it would pass them by. But after three hours at sea, both boilers lost water rapidly. Rather than turn back, the engineer attempted to repair them at sea. The crew had done this before.

Unfortunately, steam propulsion is a fairly tightly coupled system: at least one boiler must be working to supply the power to the feed pumps that fill the boilers. After repairs, the boilers cannot be tested hydrostatically unless the boilers are fired up to drive the steam-driven pumps needed for testing. But they should be tested before they are fired up. Thus, a completely separate energy source is needed to fill and test boilers—an electrically-driven pump, capable of being driven from an emergency generator. They didn't have a working pump of this type; it was broken, waiting for a part. But even if they had such a pump, they were soon to lose the emergency generator.

Through jury-rigging and hard work, the boilers were repaired and placed into operation by the next morning, without testing. Then the number one generator sustained a major casualty, and the emergency generator was pressed into service (pressed too hard, it seems, since the starter broke). The feed pumps also were clogged and various other associated problems occurred. The ship was slowed again. That evening, after repairs, the ship was up to full speed, confirming the captain's nerve and derring-do. The next morning, with the storm rising, both boilers again were losing water, and the ship was slowed. The boilers were interconnected, and because of the rolling of the ship they could not determine which boiler had the main leak, or whether one or both of them leaked. During the previous day's casualty with the generator, the starter on the emergency generator had burned out. With the boilers shut down again there was no regular power; with a broken starter for the emergency generator, there was no emergency power. The ship was blacked out in a typhoon.

They finally fabricated a hand crank and got the emergency generator going five hours later. They then jury-rigged the wash-water pump—used for bathing—with a garden hose and used it to try to fill the D.C. heater (a device for preheating the boiler water) with wash water. They secured (shut off) the port boiler so they could work on it and tried to start the starboard one, guessing that the port boiler had the worse leaks and needed the most work. After six hours of fruitless pumping, they realized that a valve had been left open and the water patiently supplied by the garden hose was going into the drain space between the double bottomed hull. Meanwhile, with the port boiler secured, they opened it to inspect the steam drum, on one end, and the "mud drum" at the other end. When they opened them they realized the problem was probably hopeless, because numerous leaky steam tubes were found. They advised the captain to call for a tow, which he did at 5 P.M. The ship wallowed helplessly through the stormy night. At 5 A.M. the next morning they received a reply saying that a tug had been dispatched and would arrive in two days!

Unfortunately, the position they gave the agents in Manila to relay to the tug was at least 80 miles off. The ship did not have a Loran navigation system, which is electronic—and inexpensive. Instead it had to use "dead reckoning" since neither the sun nor stars were visible, and had not been since they left Manila. In dead reckoning, a record is kept of the speed of the vessel and the direction, and an approximate position is charted, after correcting for wind and current effects. The typhoon was building up, and since the wind was coming one way and the current was going the other way, the captain assumed the two forces cancelled each other out. But strong winds can not only blow a ship off course but also change the currents, deflecting them, and the captain did not realize or recall this.

The vessel was now dead in the water and rolling 35 to 40 degrees (90 would put the ship on its side) in heavy seas with a force 9 wind (near hurricane force). Bilge water was sloshing through the engine room. There was enough emergency power for lighting and some ventilation. Hand tools used in the more or less continuous repair of the boilers were sliding across the deck plates in the engine room and were lost in the bilges, which were overflowing. Sea water was coming in the stacks from the wind-driven seas and soaking the men and equipment below. One of the fan motors was shorted out by this. Yet the engineers worked all day through trying to get the boiler operating.

The men opened the port boiler and crawled into the steam drum to examine the steam tubes with flashlights. To find the leaks, they filled each of the tubes with water from the garden hose, after blocking the other end in the "mud drum." This test was performed with the ship gyrating wildly. One bad leak was found, and they assumed this was the problem, plugged it, and closed up the boiler. By 8 A.M. the next morning they could begin the tedious process of filling the boiler with water through the test cock and air cock by means of that now essential safety device, the garden hose. The starboard boiler was isolated, and they filled the port one. But they inadvertently left one feedline open, and the water again drained away.

If they had had one reciprocating pump working, an engineer said, they would have been all right, because they could have filled the boilers with water from the double bottom. But the reciprocating pumps had been out of service during the voyage to Manila, and repair parts had not yet arrived on the boat. More problems occurred, and they had to refill the D.C. heater once again. An emergency line was hooked up to a condensate pump and that speeded up the filling. They hoped to have limited power for the propellers by 3 or 4 o'clock that afternoon.

At 11 that morning the sun came out briefly and they were able to get navigational bearings. While taking them, the navigator noticed breakers on the horizon, standing out from the crashing sea. Using the readings from the sun and examining the charts they realized they were four miles from Loaita Bank—reefs. They were 90 miles south of their dead reckoning position. The captain ordered a distress signal, but it took thirty minutes to get it out because the frequency was occupied by another vessel that was also in distress. The H.M.S. *Eagle,* a British aircraft carrier, responded and was the closest. The engine room crew continued to work on the boilers during the next three hours as the ship was blown toward the reef. Even after the ship struck the reef, they continued and finally began to get up some steam. An anchor had been let out, and finally grabbed, but did not hold. With each wave the ship hogged further up on the reef. They were taking water on from holes in the hull, and finally the level was so deep in the engine room it put the fires out in the boilers that were just getting up steam.

A helicopter from the *Eagle* appeared and lowered a man to the vessel. (Life boats had already been made ready and lowered partially.) It was decided to remove the crew by helicopter, and two hours later the helicopter returned and the rescue of the officers and crew of thirty-five was completed in an hour, with the captain being the last to leave.[27]

Multiple failures certainly abound in this accident, and if it hadn't been for failures B, C, D, et cetera, there might have only been an incident. But what strikes me is that a captain, without reasonable navigational equipment, two balky boilers and no reciprocating pump, with a typhoon in the area, failed to return to port when, three hours after sailing, the boilers again began to malfunction. I doubt that he loved the sea so much that he had to sail; I suspect that he would have been fined or otherwise penalized by insisting on Loran equipment, not starting until reciprocating pumps were received, waiting for the typhoon to find its steady direction or blow itself out, or delaying until the boilers were properly repaired. This was a conventional cargo ship; next, let us take a look at some quite unconventional ships where the hazards, complexity, and coupling are greatly compounded—the supertankers.

Supertankers

About half of the tonnage afloat is in the form of tankers. All of the biggest vessels are tankers carrying crude oil (or liquid natural gas—LNG tankers). The biggest, as of 1974, was the *Globtik Tokyo,* with a deadweight (maximum tonnage of cargo, fuel, stores, and ballast) of 476,292 tons (think of 50,000 10-ton trucks), a length of 1,243 feet (four football fields, or nearly one-quarter of a mile), and a draft (the depth below the water line) of 92 feet (about the height of an eight-story office building). The growth of tankers has been astonishing; at the end of World War II the largest tanker had a deadweight of 18,000 tons. The *Globtik Tokyo* is 265 times as large. Most of the large tankers, however, are in the 200,000 to 300,000 range. They don't even call them ships when they get over 200,000 deadweight tons, but VLCCs, which stands for Very Large Crude Carriers. The very name suggests a different order of sea things than we had in the past.

The *Torrey Canyon,* which grounded off England in 1967 with devastating results for the coastlines of the English Channel, was only 120,000 tons. Plans for 750,000-ton tankers were on the books in 1974.[28] Mostert, in his fascinating book *Supership,* mentions industry talk of million-ton tankers, but since that book was written in 1974, oil conservation and recession have come upon the world, and at present they are actually cutting out the midsections of supertankers and downsizing them because of reduced demand and a glut of ships. No supertanker has been built since 1979; twenty-seven were scrapped in the last two years. They are not built to last much more than fifteen or twenty years anyway.

But they are still with us. As systems they are impressive for their size, but not for much else. For economy, they generally have only one propeller, or screw, which makes maneuvering very difficult; they often have only one boiler (a passenger liner will have several), which makes a breakdown incapacitating; they are underpowered, by traditional standards, making maneuvering difficult and slow. The illustrious sailing clipper the *Cutty Sark* could beat a large tanker, since they only make 14 to 15 knots an hour. This means two and a half to three months for the roundtrip voyage from Europe and the Persian Gulf, with supplies delivered by helicopter at the Cape. In port, though, all is arranged for speed. Turnaround is twelve to eighteen hours. The crew does not get off.

When on the bridge, the captain is about 100 feet above the waterline, and has to walk 150 feet from port to starboard to see what is happening there. Since it takes three miles and twenty-one minutes to stop a 250,000-ton tanker, I guess there is no hurry. Falling off a tanker is inadvisable, because of stopping distance, and there is also the 60-foot drop from the deck to the water to consider. The crew is so far above the water that they can run over trawlers and fishing boats without ever seeing or feeling them. Maneuvering in channels or near shore is difficult. Anchors won't stop tankers; even if such a ship is only slightly underway, the chain would be wrenched away.

The draft is so deep that tankers have to stay in the middle of the channels and thus cannot turn even if they have the 2 miles to do it in. They are so long that at night smaller ships have tried to steer between the fore and aft lights, thinking there were two ships. In many areas, reports Mostert, tankers sail with a clearance of as little as 3 feet. In 1967 Shell International, the largest single operator and charterer of tankers, declared its policy was to allow a minimum of 2 feet of water under the keel. With this clearance the ships are virtually unmaneuverable because of suction effects. Even when a ship with a 50-foot draft has 20 feet under her keel, her turning circle is doubled. Most tankers in Europe use the

English Channel, but the bottom keeps changing and sand rifts as high as 20 feet can be created so that ships touch bottom there when they thought they had 20 feet of clearance, judging from the charts. VLCCs are part ships, and part submarines, Mostert notes, and submarines consider the English Channel unnavigable.

Docking tankers is a problem. Moving at a very, very slow walk of a quarter knot (25 feet per minute—try it in your living room), the contact with the dock or jetty can badly damage the ship, with the chance of an oil leak or an explosion. Recall that the captain or pilot is 100 feet up, looking at a bow almost a quarter mile away. Of course, tugs are used, and even side thrusters and sonic measuring devices.[29] This is a manageable problem. But if tankers break down there are very few places on earth that can receive them, and towing them into a dock must be quite a problem. In 1971 one 200,000-ton tanker had a collision in the Persian Gulf. She was towed to a port for repairs, but they refused to receive her because of the oil leaking from her tanks; she was towed about, leaking, for two months before they found a port that both could take her and would. Mostert gives the details of others less fortunate, those that break up at sea, spreading their balm on stormy waters.

The size of such ships makes even trivial mistakes monumental. A valve was left open by mistake for thirty minutes, and 22 miles of coastline were affected; oil coated rocks and the beaches of beautiful Bantry Bay in Ireland. In Rotterdam, the world's biggest oil port, the Dutch authorities maintain rigid antipollution controls. But mistakes will happen in the best of Dutch families. Two thousand tons of oil went into the harbor once by mistake, Mostert was told, when an inexperienced seaman opened the valve that sent the oil into the sea instead of the shore. When 16,000 tons escaped in a wreck off the Spanish Atlantic coast near Vigo, the oil caught on fire, creating a firestorm that raised hurricane winds in the vicinity of the ship. These winds broke up the oil into a fine mist and sent it aloft; it came down some days later as a black rain on the coast, damaging homes, crops, and the cattle that ate the oil-soaked clover.[30]

Collisions between even modest tankers can be frightful. Two ships of Liberian registry, a U.S. and a Greek tanker, collided in the Indian Ocean 23 miles from the coast. The explosion rocked buildings 40 miles inland. One ship vanished in four minutes, and thirty-three seamen vanished with it. Both ships were traveling at high speed through a fog so dense that one master could not see the mast of his own ship. Though they observed each other on radar, neither ship reduced speed; the Greek ship made two attempts to plot the course of the other ship, one four minutes before the collision; the U.S. ship made no attempt. The master of the U.S. ship (a Chinese captain) ordered his ship away at full speed and did not attempt to pick up survivors. He also broadcast the wrong position on his SOS call, which he discovered six hours later, but never made a correction. Survivors were picked up by a passing freighter that watched the whole thing on radar—the chief officer of the freighter watched the two dots come together, heard a terrific explosion and felt his ship shake, then saw the two dots come apart and one disappeared.[31] Mostert gives a horrendous list of other examples of incompetent seamanship, and, of course, inoperative equipment (such as one grounded tanker, where the gyrocompass, echo sounder, radar, automatic log, speed indicator, and rudder indicator were all out of order)!

The English Channel is so crowded at times (forty ships can be within it at once) that they pile one on top of another as if it were the New Jersey Turnpike. There are lanes in the Channel (and, since the early 1970s, in sixty-five other crowded parts of the vast oceans). One freighter was in the wrong lane, to save time. She hit a tanker, which exploded, breaking windows five miles away in Folkestone. The freighter went down too, but being intact, created a partially sunken hazard (the Channel is shallow), which was marked with lights. A German freighter hit her the next day, and sunk. A light ship was added to the warning buoy lights. A month later a Greek freighter added itself to the pile. A second lightship was added and more light buoys, now fifteen of them, and, of course, advisories had been going out for updating the charts. Two weeks later an unidentified tanker ignored a barrage of rockets and flashing lights from the two lightships, ran through one row of buoys, and to the surprise of all, made it through and vanished in the night. By now forty-seven had died on the spot. British coastal authorities reported that within a two-month period sixteen ships had ignored the warnings and entered the area of the wreck; it was the fastest route. When the weather cleared, they were able to demolish the wrecks, and the cowboys were safe from that hazard.[32]

The size of the modern tankers has not only created the catastrophic potential, but is beginning, belatedly, to require changes in organizational structure aboard ship. As outlined by Mostert, first came the consequences of moving all personnel to the aft portion of the ship. Before that, the ship consisted of two communities, connected by a bridge or catwalk; the crew and the engineers aft, above the machinery, and the master and the navigation officers midships, below the bridge. The VLCCs put everyone aft for economy reasons and also because a midship explosion could wipe out the bridge. The two groups, traditional rivals going back to the advent of steam power, were now thrown together; the engineers shared a table with the master and navigator. "But automation carried these changes even further," Mostert notes.[33] Now there was a third caste—the electronics engineers, a transformation of the old light bulb and fuse changer, the electrician, into a systems engineer.

The systems engineer, or electronics engineer, is responsible for the automated equipment in the engine room, the radar and collision avoidance systems, ship-to-ship radio phones, and the computers that in a few modern ships receive a ship's position from a satellite and direct it accordingly. It is no longer so clear who is in charge. One British company, reports Mostert, dropped the term "master" and calls the captain the ship's "manager," with the ship being run by a "committee," consisting of navigation, engineering, and electrical officers. Meanwhile, the owner or charterer (most tankers, at least, are chartered; the large oil companies own only about 35 percent of the fleet) directs more and more of the ship's activity from shore. The captain receives a print-out of maintenance tasks to be performed, and a detailed schedule for them. Sailing directions are frequently changed; schedules, courses, and arrival times are set up on shore and radioed to the ship. As with the rationalization of air transport, these changes are probably for the better. Weakening absolute authority in a moderately complex and fairly tightly coupled system should allow for more effective problem solving. (But as we shall see in the chapter on space missions, increasing the power of the head office can go altogether too far.)

Of course, the automation makes the ships vulnerable to small errors. An incident is described in Mostert's book in which the engine room was on automatic control and everyone was asleep. Alarms then went off; the main boiler had tripped, and the engines had automatically tripped in turn—a safety device, just as in a nuclear power plant. Emergency electrical power was produced by the residual heat—twenty minutes of it, and then blackout. The crew couldn't find the problem, so they started the boilers up anyway in order to have power to search further. It took six hours to find that a half-inch-wide rubber diaphragm had split on a reducing valve. The valve used high-pressure air to hold up a flap on a forced draft fan. The diaphragm failure caused the flap to close, but only momentarily, which told the computer that the fans had stopped, though they hadn't. Thinking the fans were stopped, the computer immediately put the boiler fires out, which stopped the engines and the ship. Fortunately, the ship was not in busy waters or close to a coast, or in a storm.[34] As Mostert aptly puts it, automation is marvelous; "it has a pretty, animated face." But automation depends upon a ship's undependable power system, itself automated.[35]

Explosions

VLCCs have not only added a measure of complexity to lumbering size, they have introduced new complexities into the explosion problem. Oil tankers have been exploding since they were first used, but the problem became especially worrisome with the arrival of the Very Large Crude Carriers. Mostert notes that from 1959 to 1974 there was an average of fourteen explosions a year on oil tankers.[36] During an eighteen-day period in 1969 three VLCCs blew up. All were new, and all were cleaning their tanks at sea at the time. Shell and other companies launched an intensive inquiry into the problem, and probably, Mostert notes, spent more trying to understand the explosions than was spent by the industry on models, experiments, and research to build the new 200,000-ton plus vessels.[37] The problem is one that the chemical industry confronts regularly: the transition from an over-rich atmosphere to a under-rich one. Here is how it works.

Oil itself does not explode, or even burn all that easily; it is the gas given off as it evaporates that creates the problem. This gas is rich in hydrocarbons, and these are explosive. After the tank is pumped out, there is still much residue in the tank. A tank, incidentally, can be as large as the inside of a glorious cathedral; a ship will have a string of these, since a 200,000 dwt (deadweight ton) ship is 1,000 feet long. A small 20,000 dwt tanker will have thirty small cargo tanks; a 250,000 dwt ship will have only fifteen cathedral-sized ones. With all that exposed surface, evaporation is rapid. When a tank is full or even one-quarter full, the air in the tank is so rich in hydrocarbons that it will not explode. If it is "empty" and scrubbed, though oil still remains, there is plenty of air to dilute the gas. It is in the intermediate state, passing from over-rich to over-lean, that a period of time is inevitably reached when the mixture is explosive. Since the air in these cathedrals does not move much, a pocket in one corner of the tank, or between two ribs of the ship, or up near the deck, can be of the proper mixture while the rest is safe. It may be a small pocket—a few square meters—and impossible to detect, since it can move about. Yet is is enough to ignite the whole tank once it goes.

A spark is required to set off the explosive mixture. A spark can be produced by a nylon shirt; a nylon rope; by a nut falling off a hose used for cleaning the tank; or by the wash water itself. When water is slammed against the steel side at high pressure, it can create static electricity sufficient to ignite an explosive mixture. Experiments by Shell led them to posit the sequence that is paraphrased below.

The oil is washed from the sides and bulkheads and other internal structural supports by automatic machines that rotate their gigantic streams of water. Water from the nozzles expands into huge "chunks" of water, which pass through thick clouds of spray. The static electricity in the spray is absorbed by the chunks of water; they become charged as a thundercloud is charged. When they hit a piece of metal they are capable of drawing a flash, somewhat like lightning.[38] The solution, now applied to new tankers, is effective but expensive: waste gas from the boiler is high in nitrogen, and not explosive. It is pumped into the tanks as the oil is drawn out, replacing the oil with an nonexplosive mixture. But there are still problems. The tanks must be inspected from time to time for cracks and open seams; the ships are so long that gigantic stresses are set up on the hulls and the bulkheads. Inspection then requires oxygen equipment and presents the risk of being poisoned. It is a well-recognized hazard, and elaborate protections are taken because of the many people overcome and killed in tanker holds.

Furthermore, there are times when inert gas cannot be used, even though the ship has the capacity. Tankers frequently blow up in our gulf and river ports when inert gas is not used, for example, because of leaks in the "ullage" covers (ullage holes allow one to drop a line into the tank to measure the contents). In one such accident, a vapor cloud from the tank collected on the deck during a windless day, and was drawn into a temporary hose, traveled a few decks down in the hose through a siphoning effect, to where some welding was going on. The welding ignited it, and the fire flashed up the hose and out on the deck where the vapor cloud exploded.[39] It was, we might say, an unexpected interaction. Or, lightning bolts from the frequent thunderstorms in the Gulf of Mexico area can find an errant leak, or an open vent, or out-of-place flamescreen. An event such as this took place in the Houston ship channel on September 1, 1979, when S.S. *Chevron Hawaii* blew up and killed three people and injured thirteen others. A heated projectile from the exploding ship, 5 by 7 feet, flew 600 feet inland and hit an ethyl alcohol tank and blew it up; the flaming oil on the water ignited several barges in the vicinity; fires burned for eighteen hours. Damage to the 70,000 dwt ship was $50 million; damage to the terminal, barges, and other vessels exceeded $27 million; the cost of clean-up of the waterway was expected to be $6 million.[40] As usual, it could have been worse; a nearby tank farm was almost set on fire. Fortunately, none of the barges were butane barges. One of those was hit in another accident by a freighter in Louisiana, and the explosion killed twelve people and the damage was estimated at $10.5 million.[41]

The Technological Fix

Extra boilers, redundant steering apparatus, inert gas, emergency generators, and the like are all useful devices aboard ships, and the Coast Guard and the NTSB call for them regularly. But the real problem is seeing. Ships have to avoid hitting the bottom, the shore, bridges, and other ships. They hit all these obstructions because they don't see them, don't see them in time, or don't see how the direction in which they are heading is going to interact with the current, the wind drift, or the other ship. If they could only see the obstacles and see the expected point of collision or impact in time they could avoid disaster. Consequently, the major technological fixes have centered on the vision problem and the relative motion problem.

One seeing device is the fathometer, which tells the ship when the water is getting shallow. Fathometers are an improvement over the lead line and have been around a long time. They can be used in conjunction with charts to navigate, in a limited way. "If they had only checked the fathometer, they would have seen that they were five miles off course and about to run aground" is a familiar refrain in the accident reports. But if you have no other reason to believe you are off course, you are not going to be particularly attentive to a fathometer. Besides, they are often inaccurate or out of adjustment. Seeing the bottom, though, is not the greatest vision problem; most groundings, I suspect, occur because the land is not seen, not the bottom.

The major breakthrough for sightless vessels has been radar, developed in World War II by the British for defense and weapons. (It is one of the great many things that has been said to have "won the war" for the Allies.) Once it achieved reasonable accuracy and reliability, it began to be used aboard merchant vessels. The initial results were dramatic. In the English Channel, at night or in a dense fog or storm, a ship equipped with radar could zip through the fleet with impunity, watching the slow progress of sightless vessels and maintaining full speed. She would know that the other vessels were not likely to turn sharply, or speed up, because they could not see. They were on predictable courses at predictable rates, blowing useless whistles for legal purposes. Passing close was no problem. But when several ships could peer at the cathode ray tube, the problems began. "Target" Y (the language is probably a holdover from the early military applications where you used radar to find targets, not to avoid them) would appear to be on a steady course at a steady speed, but if X also had radar, it could decide to change direction suddenly, or speed up, because it anticipated that Y was proceeding blindly with a lookout and would not change speed or direction. The result was what was called, in marine circles, radar-assisted collisions.

The collision rate did not go down with radar. In particular, collisions between ships where at least one had radar did not go down; they may have gone up. What certainly went up was speed, for vessels used to slow down when they were sightless.

Radar is hardly foolproof and sometimes quite ineffective. When the master of the New York harbor ferry, *American Legion,* collided with a cargo vessel in dense fog on May 6, 1981, its radar was operating. However, the master testified that the plotting of targets on a radarscope is not done, nor is it practical, for ferries. The radar operator has to "reduce contact error," that is, correct for errors in the system; this is difficult to do when the vessel is not held to a steady course or is yawing in rough seas. Furthermore, the complexity of the task of radar plotting is considerable. The TSB report gives some of the problems. The radar observer must get a compass heading from the pilot or helmsman while getting a relative bearing of a contact on the radar. To make a relative bearing projection (the target is moving, but so is the ferry), another bearing must be taken after an interval of time. Because the ferry may change its heading in the meantime, another bearing must be requested from the helmsman and the difference calculated and applied. If the target changes its heading, the process starts over. Meanwhile, the image on the radarscope is likely to be blurred; and multiple radar contacts in a place such as New York Harbor are likely, making the work time-consuming and requiring complete concentration. At the same time, the same radar is also used to determine the course to steer, by observing landmarks, and to monitor various navigational aids.

Consequently, radar's value as an anticollision device is limited. Add to this the fact that recently 57 percent of the persons examined failed the Coast Guard's radar examination and the value of this technological fix is shown to be even more restricted.[42] Finally, in one study found that when initial detection was made by radar, the vessels in the sample made as many course changes in the direction of the target as away from it![43]

Not surprisingly, efforts have been made to automate the complex task of plotting the courses of target ships and "ownship" (one's own ship) and predicting the closest point of approach. The various devices designed to do this, whose development was strongly recommended by the Coast Guard and the NTSB, are called collision avoidance systems (CAS). They process radar data, solve for the speeds of own and other ships, the courses involved, and the closest point of approach (CPA). A trial maneuver can be entered and the results determined. Alarms indicate when a new target comes into range, and some automatically enter the new target data if the crew doesn't happen to see them appear on the screen. The targets are displayed as a line or vector on the screen. An alarm indicates when a closing vessel is likely to come within a predetermined range, say one or two miles. It is a marvel of modern electronics, simplifying everyone's work.

Of course, if the other vessels, with their CAS, start changing direction, the system takes some time to determine the stable direction of each ship; there may be a delay of two minutes, which may be too long if ownship and one target are both traveling 15 miles per hour in reciprocal (converging) directions. The resulting speed is 30 miles per hour for boats that need at least a mile to significantly alter their course.[44]

An additional system proposed is the marine radar interrogation transponder (MRIT), which when asked by another ship will automatically send the ship information on speed, direction, cargo, and draft. But in a crowded passage one could get so many likely targets and warning signals that the "conning officer" (the one watching for dangers) would turn the warning device off. It happens with present CAS systems.

My pessimism about CAS and MRIT is shared by an outstanding and respected analyst of marine accidents, John Gardenier, of the Office of Research and Development of the U.S. Coast Guard. In a 1976 paper he reports on research he and his colleagues did that contradicts most of the norms of the marine engineering fraternity, and of the Coast Guard itself, I would think. He took a large number of accidents, including all the collisions of medium to large ships over several years, and asked, How many of these might have been prevented by a CAS using radar? Since CAS is designed to prevent collisions, would it do so in these cases? They gave CAS the benefit of the doubt; if it were present, it was assumed to be working, and if working, correctly interpreted (both quite generous assumptions). If all this were true they asked if it could *possibly* have prevented a collision; they did not require that it definitely would have. In only 9.6 percent of the 198 collisions might it have prevented the collision; the independent raters disagreed about another 2.5 percent and were uncertain about 1 percent. So, resolving these disagreements and uncertainties in the favor of CAS and assuming it was working and properly used, we still would have only 13.1 percent avoidance of collisions by a collision avoidance system. A 9 or 10 percent figure is probably a more realistic one.

One might require ships to have the equipment even if only 9 percent of the collisions might be prevented; collisions are expensive. But this result strongly suggests that the overwhelming causes of collisions are not due to lack of information about the relative positions and headings of the two ships, which would be provided by CAS. In 68 percent of the collisions, this kind of information was not a problem (which means the ships were staring at each other and neither would change course). In 13 percent of the cases, CAS might have made a difference, as we have seen; that leaves 19 percent remaining where the information might have been relevant, but could not prevent a collision. Why? There were two major reasons: for almost half of these cases the maneuvering of the ships would have made it impossible to determine useful course projections or the closest point of encounter; the system requires a stable course on the part of the target and ownship. For almost half of the rest, the radar wasn't working well for such reasons as heavy rain. The remaining cases involved lack of vessel control, malfunctioning radar, and so on. Still, a 9 percent reduction—if working, used properly—might be worth it, if it didn't just encourage higher speeds.

Gardenier and his associates also looked into the case of bridge-to-bridge radio/telephone communication. They examined the percent of collisions potentially preventable by bridge-to-bridge radio. It was initially high—averaging 45 percent from 1964 through 1969, the period of the study when there were few such radios. By the time they were in widespread use, the potentially preventable collisions fell to an average of 19 percent (1971 to 1974). Thus, since more ships in the 1971–74 period had the equipment, there were fewer accidents the researchers could say might have been prevented if ships did have bridge-to-bridge radios. But even as they "worked" the problems with their use rose steadily—such as wrong channels used, too much traffic on the channels, misunderstandings of what was said, and mistaking the identity of the vessel one is talking to. By 1974, 18 percent of the collisions disclosed such problems; the equipment was there but could not be put to effective use. It is like radar; the initial impact for a few ships is probably great, but when sys-

TABLE 6.1
Frequency of Collisions

Cause of Collisions	*Total Percent*
Deliberate violations of rules of road	55.6
Judgment errors	50.0
Environment	46.5
Vessel design/waterway design	31.3
Late detection	30.0
Multiple vessels	9.5
Mechanical failures	8.0

SOURCE: From Gardenier, John S., "Toward a Science of Marine Safety," Symposium on Marine Traffic Safety, The Hague, Netherlands, April 1976.

tem effects take over, the advantage is reduced. Heavy traffic on the channel is one such "system effect."

Gardenier and his colleagues also looked into the causes of collisions, and found that mechanical problems, such as broken steering gear, have dropped considerably from 1964 to 1974; in the 1970–74 period they accounted for only 8 percent of the collisions. The major categories for collisions where at least one of the vessels was over 10,000 tons during the 1970–74 period are given in Table 6.1. (Since there may be multiple causes, the percentages will not add to 100.) Note the high role of "human factors"; either or both of the first two types of errors, violations or judgment errors, were found in 89.4 percent of the collisions. In a separate listing of violations, excessive speed was the most frequent (and is the one most closely linked to the economic concerns of owners). It was followed by not staying on correct or agreed-upon side of channel, improper lookout, and improper interpretation of the rules regarding which is the privileged boat and the boat most burdened. Excessive speed in restricted visibility accounted for almost half of the violations.

We are led back to production pressures again. Even if CAS and radiotelephone communications (and inertial guidance systems, et cetera) would appear to reduce accidents in these studies, they also appear to increase speed and risk-taking, because the accident rate is growing steadily. Production pressures defeat the safety ends of safety devices and increase the pressures to use the devices to reduce operating expenses by going faster, or straighter; this makes the maritime system as a whole more complex (proximity, limited understanding) and tightly coupled (time dependent functions, limited slack).

If we grant that production pressures are important, and the system induces errors because of authoritarian structures aboard ships, weather problems, seeing problems, and so on, we still have one remaining problem. Why do ships turn at the last minute and collide with each other when they were not on a collision course? How can we explain noncollision-course collisions?

Noncollision-Course Collisions

Collisions only account for 10 percent of ship accidents, far behind foundering (40 percent), wrecks (32), and fire and explosion (18).[45] Yet they are the most baffling, because they would seem to be quite avoidable. A large part of the technological fixes in the shipping industry are concerned with collision avoidance systems, possibly because a collision would appear to be an accident that should not happen and could be prevented. It should not happen because there is evidence that most collisions have required energetic action to make them happen.

When we think of collisions we generally think of two ships on intersecting courses that collide; in the last minute or two they may see the imminent collision, but it is too late to make an effective course change. This is diagrammed in Figure 6.1. Remarkably enough, this appears to be a rare event. Most collisions that I could find suitable accounts of involved ships that were not on a collision course, but one or both managed to change course after becoming aware of the other in such a way as to effect a collision. An organization concerned with marine safety, the Chamber of Shipping of the United Kingdom, published a compilation of fifty accidents in 1972 with accompanying charts where these were

FIGURE 6.1
A Hypothetical Collision

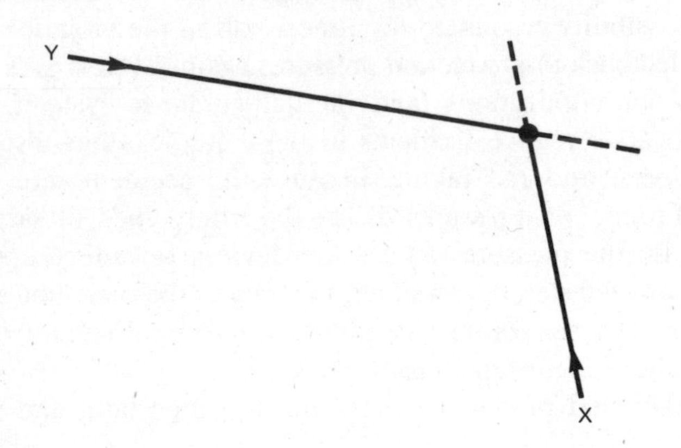

relevant (and kindly allowed me to reproduce some here).[46] Eight are presented in Figures 6.2, 6.3, 6.4, and 6.5.

Of the twenty-six collisions in this compilation of fifty accidents, only two represented collisions where neither ship changed its course. One of these involved one ship ramming an anchored ship, the other a near head-on collision in a channel. In five other cases one or both captains made an avoidance maneuver between one and two minutes prior to the collision; the maneuver might have been successful—we can't easily tell—if it had been made earlier. Thus, between two and seven of the twenty-six collisions represented "collision-course collisions." The other nineteen (and possibly as many as twenty-four) were noncollision-course collisions. In these nineteen cases there was a course change by at least one of the two ships that was intended to avoid a collision but that ended up bringing about a collision that otherwise would not have occurred.

What on earth could possess people in charge of huge ships to make the sometimes elaborate last-minute course changes that would bring them into collision? Most of these collisions were in the open sea, without constraining coastal obstacles, and for most, the presence of other ships was not a factor. If ships tried to strike one another, it would require a great deal of coordination to manage to do as well as some of the eight collisions reproduced in Figures 6.2 and 6.3. We know they do not try to collide, so the mystery deepens. Unfortunately, the accounts in the publication do not give enough details to allow us to explain these events. Sometimes the officers of one or both ships did not survive to explain their actions; evidence in the engine and chart room will go down with the ship if it sinks; officers may refuse to testify; they may give explanations that seem patently unreasonable.

Fortunately, we have a few documented accounts in the NTSB studies that will shed some light on the question of why ships that were about to pass in the night, or clear daylight, managed to collide against rather large odds. But first, let me remind you of the conclusion of the last section, by again turning to John Gardenier. In a recent article he states:

> The vast majority of collisions [involving at least one U.S. vessel] occur in inland waters in clear weather with a local pilot on board. Often the radar is not even turned on. If it were on, the alerting in function would almost certainly be turned off or ignored because of the large number of objects that routinely approach within collision threat detection range.[47]

We will examine some collisions in the fog, at sea as well as in inland waters, but even with poor visibility the evidence is that the technological safety aids were not being used; or if used, were not useful, and were

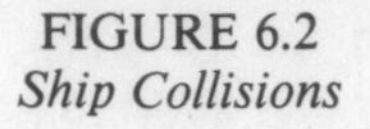

FIGURE 6.2
Ship Collisions

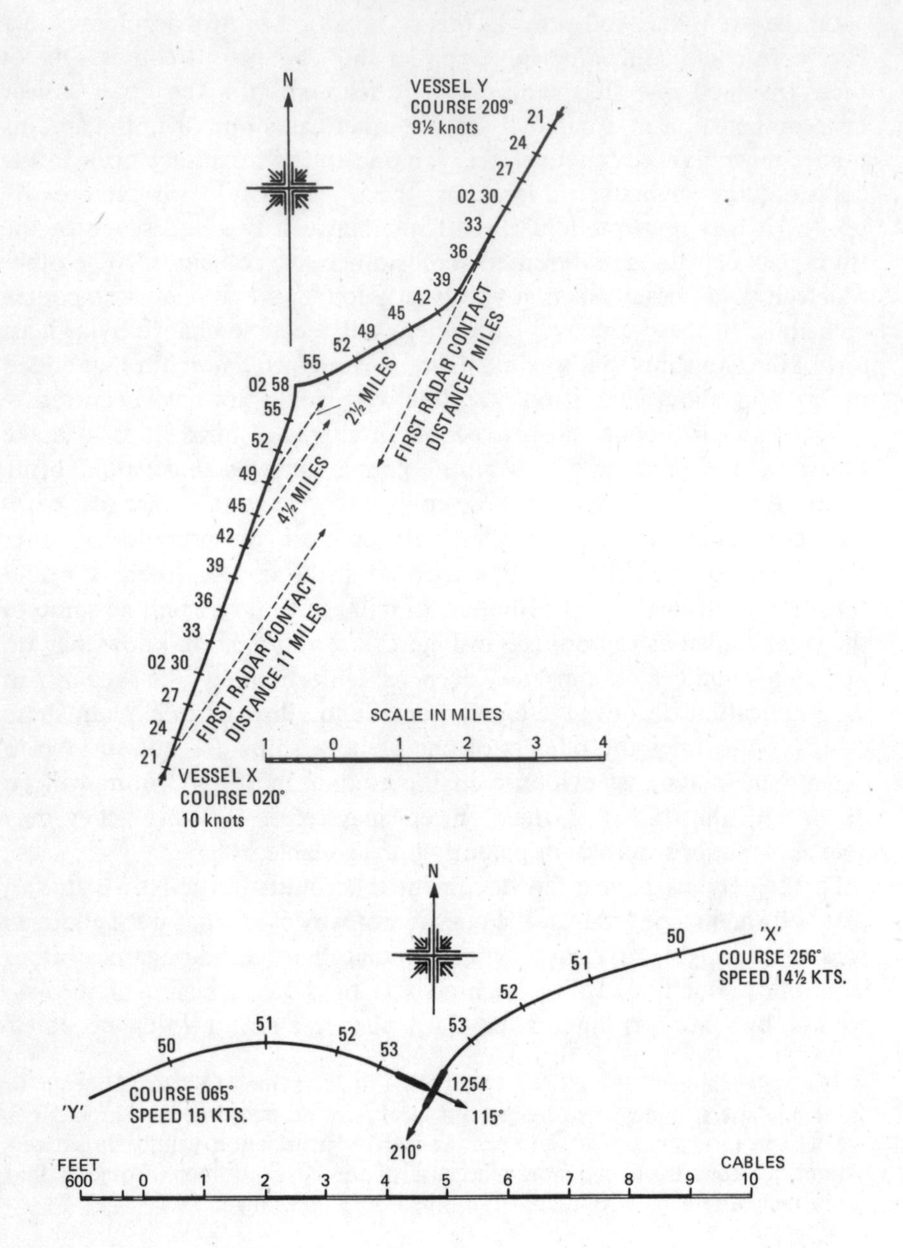

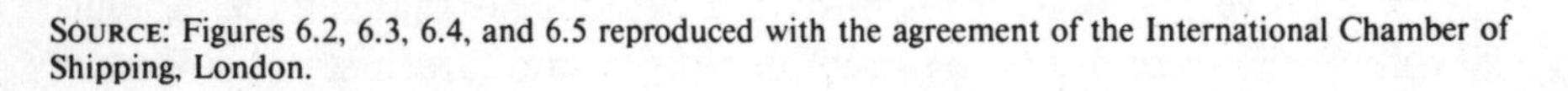

SOURCE: Figures 6.2, 6.3, 6.4, and 6.5 reproduced with the agreement of the International Chamber of Shipping, London.

FIGURE 6.3
Ship Collisions

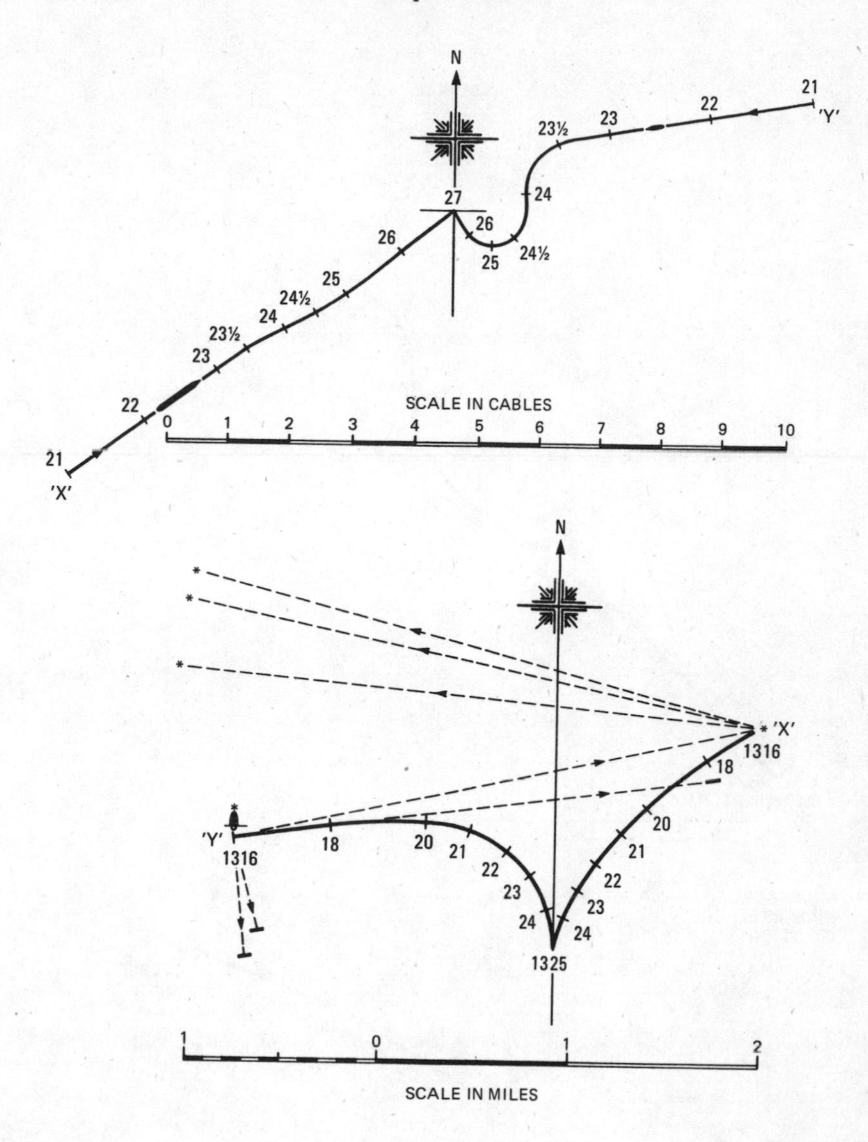

FIGURE 6.4
Ship Collisions

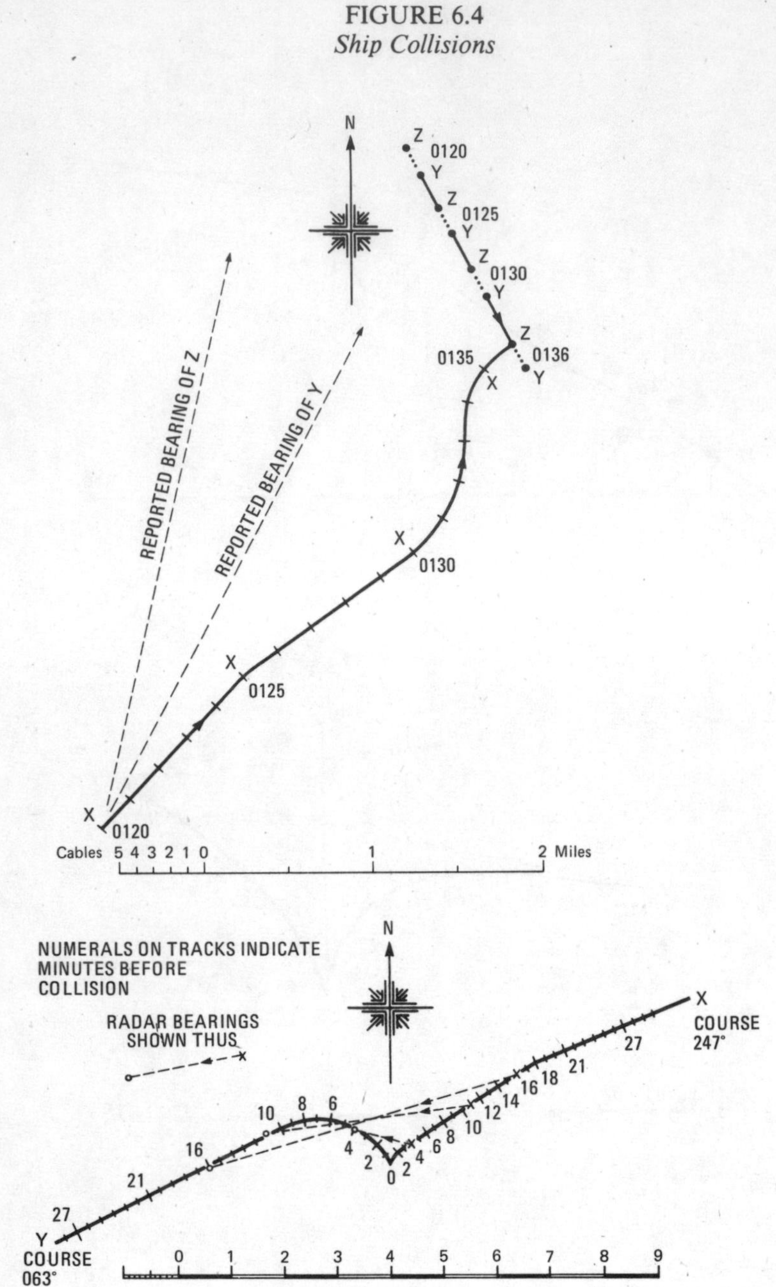

FIGURE 6.5
Ship Collisions

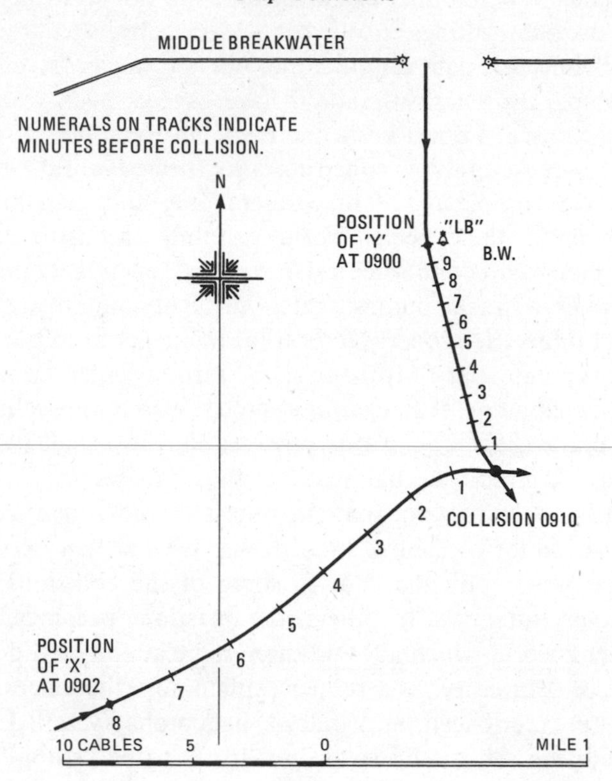

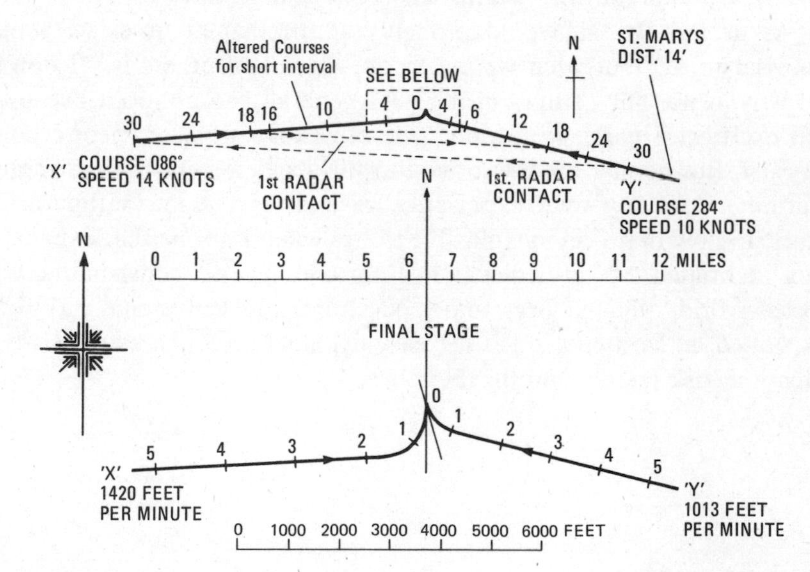

even deceiving; or were not needed because the weather was clear and passing agreements were either established or did not seem necessary.

When we do dumb things in our car occasionally, we get an insight into how deck officers might do the same. Why do we, as drivers, or deck officers on ships, zig when we should have zagged, even when we are attentive and can see? I don't know the many answers, but the following material will suggest that we construct an expected world because we can't handle the complexity of the present one, and then process the information that fits the expected world, and find reasons to exclude the information that might contradict it. Unexpected or unlikely interactions are ignored when we make our construction. Tight coupling also inhibits us; we cannot intervene properly to prevent incidents from becoming an accident. The operators at TMI made such limiting constructions of reality. In the last chapter we shall examine some research in psychology that attempts to make sense out of this process. Here we shall just present some fleeting and circumstantial evidence that it exists.

Note what is *not* being said. Inattentiveness is sometimes given as the major explanation for collisions. Indeed, that is what mariners reported in a poll that was conducted. But in some of the collisions, at least, people were very attentive. In others, the questions becomes, Why did they construct a world which allowed them to be attentive? I don't think inexperience or "stupidity" is a proper explanation; the officers in these collisions were experienced, and indeed, had probably sailed for years without a collision. Ship officers are not likely to be "stupid"; in fact, very few people in any walk of life deserve that appellation. Nor does the attribute "risk taker" help us much. It is not exciting to suddenly turn to the starboard and run into a ship when one should have turned to the port. As drivers, we all would probably admit that at times we took unnecessary risks; but what we say to ourselves and others is, "I don't know why; it was silly, stupid of me." We generally do not do it because it was exciting. Finally, we cannot rule out exhaustion, or inebriation. Both exist. But neither are mentioned in the accident reports, and more important, from my own experience as a driver, skier, sailor, and climber, I know that I do inexplicable things when I am neither exhausted nor inebriated. So, in conclusion, I am arguing that constructing an expected world, while it begs many questions and leaves many things unexplained, at least challenges the easy explanations such as stupidity, inattention, risk taking, and inexperience.

Marine Accidents

Some Collisions Explained

On a beautiful night in October 1978, in the Chesapeake Bay, two vessels sighted one another visually and on radar. On one of them, the Coast Guard cutter training vessel *Cuyahoga,* the captain (a chief warrant officer) saw the other ship up ahead as a small object on the radar, and visually he saw two lights, indicating that it was proceeding in the same direction as his own ship. He thought it possibly was a fishing vessel. The first mate saw the lights, but saw three, and estimated (correctly) that it was a ship proceeding toward them. He had no responsibility to inform the captain, nor did he think he needed to. Since the two ships drew together so rapidly, the captain decided that it must be a very slow fishing boat that he was about to overtake. This reinforced his incorrect interpretation. The lookout knew the captain was aware of the ship, so did not comment further as it got quite close and seemed to be nearly on a collision course. Since both ships were traveling full speed, the closing came fast. The other ship, a large cargo ship, did not establish any bridge-to-bridge communication, because the passing was routine. But at the last moment the captain of the *Cuyahoga* realized that in overtaking the supposed fishing boat, which he assumed was on a near-parallel course, he would cut off that boat's ability to turn as both of them approached the Potomac River. So he ordered a turn to the port. This brought him directly in the path of the oncoming freighter, which hit the cutter. Eleven coastguardsmen perished. (See Figure 6.6).

As with most accidents that receive extensive investigation, all sorts of imperfections and errors were discovered. The captain was myopic and not wearing eyeglasses, and had asthma complicated by aspergillosis, which might have affected his vision. Though experienced, he had never received the recommended shore training, and was operating what previous Coast Guard inspections had revealed to be a grossly understaffed vessel with an overworked crew. A second radar screen near the captain's station that might have prevented the accident, thought I doubt it, was due to be installed in just five days (it was the only Coast Guard cutter without one). The pilot on the cargo ship did not sound the five-blast emergency whistle at the initial sign of danger, but an ambiguous one-blast warning signal; the pilot believed that he was the "privileged" vessel and the other one was "burdened," according to navigation rules, but he was wrong according to the Coast Guard inquiry. The loss of life might have been less if two watertight doors were not standing open on

FIGURE 6.6
Ship Collision in Cheasapeake Bay

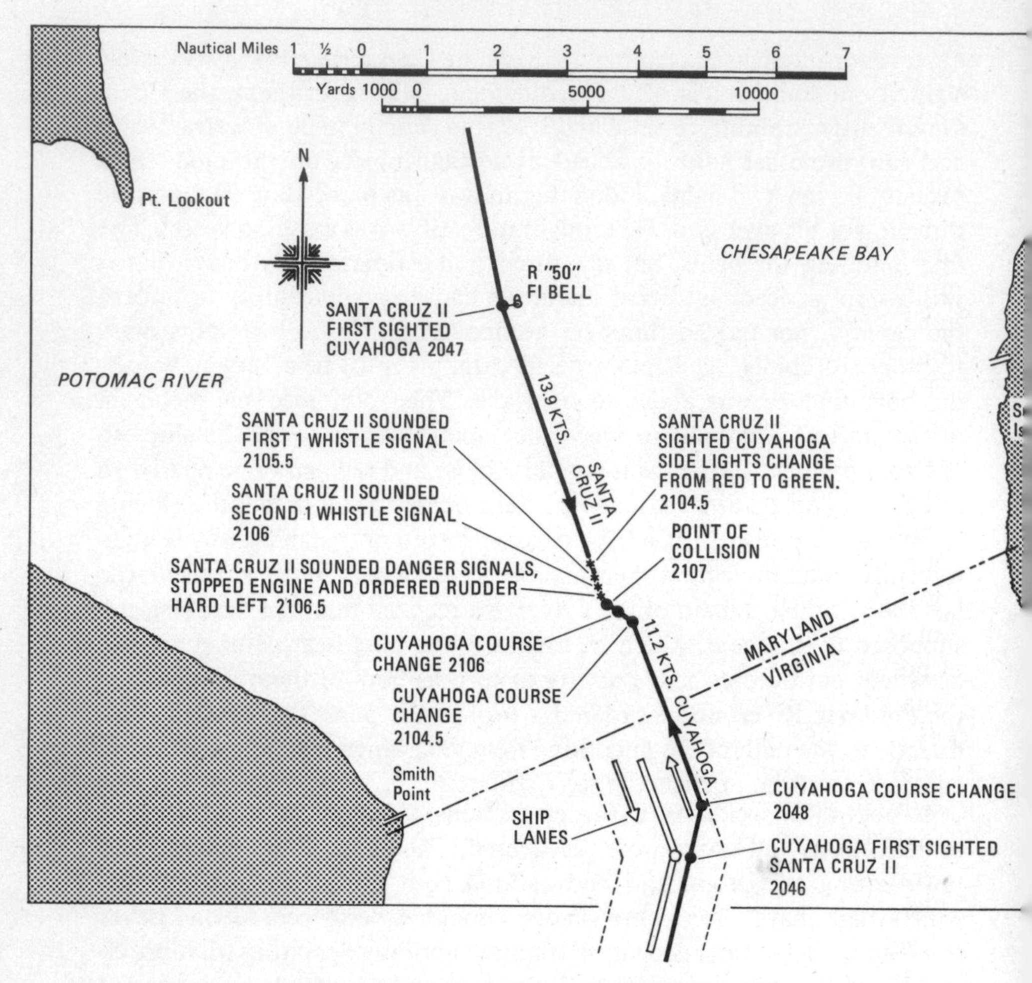

SOURCE: USCG/NTSB, Marine Casualty Report, No. 16732/92368 31 July 1979.

the warm autumn evening and if clothes and other personal gear had been properly stored below.

All true. This is the list one will find in any careful accident investigation. One would also find such a list, perhaps even a more lengthy one, for all those close calls that never resulted in accidents, and for all the routine operations that do not even have close calls. The Marine Board of Investigation is right to cite them, since every little bit of vigilance-stirring is useful. But what strikes me about this event is how well it exemplifies the easy way in which we can construct an interpretation of an ambiguous situation, process new information in the light of that interpretation—thus making the situation conform to our expectations—and, when distracted by other duties, make a last minute "correction" that fits with the private reality that no one else shares.

Further, it is interesting to note two minor events that conspired to prevent a new construction from emerging. Since the lookout knew the captain had seen the lights, it would have been redundant (and quite outside the authoritarian structure) to say, "Hey, don't you see that ship coming our way?" Later, as the collision neared, the lookout and another seaman discussed the situation and decided that perhaps the approaching ship actually should be reported again. But at that moment the captain saw it and blew his whistle. There then did not seem to be any point in notifying the captain of the obvious. But they did not know that the captain still assumed that the ship was going in the same direction as they were; thus, they did not contradict his sense of reality. Of such trivial events can accidents be born.

Why would a ship in a safe passing situation suddenly turn and be impaled by a cargo ship four times its length? For the same reasons the operators of the TMI plant cut back on high pressure injection and uncovered the core. Confronted with ambiguous signals, the safest reality was constructed. In the above marine case this view of reality assumed that the other ship was not a head-on collision threat. That this accident was not a freak or highly unusual event was attested to by an expert witness at the hearings, a pilot with over twenty years experience in those waters. The report notes that he testified "that he had experienced a number of occasions when other vessels took sudden and unanticipated actions, changing proper and safe situations to hazardous situations." The report itself notes, "The case books are replete with collision cases in which one of two vessels took a sudden and unexpected action precipitating collision."[48] Something other than myopia, or, as the court-martial charge initiated against the skipper put it in the inelegant language of sea lawyers, "negligent hazarding of a vessel" would appear to be present in this and other noncollision course-collisions.[49]

Figure 6.7 shows the track of two ships about to pass without incident. Both of them, however, executed sharp turns that brought about an expensive ($6.8 million) collision. On the face of it, one is tempted to call the pilots insane. The cause is human error, but a close review will indicate how human it is to err on the Mississippi River. The *Pisces* was heading downstream in the river, and the *Trade Master* upstream. Both were small ships, 600 feet long, and 24,000 and 33,000 deadweight tons respectively. One carried bauxite, the other fuel oil. They established a port-to-port meeting agreement before they even saw each other. As the *Trade Master* rounded a bend going upstream, the pilot saw that he had to have a passing with a tug and its tow. This would bring him to the port side of the channel, so he called the *Pisces* and another boat that was between both of them and arranged it all; the *Pisces* agreed to a starboard-to-starboard meeting.

But a *Pisces's* crew member discovered a tug with a tow on its port aft quarter starting to overtake the *Pisces*. (This is possibly the first faulty reality construction; a watch officer saw something, but later no one could identify the existence of the tug and tow, although it is also quite possible there was one.) *Pisces* had passed one before, then had slowed up for a disabled vessel in the channel, been passed by the tug, and then passed it again. The pilot assumed it was the same one. He did not try to call the tug on the radio-telephone because it had signaled that it was going off the air. (The captain of the tug the *Pisces* had passed before explained later he had not meant that he was going completely off the air, but just was turning down his set because of all the chatter on it. This was another small error that might have prevented the $6.8 million collision.) Actually, if there was such a tug and tow on the port quarter, it was not the one *Pisces* had passed and which had turned down its radio; that was later found to have been three-quarters of a mile behind. But it was not unreasonable to expect that the *Pisces* was being overtaken again, and could not call the tug. Tugs with tows are long, unmaneuverable objects in the river, hard to get around.

The *Pisces* pilot figured that if he stayed to the port in order to have the agreed starboard-to-starboard passing with the *Trade Master,* he would collide with the overtaking tug and tow, and possibly be so close to rafts of barges moored alongside of the river that they would be pulled loose of their moorings by the suction—which happens in the river. (Having constructed a reality—which may or may not have been correct—he acted in terms of it, and not unreasonably, for the passing agree-

FIGURE 6.7
Tracklines of Pisces *and* Trade Master

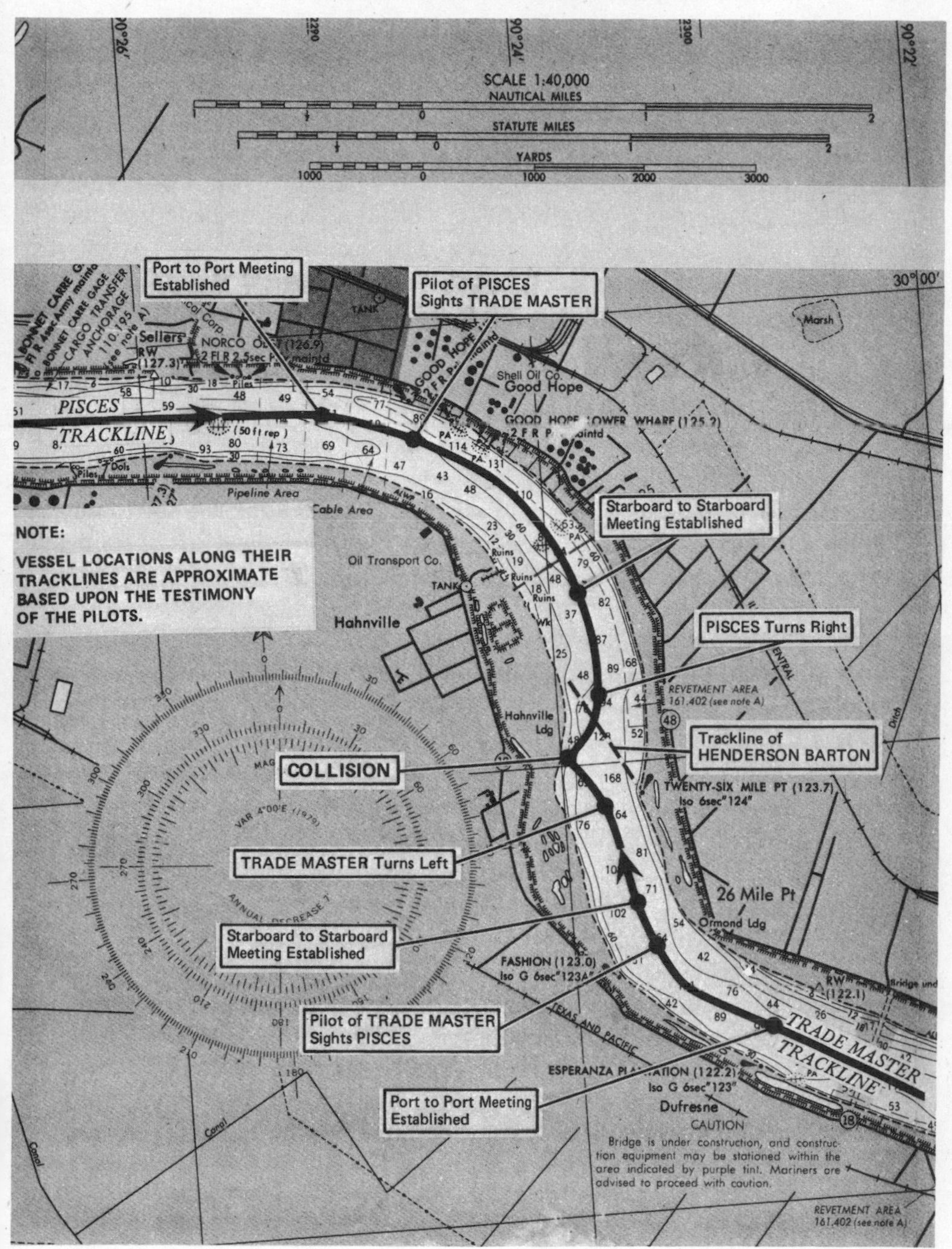

SOURCE: NTSB, MAR-82-39, February 1982.

ment could be changed.) So he blew his whistle and called on the radio for a port-to-port passing, instead of the starboard one, and turned sharply to the right (starboard) at the same time. The *Trade Master* saw the maneuver, but said it did not hear any broadcast or whistle, and assumed it was a mistaken maneuver that would be immediately corrected. (Here we have the second construction of reality: the other fellow made a mistake, and would correct it.) *Trade Master* did not turn right herself, because the pilot believed the *Pisces* would correct the error; turning right would put the two on a collision course. When the *Pisces* kept turning right, the *Trade Master* tried to call on the radio but only got a garbled message in return, perhaps from a different boat.

By now they were less than a half a mile apart, and would collide in one and a half minutes. The pilot of the *Trade Master* did not turn right, to turn away from the *Pisces,* but left, because he feared that he would hit the tug and tow that had been close behind the *Pisces*, or the barges on the far side of the river (all of which would have been better than hitting the *Pisces*, but he thought he could avoid this). So he turned sharply, hoping to cut a circle so close as to miss the *Pisces* and plow into the soft river bank closeby. (The navigable channel was less than a half-mile wide here; end to end, the two ships would take up half of it.) The *Pisces* turned even more sharply and backed her engines in order to avoid hitting the other boat directly amidships, and they collided. The *Pisces'* bow rammed into the front portion of the *Trade Master*. No one was hurt.

Having decided what vessels were in their paths, and having committed themselves to avoiding the tug and tow, both boats effected a collision. I think it is likely, though one can never know, that if both boats had decided at the last minute to risk hitting the tug and tow and sweeping close to the barges on the far side, they would have had collided anyway, on the opposite side of the *Trade Master*. If one had decided this, and the other not, the accident would have been avoided. But in terms of probability, a right-left or a left-right decision (avoiding an accident) is no more likely than a right-right or left-left decision (the actual one). In terms of commitment to a version of the world, the former version, retrospectively the safe one, was far less likely to be embraced; one or the other would have to change their views about the tug and tow, and change the heading of their ship. The course both took was the most "economical" and "consistent" one.

The investigating board points out that the *Pisces* had less to fear from the tug and tow and the barges than from changing an agreed passing agreement and cutting across the bow of the *Trade Master*. This is true, but only if one assumes that one has a choice of accidents; it is irrelevant if the *Pisces* believed the chances of any accident at all were less if it turned sharply to avoid the other ship. What the *Pisces* did not know was that the other ship did not hear it change the agreement, and would also turn sharply. The same happened with the *Trade Master;* it feared hitting the mysterious tug and tow, and believed it could avoid *Pisces.* But, unfortunately, *Pisces* kept turning. Thus, we may view this accident in terms of faulty reality construction, "explaining" the apparently irrational event.

The NTSB goes into some other details that we can learn from. The Board appropriately points out that the *Trade Master* did not reduce speed when the pilot first saw the *Pisces* make the unexpected turn; and the *Pisces* did not reduce speed when she first divined trouble. Thus, neither pilot behaved with prudence. This is true; prudence is sometimes in short supply in marineland. The Board also faults the *Pisces* on the initial right turn after a different passing had been agreed on, thus saying, in effect, it should have discounted the tug-tow-anchored barge.

But most of all, the Board faults the condition of the radio communication frequency at the time of the accident. This is a matter we discussed earlier—the unanticipated consequences of new safety devices. The requirement that all ships should have radio-telephone equipment, and use it in passing, is supposed to reduce accidents, the Board ruefully notes. Unfortunately, it says, a new set of problems arose with the radios —towboats put theirs on at high power (despite regulations) blocking out nearby radios, which are generally low-power, hand-held sets on large vessels. Furthermore, tugs and other ships abuse the airwaves by playing music on the emergency channel, engaging in much gossip, and doing a lot of cursing. One pilot on the river at the time of this accident remarked that the foul language was lengthy and incredible. In the lower Mississippi River, foul language seems to be a specialty; four different vessels associated with this collision testified to the routine use of the emergency channel for this purpose, or for country music. Because of budget cuts in 1980, neither the Federal Communication Commission nor the Coast Guard (the two share policing responsibilities) could do much to police the abuse. Four recent collisions, costing sixteen lives, one hundred injuries, and damages of over $12 million were associated with the abuse of the designated broadcast channel, the NTSB report notes.[50] Life on the Mississippi apparently needs the distractions of music, chatter, and swearing. It would probably be easier to change the technology than the culture.

On May 6, 1981, the *Lash Atlantico,* a U.S. container carrier, 820 feet long with a deadweight tonnage of 30,000, was sailing south in the Atlantic off of Kitty Hawk, and the *Hellenic Carrier,* a Greek freighter of half its size, was sailing north. Intermittent fog set in during the early morning hours. The *Lash* saw the *Hellenic* on its radar several miles away, and judged that she was on the port (left) side, when in fact she was on the starboard side. Why this error was made is not clear to the investigating board. The radarscope may have been misaligned; it shows a line that indicates the heading of ownship, and this may have been off enough to show the *Hellenic* to the left of the line, when in fact she was on the right. The *Lash* estimated they would pass within one mile of each other at the closest. The *Hellenic* also found the other ship on its radar, but correctly saw that the *Lash* was to its starboard, and the passing would be within one to two miles.

As the ships neared each other neither reduced speed, thinking all was fine. The fog settled in. Then the *Lash* turned right a bit to increase the passing difference, or so the crew thought. The distance did not increase on the radar, so when the ships were only about 2 nautical miles apart, a more radical course change was ordered on the *Lash.* Perhaps unfortunately, the master of the *Hellenic* noted this sharp turn on the radar, and ordered hard left rudder, just one minute before the collision. But it is not clear whether it made any difference whether he went hard right, left, or did nothing.

The captain of the *Hellenic* ordered the ship to be abandoned, and the captain of the *Lash,* which was not as seriously damaged, managed to call for help through a satellite communication system, since both the main and emergency antennas on the *Lash* were disabled by the collision. Help arrived quickly, thanks to the presence of the satellite communication system, and no hands were lost. The damage, including oil cleanups on the beaches, was about $8.5 million.

The Safety Board had its usual list of criticisms and recommendations: traveling too fast in fog (but ships with radar, Loran, etc., should be able to); failure of the *Hellenic* to sound fog signals (but those of the other ship were not heard by the *Hellenic*); failure to carry out a closest point of approach analysis of the radar data, complicated and time-consuming, which would take into account the relative movements of each (but there seemed to be no need, since neither ship expected any problem with a routine passing, until the last minute); failure to use VHF-FM radiotelephones to establish a passing agreement (it is not mandatory on the open seas; despite much effort the Coast Guard has not been able to get the international shipping community to agree to make this required). This list would have helped, but again, the meeting was assumed to be routine.

There is a hint that the *Lash* might have fiddled with its course recording device after the accident, and the captain gave a quite unconvincing reason for not having both radars operating, one of the 3-mile and one of the 6-mile range. The *Lash* did not use the radio because the captain would have had to leave his lookout position too long, but the *Hellenic* did not even have its radio turned on. As usual, it is the familiar tangle of "if only . . ." But what is important about this accident, where two ships would have passed in the fog if one had not wheeled around to strike the other amidships, is the initial assumption, or construction of reality, in this case shared by the captain and the mate. Once having assumed the other ship was on the left side, each step taken to avoid hitting it made it worse. The technological fixes were there; they either were irrelevant or not being used because there seemed to be no danger. It is true that the fog was dense, but that is a commonplace at sea.

Our last example concerns two ships that collided in the Mississippi River after a seven-day incapacitating fog which was just beginning to lift. The fog was impenetrable from the ground up to 30 feet; the air was clear above that. The downcoming ship, *Keytrader,* assumed the *Baune,* when she was sighted on radar, was pulling out of an anchorage on the *Keytrader's* right side. So the *Keytrader* called and whistled for a starboard-to-starboard crossing (a port-to-port crossing is the expected one, and the *Baune* was actually on the far side of the river, where port-to-port would have been the proper crossing). The *Baune,* having just transferred from a hand-held radio to the ship radio as the pilot was about to be relieved, had the set on low and did not hear it; there was also a lot of traffic on the radio. The whistles were not heard since the wind was such as to create a "shadow zone" in front of the *Keytrader* where sound is considerably reduced. The *Baune* did not have a watch forward since she did not expect downriver traffic, having participated in an informal agreement among various pilot associations and traffic stations (yes, in some parts of the river, there are even traffic lights) to only move ships up the river as the seven-day fog cleared; the pilot of the *Keytrader,* however, was a federal pilot, and had not known of this decision.

Once the ships could see each other's superstructures, it was still difficult to judge positions and directions of movement, because they could not see the shores, only masts and superstructures of various anchored ships. The *Keytrader* was going about 20 knots over the ground, but only 7 knots in the water, which she claimed she needed for maneuvering.

Nevertheless the inquiry boards (the Coast Guard and the NTSB) censured her for traveling too fast. The boards censured the *Baune* for not maintaining a lookout at the bow, but with the fog 30 feet high, it is not clear he would have been able to see much; the bridge is much higher. Thus, all in all, a variety of small failures and special circumstances led both shops to be totally surprised by the position of the other. Fortunately, the ships did not ground near docks or tank farms. Unfortunately, the *Keytrader* was loaded with gasoline; it ignited, and surrounded the *Baune,* which was downstream, with a 15-foot wall of flame. Sixteen persons died and three were injured; the fire burned for fifty hours.

The Larger System

Though 80 percent of marine accidents are attributed to human error, generally excessive speed in poor weather and errors in navigation, ships are certainly not free of equipment and design failures. A ship is a fairly complicated engineering system, as some accidents show. The following example, our last marine accident, involving the loss of the tankship S.S. *Transhuron* in the Arabian sea in 1974, indicates this complexity. But more important, it gives us a glimpse of the systems a ship on the high seas is connected with—the operating company in New York, the Indian government, and passing ships. At this level we are less concerned with complexity and coupling as an explanation of a particular accident (though that is apparent) than with the successive links with even larger systems, which add their own sources of failure. I will draw on a Marine Casualty Report of the National Transportation Safety board, supplemented with a telephone interview with the operating company, which was very helpful.[51]

When the *Transhuron* was reconditioned, air conditioning was installed. It was put on the level that was directly under the propulsion switchboard. This occasioned no comment from the Coast Guard inspector, because while piping should not be "in the vicinity" of the switchboard, this piping was separated by a steel floor from the switchboard, and ran to a nearby condensor.

After installation, engineers found that they needed a by-pass valve installed so that they could use the cold water system when the cooling pump needed repair. An iron nipple was installed on the bronze condensor head, to hold a gauge, and the dissimilarity in metals slowly created corrosion. Unfortunately, when the unit was cleaned a few years later, this obscure addition was neglected. At sea, it failed, and sprayed water into the propulsion switchboard 6 feet above it through an opening in the deck through which cables from the switchboard passed, and shorted the switchboard out. Since the system had, at this point, 2,300 volts and 1,000 amperes, it was a big short, and it started a large fire. The crew failed to disengage another system on the panel, and that system also failed.

The crew tried two methods to shut the main system down, but both were defeated by arcing and shorting; they did not try the method the inquiry board sternly recommended with all the assurance of hindsight. (One is reminded of the hapless operators at TMI and the assurance of the pro-industry members of the President's Commission, who argued that operator error was the cause.) The *Transhuron* tried to turn on the various CO_2 (carbon dioxide) fire suppressing systems, but they malfunctioned. The fire was put out with hand extinguishers.

With the fire finally out, the master sent an urgent message to his New York company for a tug; the ship had no propulsion system and was drifting in the Arabian Sea. The message was sent through the nearest relay, Cochin Radio, in India. No reply was received, so another was sent six hours later; then three hours after that; then seven hours after that. The next morning, thirty hours after the first call, the *Transhuron* radioed they were drifting in heavy 10-foot seas through frequent rain-squalls, 23 miles from an island, with only emergency generator power. On this morning, September 25, the radio operator at Cochin, India, told the ship that his station did not recognize urgent messages; all messages went in and out on a routine, first-in, first-out basis. The word "urgent" that headed all his messages was simply deleted by the operator.

The company finally replied thirty-one hours after the original message, close to a day and a half later. They had received at least the first two messages, since the second one from the ship gave more details about the accident, noting that the plug blew out of the air-conditioning condensor and listing the systems that were completely destroyed. But the operating company, the Hudson Waterways Corporation, of Manhattan, was not to be rattled. "Keep advised your position," they cabled. "Can possible emergency repairs be made. Did salt/water air condition condenser cause fire. Did control desk cubicle short and blow out what parts. Did control desk cubicle parts and wiring burn with fire and for how long . . . " and so on for some length. It concluded, "Answer immediately and start message with urgent." [52] They said nothing about getting a tug, or any difficulty in getting one; had they done so, this might have led the master to call for assistance from a passing ship.

We can imagine what the master wanted to reply to this engineering quiz, with 10-foot seas, rain squalls, no propulsion, and drifting toward an island, but his response was temperate. "Bombay FCC disallows my sending urgent messages. Position . . . Fifth message since breakdown. Require assistance." Meanwhile passing ships had offered assistance, but the captain was waiting for a tug to be dispatched by the home company, an unfortunate decision given his condition. Finally he sent Hudson Waterways a message saying that if he did not get an immediate response, "will take personal action." Again, it is curious that he needed authorization; according to the policy of the company (personal communication) and all or at least most companies, the captain is in charge and can do what he needs to do for the ship and the safety of the crew (though of course he must answer for his decision later). This message was sent at 8 A.M. on September 26. At 11 A.M. he sent out a distress signal.

The distress signal was acknowledged by fifteen vessels, but the closest was 110 miles away. The ship then radioed Manhattan again asking for reef and shoal details for the south side of Kiltan Island, so they could try to anchor as they drew near. They had missed Chetlat Island. These were the only two islands for hundreds of miles. The captain had passed over an anchorage near Chetlat Island, the company said later, but unfortunately had not let his anchor hang in the water until it caught on the bottom. The *Transhuron* then sent out a more urgent distress signal, and found a ship about 45 miles away. It sped toward them.

But nothing is certain at sea. The ship, the S.S. *Toshima Maru,* prepared to take them in tow, came close and fired their line-throwing gun—but it fell short. The *Toshima Maru* then signaled that their third officer had been injured when the gun was fired, and they were going to the nearest port to take him to a hospital. The master of the *Transhuron* was probably beside himself; they were two miles from an island with reefs and rocks, and being blown toward it, with high seas. He implored the *Toshima Maru* to give a tow for a mile, in order to clear the island. The *Toshima Maru* refused. The *Transhuron* then pleaded for just a half-mile tow, since the lives of thirty-five men were at stake. There was a long delay, perhaps fifteen minutes, and the *Toshima Maru* replied they would try once more, but their line-throwing gun was broken so the *Transhuron* should fire a line to them. The crew of the floundering ship were shifting their gun to the stern to get a proper shot when the ship struck bottom. They "walked out" the port anchor and called for the rescue vessel to come closer. But the *Transhuron* appeared to be breaking up in the large swells and was taking on water. Its cargo of special fuel oil for the Navy in the Philippines began oozing out.

The captain gave the crew the choice of trying to get to shore or to the *Toshima Maru;* they chose the former, because of the rough seas, though the island appeared deserted. The *Toshima Maru* left. An hour and twenty minutes later two boats were lowered, and all but the captain and four other officers left. The officers radioed the operating company through Cochin yet again. No answer. The next morning the five officers again radioed, asking for assurance that the crew would be rescued and "repatriated" from what looked like an uninhabited island. It wasn't unhabitated; the crew was safe, the master soon learned, but under arrest, and about to be transported by an Indian naval ship to Cochin.

Late that afternoon, three and a half days since the fire, the master finally received a second message from Hudson Waterways. They had acted, and were "doing the utmost." The message read: "Tug Challenger leaving Bombay early Friday morning and will arrive your position within forty-eight hours reverting with tugs callsigns etc. Send your position every 12 hours. No other tug available from Singapore to Persian Gulf. We are doing utmost expedite tug assistance."[53] That utmost effort would appear to bring the tug to the grounded vessel in two and a half days. But the company did not yet know the vessel was aground when they sent the message.

Meanwhile an Indian naval vessel appeared and boarded the *Transhuron* and made an underwater survey. All was lost; the ship had a hopeless crack most of the length of the bottom. For reasons that I do not understand—they probably concern insurance and salvage matters, but there may be other legal problems—the master and the four officers did not leave the ship for the naval vessel, but stayed on board. The next day the tug arrived, though the master had told Manhattan to cancel it. The emergency generator ran out of fuel, and the *Transhuron* started to move in the heavy seas. The officers prepared to abandon ship. They had to use an "oar-propelled" lifeboat because, taking precautions, they had tried to start the engine of a motorized lifeboat before the grounding to make sure it worked, and it had blown a seal ring in the hand pump and was inoperative. The swells were 12 to 15 feet.

Launching lifeboats in stormy waters (the time when you most have need to launch them) is quite difficult, as many accounts of marine accidents make clear. Even the ultra-sophisticated, totally enclosed, circular survival boats used on off-shore drilling rigs, with every imaginable safety device, flip over and leave their imprisoned workers to perish (eighteen in one tragic case). Inflatable liferafts are forever failing to fill, or having filled, blow away in gales, or fall on the swimmers awaiting them, injuring them. In one endless and awful marine tragedy in 1969, the crew of a naval vessel were trying to castoff the lifeboat they were in when a 2,000 pound bomb broke loose on a cargo deck; the "loose gun" smashed through the side of the vessel, and dropped into the lifeboat, smashing it in two.[54]

On the *Transhuron,* there was a sophisticated launching device, a Rottmer Releasing Gear, which releases both ends of the "falls" at the same time. The falls are devices that handle the rise and fall of the boat in the swells, preventing it from coming loose prematurely. The radio operator released the falls at the stern, but the man in the bow was unsuccessful. In a particularly large wave, the boat, still attached at the bow, assumed a nearly vertical position and slammed against the hull. The proper procedure, the Marine Board notes, is to release both falls at once, from a midships lever. In high, oil-covered seas, this might be difficult, but certainly not impossible. We do not know why the captain and officers did not try, or even whether any of the four men were amidships in the 24-foot boat. Upon dropping from the vertical position, the fall mechanism ripped out and the boat was free. The men managed to row it away and a motorboat from the tug picked them up.

But still on board was the first assistant engineer, who had lowered the lifeboat and was to join the four others in it. He ran to get a fire axe to release the forward fall and was chopping at the rope when it tore free. The lifeboat was immediately too far away for him to reach so he went to the upper bridge and launched the inflatable liferaft by rolling the container over the side. When the container was in the water he pulled the painter, which secured the raft to the vessel, to trip the cylinder to inflate it. It worked, but the wind caught the big balloon and the line ripped out and the raft sailed away without him. Manufactured by the Switlik Parachute Company of Trenton, New Jersey, the painter was supposed to withstand a tug of 3,000 pounds, but the Marine Board notes ominously, and without further comment, "Reports have been received of painters found severed when rafts are opened for servicing."[55]

The engineer then ran to the bow, where there was the least oil slick, and the boat from the tug managed to get about 30 feet from the bow. He jumped over and was saved. The officers and the men on the island were held in Cochin for three weeks under house arrest by the Indian government, and finally repatriated. The Indian government had the rest of the oil pumped from the ship to avoid further contamination.

This accident, or rather series of accidents, illustrates: design failures (the location of the gauge, materials used), a procedure failure (gauge and nipple not serviced), initial operator error (the damage would have been minimized somewhat—but not much—by deenergizing the control deck immediately), equipment failure (the CO_2 system), later operator error (failure of the captain to request a tow from a passing ship sooner; failure of the operators in Manhattan to respond promptly and inform the captain of difficulties in getting a tow, although perhaps they had to get clearance from the U. S. Department of Commerce, which owned the ship and used it for Navy business). The failures of the rescue attempt were multiple—the injury to the officer and failure of *Toshima Maru*'s gun, the delay in moving the *Transhuron*'s gun, lifeboat and liferaft problems, and so on. This is a particularly valuable story in that we are able to see a variety of systems interacting with the ship—the operators in Manhattan, the Indian radio and government, the tug assistance problem, passing ships, and ill-placed islands. Ordinarily, none of these problems are serious in themselves; it is only when they come together in a quite unique situation that we have an accident. Fortunately, it was not a very serious one. No lives were lost and the ship's loss would hardly make a dent in the budget of its owner, the U. S. Department of Commerce.

Conclusions

As complicated as nuclear power and chemical plants are, the complexities are contained within the hardware and the human-machine interface. We may not grasp the functions of superdeheaters, but we know where they are and that they must function on call. In the aircraft and the airways systems, we let more of the environment in. It complicates the situation a bit. The operating envelope of air becomes a problem, as do thunderclouds, windshears, and white-outs. There are also other systems in the air and moving about on the runway. Two government agencies are concerned with air transport, where only one is important for nuclear plants (the chemical industry has several, but the impact is minor). With the marine chapter we encountered a still more complex system, which required a more extended analysis. The ship itself, with its power plant, explosive mixtures, steering apparatus, and draft in shallow channels is important, but so are other ships, the insurance industry, the fragmented shipping industry, attempts are regulation, rules of the road, dangerous cargoes, national jealousies and interests, and, of course, the horrendous environmental problems of fog, ice, and storms. We had to look much more carefully than in other chapters at risk aversion and risk taking, casualty probabilities, production pressures, and the organization of the operating unit.

Marine transport appears to be an error-inducing system, where perverse interconnections defeat safety goals as well as operating efficiencies. Technological improvements did increase output but probably have helped increase accidents; with radar, the ship can go faster; when two ships have radar, they are even more likely to collide. The equivalent of CDTI (cockpit display of traffic information), which is to be introduced into the airliners, already exists at sea, and is useful only a small percent of the time, and may sometimes be counterproductive. Anyway, despite the increasingly sophisticated equipment, captains still inexplicably turn at the last minute and ram each other. We hypothesized that they built perfectly reasonable mental models of world, which work almost all the time, but occasionally turn out to be almost an inversion of what really exists. The authoritarian structure aboard ship, perhaps functional for simpler times, appears to be inappropriate for complex ships in complex situations. Yet it may be sustained by the shipping industry and the insurance industry who need to determine liability almost as much as they need to stem the increase in accidents. It is reinforced by the "technological fix" which says, "Just give the leader more information, more accurately and faster."

In Europe, Michael Gaffney notes, the technological imperative has been moderated by the social imperative.[56] There, the captain and other officers are trained to work as a team, and the equipment is being designed to sustain teamwork where a lowly helmsman or lookout is expected to contradict the mate or captain if necessary, where all are expected to share and check their mental models, and all share the responsibility.[57] This is the first sign I have seen of an exception to this error-inducing system. It is, as yet, a very small beginning, but it might unravel a bit of the tangled, self-defeating interconnections.

In the next chapter we move on to a much more prosaic world, firmly grounded in dams, earthquakes, mines, and lakes. The reduction in relevant system elements is substantial. Mining appears to share some of the error-inducing characteristics of the maritime world, but they could probably be corrected with much greater ease. We will see organizational failures in the case of the tragic Teton Dam failure, but no complex technological analysis is needed of that commonplace failure. Only in earthquakes and lakes do we find substantial potential for enlarging the system, but it is an interesting one, introducing the notion of eco-system accidents, something that becomes frightening when we come to recombinant DNA research in Chapter 8.

CHAPTER 7

Earthbound Systems: Dams, Quakes, Mines, and Lakes

In this chapter we will deal with the movement of large quantities of earth or water, whether deliberately or by accident. The production systems are primitive, compared to nuclear and chemical plants, and there are few unanticipated interactions in mines and virtually none in dams. Why, then, be concerned with these systems? First, it is useful to have some contrast to complex, tightly coupled systems to explicate and illustrate the value of our basic concepts. Dams have catastrophic potential, a matter of interest to us, but dam failures are not system accidents. The system is tightly coupled but very linear. Mining is a death-dealing activity, but the accidents are the prosaic ones of a largely linear system with a fair bit of loose coupling. Accidents in both systems could readily be reduced; the fatal combination of complexity and tight coupling is not present. Second, two themes have been emerging from the other systems that need more explicit discussion: organizational failures, and forced operator error. The Teton Dam disaster was not so much an engineering failure as an organization one. It is a theme that will grow in importance in this book, and this case study outlines it well. Forced operator error is flagrantly apparent in the mining industry, and our brief discussion of it will raise doubts about accident investigation and safety programs in all industry. Third, despite the linear nature of these systems, there is occasion for system accidents when they are linked to other systems in unanticipated ways. Earthquakes not only stem from "the restless earth," as Nigel Calder calls it, but from restless humans who think on a small scale in an ecology that is large scale. With some dams an unanticipated expansion of system boundaries creates what we shall call an "eco-system" accident, a concept relevant to the toxic waste problem as well as earthquakes. It is also the only way to explain one hilarious example, the loss of a large lake in a few hours.

Dams

The Grand Teton

When construction of the Grand Teton Dam began in 1972, there was little reason for concern about its safety. It was being built by the Bureau of Reclamation, one of eight federal agencies that construct dams, and there had never been a failure of a dam built by any of these agencies. In fact, dam failures are quite rare, and catastrophic, or even serious consequences are much rarer. One study concluded on the basis of U. S. data that there was only one chance in 10,000 of a dam failing in any one year (a 10^{-4} probability); only twelve dam failures were identified in the United States between 1940 and 1972, about one every two and a half years, and most of the failed dams were quite small. The most common failure is in the first year of operation—indeed, when the dam is first being filled; half of the failures occurred in the first five years.[1] Another study of dams constructed over a sixty-year period in the United States found that only thirty failed, a rate of one every two years, or less than two percent over the whole period.[2] Small dams built by industrial concerns and municipalities, generally with little catastrophic potential, are more likely to fail. We should not forget the small industrial dam that destroyed the whole community of Buffalo Creek in West Virginia in 1972. The fascinating sociological classic, *Everything in its Path,* by Kai Erikson, tells the story of the history of the community, the disaster, and the poignant attempt of the survivors to establish new lives when the social web of their community was destroyed.[3] Not insignificantly, Erikson gathered his data while helping in the preparation of a lawsuit against the mining company, which had ample warning of the danger they had created; the suit was successful.

In eastern Idaho, there is a plain of low population density, watered by the Snake River and farmed. There had long been an interest in building a dam in a mountain pass, to control the runoff during the spring and nourish the land during the hot summer. As a result of severe weather, the area was declared a drought disaster area in the summer of 1961, only to be declared a flood disaster area six months later in 1962. The next year the Bureau of Reclamation proposed a dam. The year after that a bill passed Congress with no opposition. The dam was to be on a tributary of the Snake River, and would be an earth-fill structure 310 feet high and about six-tenths of a mile long, creating a large lake 17 miles long in a 22-mile-long canyon. It would include a hydroelectric plant. An environmental impact statement for such enterprises was required by a law passed in 1969, and when one was issued for the dam in 1971 there were environmental objections raised, but there was no concern about a dam failure.[4]

Construction began early in 1972. In December of that year a group of geologists from the U. S. Geological Survey were working in the area, and became concerned about the dam since they had evidence that the area was seismically active—that is, had recently had earthquakes. One of the geologists drafted a memorandum intended to alert their superiors in the Geological Survey and officials in the Bureau of Reclamation of the danger. The memo urged that the Bureau be informed as soon as possible, "certainly within a month or two" (which is very soon in such circles) that major destructive earthquakes could occur in the area. In fact, there had been five earthquakes within 30 miles of the dam within the last five years, they noted, and two were of substantial magnitude. They also wanted to remind the Bureau that there was evidence that reservoirs actually *cause* earthquakes.

For example, in 1935 the Colorado River was dammed, creating the large reservoir called Lake Mead. In the next ten years, 6,000 minor earthquakes occurred in what was previously an earthquake-free area. The underlying rocks—actually a thin skin in terms of geologic proportions—had 10 cubic miles of water set on top of them. More violent disturbances were created when the Kariba Dam in Africa was built in 1963 through 1966. When the Koyna Dam in India was being filled with water, there was a violent earthquake which cracked the dam and killed 177 people living nearby.[5]

The geologists were sufficiently alarmed to note at the end of their draft memo that a failure would cause an enormous flood. Further, they noted satirically: "Since such a flood could be anticipated, we might consider a series of strategically-placed motion picture cameras to document the process of catastrophic flooding." [6]

The original memo was revised and expanded to seventeen pages, detailing the problem and concluding that their observations and other evidence "bear upon the possible safety" of the proposed dam and should be made available as promptly as possible. But the officials in the regional office of the Survey in Denver and the national headquarters in Washington, D. C., objected; there was too much "emotion" in the memo. It was redrafted several times. The team members finally became disgusted with the process of repeatedly redrafting the memo to meet the criticisms of Denver and Washington; indeed, as other information came in, they did not bother to report it because of the difficulty they had in trying to communicate the early warning.

The memo finally reached the Bureau of Reclamation six months after it was first drafted. By then, the urgency was sapped from the carefully worded memo. While all the facts remained, the implications now drawn were as bland as could be contrived: "We believe that the geologic and seismic observations, though preliminary, bear on the geologic setting of the Teton Basin Project." [7]

Though there is no evidence in the House Committee report relating to this, I think the Geological Survey's action (or lack of it) is readily understood, though it is deplorable because of the risks they imposed upon the public. The Bureau of Reclamation had by then spent $4,575,000 on the designs and the initial construction of a dam in a particular spot. An agency would not normally run to an adjoining and cooperating agency, and say, "A mistake has been made. Four and a half million dollars have been wasted because of it. Find another spot, or redo your work even though it may double your costs." Agencies have a long-term interest in cooperation, and will not make hasty charges such as this. Nor was the Bureau likely to readily abandon construction on such grounds. Even if there were more earthquakes, the dam might not be damaged. The Nuclear Regulatory Commission did not order a halt to the construction of the Diablo Canyon nuclear plant after it was discovered that it was a few miles from a major fault. It was also very reluctant to refuse a permit, though it finally did, to the Westinghouse Corporation to build a nuclear plant in Taiwan in an area that had had severe earthquakes (and tidal waves, and four nearby active volcanoes). Some calculated risks must be run, the reasoning seems to be, or nothing worthwhile will be built.

In July of 1973 the revised memorandum was delivered to the Bureau of Reclamation. There it appears to have had no impact. One of their geologists discussed it with one of the authors of the memo, but then dropped the matter. Another Reclamation Bureau geologist disclosed the attitude of the agency: he made a marginal note that they should prepare some "constructive criticisms" of the Geological Survey's memo. The House Committee's report editorializes strongly at this point that a serious warning by professionals that the dam might not be safe was ignored by the Bureau that was constructing it. Construction should have been halted until the issues were resolved, it said.[8]

But in any case it wasn't an earthquake that caused the dam to tumble. It was far more prosaic. The Bureau ignored its own data that rocks in the area were full of fissures, and in addition they filled the dam too fast. The cracks in the rocks were discovered as early as 1970 by a Bureau geologist; he was concerned, but concluded that the widest fissures were only 1.7 inches, so they could be grouted—filled in with a cement. But in 1973, with the dam half-built (that is probably the important point), the Bureau found that on the right side of the dam the cracks were caves, large enough for a person to walk through. The amount of grouting eventually used in the dam was over twice as much as had been estimated, and most of the grouting was needed in the right abutment, where the caves were. All it takes to bring a dam down is one crack, if that crack wets the soil within the interior portions of the dam, turning it into a quagmire. Grouting is not an exact science, and cannot be guaranteed to seal all cracks. When cracks of that size are found, the House Committee felt, the confidence of the Bureau in the grouting was no longer warranted.

Even after the dam had failed, however, the Bureau publicly insisted on its confidence in its grouting. The House Committee was incredulous about their stubborn insistence. But this may just mean that a stone wall that invites incredulity may be more attractive than an admission that could result in reorganization or replacement of key personnel.

We now have a warning about earthquakes, including the argument from geologists that even a small one could turn the rock and soil to jelly because of groundwater seepage, and a warning about large fissures into which 118,000 linear feet of grout was pumped under high pressure. One might think this would heighten the perception of risk and encourage future caution. In fact, the fissures did bother the Bureau; a memorandum spoke of the potential hazards of "fissures and other foundation voids in close proximity to the embankment."[9] Seepage could be a serious problem, the memo noted. But they went ahead, filling the reservoir at the standard rate of one foot a day as set in the original design.

But five months later, while the reservoir was still filling, the project construction engineer asked for permission to fill it *twice as rapidly*. There were higher runoffs from the heavy winter snows than had been expected. Besides, he said, this would provide an opportunity to test the grouting (but he did not indicate what would be done if the test failed catastrophically), and would provide full-power generation sooner, and finally, provide recreational benefits. They would continue to have daily inspection for leaks, of course, and continue to monitor the groundwater in the nearby wells, which would provide evidence of dangerous saturation of the area.

A month later a memo indicated that the monitoring effort was faulty, and that conditions underground were not favorable. Some data in this memo were six months late, due to winter conditions; three of seventeen other monitors were malfunctioning. Most important, the monitors that were functioning indicated that the rate at which groundwater was flowing was 1,000 times that which had been anticipated. They continued, nevertheless, to fill the dam. Indeed, they increased the speed to *four times* the normal rate, for recreational, power, and "test" reasons.

Two months after this alarming report, on June 3, 1976, two leaks were located downstream of the dam, and a third was found the next day. The project engineer, Mr. Robinson, said he was not worried; the leaks were running clear, and clear leaks are common in earth dams. Two more appeared the next day, and there was no question about the seriousness of these leaks. One of these was 132 feet below the top of the dam, where the right abutment joined the actual dam—the area of the caves—and the other leak was below the first, at the foot of the dam. This last one leaked 22,000 gallons a minute. An hour and a half later, the final leak appeared, in the same area. As it grew, it sucked material from the embankment through an ever-widening hole. Earth-moving crews tried to fill in the hole, but a whirlpool grew ever larger, and just after they abandoned their equipment and fled, it sucked the equipment in. Warnings were flashed to the people below the dam. Mr. Robinson, one assumes, was now worried. At 11:57 A.M. on June 5, 1976, the dam was breached.

Congressman Leo Ryan of California headed up the subcommittee that conducted the investigation and wrote the report. Here is how he described what happened next at the subcommittee hearings:

> The huge reservoir covering 17 miles of the upstream river and holding 80 billion gallons of water burst through the collapsing wall and tore into the Idaho countryside with an almost unbelievable force. It stripped top-soil from fields; it tore the pavement from roads; it twisted railroad tracks from their beds; it lifted houses and barns off their foundations; it uprooted trees; and it swept thousands of livestock along in its flood.
>
> In the end, eleven people were dead. Thousands were homeless. Whole towns were destroyed—literally ripped apart by the force of the water—then coated and littered with tons of mud and silt and debris.[10]

More than 100,000 acres of farmland were destroyed and 16,000 head of livestock were lost. Total property damage was estimated at the time to be over $1 billion. Fortunately, there was enough warning to permit evacuation of some areas, and it happened during daylight hours. No movie cameras were present to record the catastrophe, as the geologists had recommended with tongue in cheek, but some stunning photographs by tourists made the covers of news magazines.

An Ancient but Inexact Science

The ancient Romans built fine dams that still stand—but they were small. We have had a lot of experience with dams, including ones as large as the Teton. Yet two observations are relevant: we do not take dam-building seriously enough to have adequate inspection of dams and designs; and we really don't know that much about building them after all. Congressman Ryan's subcommittee noted that recommendations to have more adequate inspection never seem to get past the Office of Management and Budget. In 1976 it was estimated that of the 49,329 dams in the United States, about "20,000 were so located that any failure or misoperation could result in loss of life and significant property damage." (We have had a lot of experience with dams and their failures, however, which indicates that warnings are often available, limiting the loss of lives.) Dam safety programs were either nonexistent or inadequate. Eight federal agencies build dams, but only three of them were considered to have adequate safety inspection programs, although the Bureau of Reclamation is ironically one of these three.[11]

As desirable as it would be to have more inspection programs, it is not clear that it would make a great deal of difference. The lessons from the Teton Dam failure is that the heads of the project simply refused to believe that it would fail, even after their own alarming inspection reports. Indeed, shortly after the disaster, the Bureau circulated a publicity release. In it they asked, "With the benefit of hindsight, was there anything that Reclamation might have done to prevent this disaster?" And their answer to their own question was, simply, "Nothing."[12]

If the officials were embarrased by the failure and the eleven deaths, they never indicated it. None lost their jobs, or were sued. In Europe life is not as easy for engineers. When the Malpasset Dam failed in France in 1959 the chief engineer was legally charged with negligence and homicide. A similar charge was filed in Italy when the worst dam disaster in history occurred.

It happened on October 9, 1963, at the Vaiont Dam in Italy, an 875-foot, highly engineered, extraordinarily large arch dam. When canyons are dammed, the sides of the canyon that are underwater are subject to some "lifting," because the water makes the rocks and soil lighter. If there are long dormant slide areas in the canyon, the lifting may reduce the pressure enough to disturb the equilibrium. This may cause a landslide, especially if heavy rains encourage it. At the Vaiont Dam a heavy landslide dumped enough material in the reservoir to produce a 330-foot wave (over one-third as tall as the dam itself; that is like dropping a golf ball in a cup of coffee). The wave swept over the top of the dam and killed 3,000 people in the narrow canyon and valley below. The Italian government, acting on findings by a technical board, held that the disaster was a direct result of "bureaucratic inefficiency, muddling, withholding of alarming information, lack of judgment and evaluation and lack of serious individual and collective consultation." (It is a conclusion reminiscent of the congressional inquiry into the Teton disaster.) Fourteen engineers were suspended and prosecuted for manslaughter, an extraordinary event, the likes of which I do not believe U.S. society has ever seen.

Naturally the engineering and geological profession was alarmed by the court's findings. It is not a precedent one likes to have set. The sentence was said to have been premature by subsequent professionals. One said, "The sliding could not possibly be foreseen by anybody in the form in which it actually took place."[13] But though the form of the sliding could not be predicted, the possibility of some sliding on the steep canyon walls was considered during design and construction. A similar fate may lie in wait for another high, thin dam, this one in Peru, 250 kilometers upstream from Lima, at Tablachaca. After filling, a prehistoric landslide was discovered, now sliding into the dam at the rate of one meter a year in the last two years. A large slide could fill the reservoir in a few seconds.

In the case of the Tablachaca Dam, the Peruvian government is forewarned and has called in experts. It is not a mysterious interaction of the dam and the ancient earth it has disturbed. Most dam failures are no more mysterious. Still, once a failure has taken place, it is sometimes a matter for intense debate as to just what happened. This is true of the Teton failure. Did the dam sink, or did it float? One geologist noted that the dam was not built on bedrock but on a brittle, cracked surface unit—which revealed, he said, a "colossal misunderstanding of the fracture pattern of volcanic rocks and a serious misconception of groundwater flow systems" on the part of Reclamation's geologists. When the dam was loaded, there may have been differential compaction, such that one side of it sank. Sinking only a matter of centimeters would exert tremendous pressure.[14]

Another expert also faulted the Bureau, but advanced a floating theory. The ground water underneath the dam built up pressure as a result of the unusually high runoff in April and May; this may have caused some lifting of the reservoir basin following a shear zone along the right abutment. The dam may have floated under the pressure.[15]

One ingenious experiment tested the knowledge of geotechnical engineers on a very prosaic, commonplace, but practical matter—how high can you build an embankment before it collapses? The results demonstrated that the experts differed greatly among themselves, and were all wide of the mark.

Geotechnical engineers from the Massachusetts Institute of Technology had been experimenting with the stability of clay foundations as a stretch of highway was being built through some marshy land. The highway construction was abandoned, so they had a chance to conduct a test. A stretch of the embankment was about 38 feet high, and various monitors were embedded in it and the sand and gravel working mat. The engineers proceeded to add more fill to the embankment to see at what point it would collapse. It did so after the additional fill had raised the height another 18.7 feet, for a total of 56.6 feet.

They then gave all the relevant data on the clay foundation, the mat, the fill, the width, et cetera, to seven experts, and asked them to predict how much higher the fill would go beyond the 38 feet before the embankment collapsed, and to indicate how confident they were of their predictions. Five predictions were substantially below the actual amount, two others substantially above it, with the range running from 8 to 27 feet (true value, 18.7). Furthermore, the experts indicated their interquartile ranges for the added height to failure, giving us an answer to the question: What is the lowest and highest height at which the embankment will collapse in order to be right at least 50 percent of the time? There is a 50–50 chance that the actual value will lie in this range.

Not one of the seven experts thought that the observed value of 18.7 feet would have a 50–50 chance of falling within their range; that is, all the interquartile ranges were either below this actual figure or above it. They also had high confidence in their estimates.[16]

Dam building, then, is an old but still inexact science; even the prosaic matter of low embankments, which exist all over the country and the world, is imperfectly understood. No political or organizational or economic problems complicated this simple test.

Radioactive Dams

Dams, unfortunately, do not only hold water; several in the Southwest hold radioactive mill "tailings" (waste sand and clay) from uranium mines, from which 15 percent of the uranium has been removed, leaving about 85 percent of the uranium ore still in this dirt. Of the fifteen accidental releases of tailing slurry from 1959 to 1977, seven have been from dam failures.[17]

On June 16, 1979, one of these dams broke at Church Rock, New Mexico, and released 93 million gallons of contaminated liquid and 1,100 tons of hazardous solid waste into an arroyo. The toxic materials then flowed through an Indian Reservation, on to Gallup, New Mexico, and then on into Arizona, where minute amounts of it probably ended up in Lake Mead, a water source for Southern California. Measured contamination extended over 100 miles of river bottom beyond Church Rock. But some experts testified during a congressional hearing that unmeasured contamination will extend further and eventually contaminate ground and lake waters. Some of the contamination sank 30 feet into the soil and eventually is expected to reach the food chain.[18]

Cleanup by United Nuclear, which built the dam, started immediately, and at the time of hearings, three months after the burst, 3,500 tons of streambed material had been removed from the first 10 miles of the stream, by workers using shovels, buckets, and 55-gallon drums. (The streambed was too soft for large equipment, and the company said diverting the stream and letting it dry hard enough to hold heavy equipment would be too expensive.) Company officials asserted that there did not seem to be any health problems as a result of the uranium concentrations.

United Nuclear claimed that the failure was due to a unique and unexpected outcropping of rock that caused differential settling. A previous crack that occurred was unrelated to the failure, they said. The company testified that, unfortunately, the liquid in the dam "temporarily exceeded" the sand beach buffer and got to the dam; because of the settling, this liquid caused internal erosion, leading to the failure.[19]

However, documents from the Corps of Engineers, an independent engineering firm called in at the time of the accident, officials from the State of New Mexico and the U.S. Environmental Protection Agency, United Nuclear's own records, and other testimony told a different story. The company's own geologists had warned of bedrock problems and the need for continuous inspection; the company agreed to put in sensors but didn't; the company's design called for buttressing the dam structure, but they did not do it; some cracking in 1977 occurred and should have given warning, but the State Engineering Office was not even informed, contrary to an agreement; and the dam did not incorporate all the necessary protections that the company's engineering consultant had advised. Nor was the proper compaction of the dam material performed. A Nuclear Regulatory Commission official testified that cracks appeared in January of 1979, a few months before the failure, giving "significant" warning. Drain zones were not built, though they were in the plans. The freeboard (distance between the top of the dam and the top of the liquid) was supposed to be 5 feet but it was only 20 inches; the dam was overloaded 50 percent beyond its design capacity. The cleanup had only six to ten people with shovels involved for the first month, until the state wrote a letter saying, "Get on with it."[20] This was not a system accident; components failed because of improper management, which included taking calculated risks.

Quake Making

The Teton disaster was a component failure accident of design and operator (management) error. But disturbing the earth's surface on a grand scale is always a bit risky; a geologic and geothermal system is being altered. The disturbances may create system accidents rather than the much more preventable component failure accidents. Increasingly we are going into the thin mantle that separates us from the core of the earth, in search of minerals, water, garbage cans, or containers for explosions.

In the process, we have created earthquakes in more mysterious ways than through the filling of dams. Some of these are worth a brief examination. Underground nuclear explosions have set off earthquakes. For example, in 1968 an underground explosion in a Nevada test site gave rise to thirty small earthquakes within three days. Apparently, an old fault was reactivated by the explosion. Minor tremors continued for weeks.[21]

Such concerns led to protests by Canada and Japan about a series of underground nuclear tests along the Pacific tectonic plate. The United States set off explosions in the Alaskan Aleutian Islands, but Canada and Japan were concerned because the plate runs from Vancouver, British Columbia, to Japan. Fortunately, no disturbance of the plate was apparent.

Denver, Colorado, had a mild earthquake in April 1963. It was a surprise, since there had not been an earthquake in the area in eighty-one years. Small ones continued for several years; one in 1967 did a little damage to the city. It turned out that the army caused them.

The army's Rocky Mountain Arsenal is 10 miles from Denver. It manufactures toxic materials, such as nerve gas, and had to get rid of large amounts of contaminated water. For a time they just put it into holding ponds, but this led to the death of crops, livestock, and wild life. So they dug a well, 2 miles deep, and forced the garbage into it under high pressure. Six weeks later there was the first earthquake, and then an almost daily series of minor tremors. The source of the earthquake was suspected within a year, but the army denied it could happen and went on pumping. The water, under high pressure, forced the old cracks in very old rocks to grow, and this allowed the rocks, under pressure from tectonic movements, to slide in jerky movements over one another. Even after the pumping stopped, for a time the pressurized water continued to force open the cracks. About two years after the army finally stopped the practice, the earthquakes also stopped.

Similar mishaps occurred in another part of the state when an oil company forced water into drill holes to increase the flow of oil; earthquakes appeared, and stopped a while after the practice ceased. The National Center for Earthquake Research took over a part of the field for deliberate experimentation. When they pumped water in, earthquakes occurred; when they pumped the water back out, they stopped. This led to a recommendation that part of the San Andreas fault near San Francisco be locked into place in a series of 500 operations. In each operation they would first withdraw water from two bore holes about 1,000 yards apart, then force water into a third hole in between the others. The withdrawal would lock the ends of the segment in place, the insertion of water in the middle would create minor earthquakes as the pressure in the fault caused the rocks to slide over one another. The result would be pressure relief along that part of the fault. Since the Pacific plate moves northward at about 2.5 inches a year, it builds up tremendous pressure. In some places near San Francisco the rocks are estimated to be 13 feet out of adjustment. Working like a chiropractor on the earth's spine, the scientist would create controlled adjustments along the spine of the fault where layers of rock have to slip under one another. Fortunately, the fault is close to the surface here, and appears to be self-lubricating below a depth of about 12 miles.[22]

The risk, of course, is that it might not cause a minor adjustment, but a catastrophic one. It would be tried out in an unpopulated area first, presumably, but even if it worked there, who can tell how the San Andreas fault would behave? Isn't it possible that a large earthquake could be precipitated, one that might not have occurred for another century? Or five centuries? How much do we know about what goes on under the earth? There are few occasions to test such devices as filling or emptying boreholes. In fact, in massive enterprises such as dams or burying hot nuclear wastes, an experiment is not possible. We cannot carry out destructive testing of large dams. We are even wary of seeding hurricanes, on the off chance that the storm might then change direction or get worse. Perhaps for this reason—an awareness of unpredictable eco-system accidents, rather than the economic costs—the chiropractic correction has not been considered for several years. We will return to the question of eco-system accidents in the case of a lake, and again, more fully in the conclusion of the chapter.

Mining

Subsurface mining in the United States is clearly a dangerous occupation. But the vast majority of fatal accidents are single person deaths; all of them are first-party victims, in our terminology. The mine, as a system, is rarely visited by system accidents, as far as my search efforts indicate. Graduate students and I examined the files of all accidents reported to the Canadian equivalent to our Occupational Safety and Health Administration (OSHA) for 1980, and searched the mine safety journals and OSHA statistics for the United States. What we found in mining was: (1) an inherently risky task; (2) frequent equipment failures, failures of the environment (such as rock explosions), and operator error;

(3) disasters that were generally avoidable and had ample warnings; and (4) only a few indications of possible system accidents.

On the Interaction/Coupling Chart in Chapter 3, mining is seen as fairly loosely coupled, and a bit more complex than linear. The argument for loose coupling is that when failures occur there is generally room for recovery because affected areas can be segregated; alternative sequences can be used for a time; some slack resources exist, and indigenous substitutions can be made in many cases. Yet it is not as loosely coupled a system as, for example, manufacturing firms, governmental agencies, or universities. There are time-dependent processes, there is only one way to perform certain activities, and the nature of the physical site eliminates some slack in resources.

The production process itself is linear—loosen the material, prop up the roof, remove the material, with few transformation processes involved, other than explosives. But the setting can create unexpected, unplanned, and invisible interactions, largely in the matter of the flow of gases and pressures in the complex web of shafts and tunnels and ventilation holes. Some of these are probably unavoidable; there are simply too many interacting factors to anticipate just where an explosive mixture will be formed, and where it will explode; which doors will be blown out, forming unexpected paths for the force of the explosion and the dangerous gases to travel. Mines, then, partake of some aspects of both complex and linear systems. It appears that some of the complexity is irreducible, though not all of it. A good bit of the tight coupling has been reduced over the years, and more could still be done in that respect. Most clearly, more could be done to reduce the occasions for failures of parts and units; incidents abound in these systems, though they rarely lead to subsystem or system failures. In this discussion of mining I wish to demonstrate that inherently dangerous activities are not necessarily prone to system accidents. In doing so, we shall also learn more about what passes for "safety" investigations in the United States, and the prevalence of "blaming the victim."

Inherent Danger

Mining is inherently difficult and dangerous. Explosives are used. Tunnels are dug, and can collapse. The lighting is poor; it is virtually impossible to eliminate deep shadows and glare because of turns and bends, and large pieces of moving equipment. Communication is difficult because of the noise and the remoteness of parts of the mine. Ventilation is difficult to control; a great deal of air needs to circulate through a mine to remove toxic and explosive gases and coal dust in coal mines. A continuous strong wind is blowing in most tunnels and shafts to clear out the fumes and explosive dust. The work requires large pieces of moving equipment and large power tools. It is performed in exceedingly cramped spaces, and it is slow and difficult to exit from the system.

In such an environment, component failures are frequent. Looking at the accident reports, one is struck by the frequency with which reports are littered with failed clutches, broken drills, broken warning devices, electrical or power failures, and so on. It seems that machinery is simply prone to failure, or, more likely, is poorly maintained and forced to its limits in this environment. A frequent "component" failure is the personnel. We do not know the extent to which personnel are trapped into "forced errors" through production schedule demands or long shifts below the ground (eleven-hour shifts are mentioned), or, on the other hand, the extent to which there is a "macho" culture that provides psychic rewards for risk taking. I am sure that the first exceeds the second; a risk-taking, macho culture has probably developed to make sense out of what is a fairly inhuman activity, a view of this world that makes it conceivable that one could function in it. But whatever the cause, operator error, forced or unforced, is common.

Finally, there is the obvious failure of the "environment." It is simply not possible to judge where faults will occur in the roofs or sides of the shafts and tunnels. Falling rock from the ceiling or the sides of the tunnel is a common problem, but explosions of rock, called "destressing," are more fearsome. These occur when rock, which is inevitably under tremendous pressure a mile underground, has the pressure on one side of it changed; the pressures from the other sides cause it literally to explode. It is difficult to predict when this might occur. Destressing is sometimes engineered, or planned, in order to make the face that is being mined safer.

To give us some idea of the range of accidents, here are a few brief descriptions from an article in a safety journal designed to show that the risks are especially high for inexperienced miners. In the first six months of 1979, fifty-three persons were killed in metal and nonmetal mining activities (not all of them subsurface), and the article gives some details about a few of the accidents. (The article shows that half of the deceased had less than one year's experience at the job they were doing, but the statistics tell us nothing about the role of experience since we do not know the "base rate"—proportion of all miners that are inexperienced.) Treating it as a typical sample of accident investigation in this industry, we can learn not only about working conditions, but also about the basic assumptions of accident investigators and reports.[23]

An employee in a Texas limestone quarry was working his usual eleven-hour shift, which consisted of shoveling away the crushed limestone that had fallen underneath a 4-foot wide conveyor belt. The material apparently was hard to remove from under the belt, and there were no provisions for stopping the belt. The safety journal reports that the Mine Safety and Health Administration (MSHA) investigators concluded that "the victim had crawled under the conveyor belt while it was running for unknown reasons." (An obvious known reason would be that doing so was the only way to keep the materials from building up to where it would damage the conveyor belt—which was, after all, his job.) The shovel he was carrying apparently got caught between the return idler (a spinning pulley) and the belt, jerked his arm and body and pulled his hand and arm in between the pulley and the belt before he could let go. When his head struck the idler, his jaw and neck were broken.

The causes of the accident, according to the investigator's report, were the "victim crawling into an unsafe position under a running belt conveyor and management's failure to guard the return idler at the pinch point. A contributing factor was poor equipment design which created spillage and buildup on the return idler under the belt conveyor making the cleanup necessary."[24] Note that the investigators agree the cleanup around the idler was necessary because of poor design, and that the idler was not guarded; but they still insist that the operator crawled into there for "unknown reasons." It is a small incident, and could have occurred in a manufacturing plant, but it gives us some idea of what most people in the accident business mean when they classify such accidents as "operator error." One might truly say that this was a "forced error," as we described them in our discussion of TMI; if the operator did not keep the idler area clean, one strongly suspects, he would be forced to look for another job.

MHSA blamed the company for poor signs and practices in the next accident. The investigation hardly seems to have been worth the effort. Two men were working with "slusher buckets," dredging an area after a blast. The area was unsafe; a danger sign was posted. The men had to drag the buckets around a corner, making the work especially difficult. About twenty minutes before the accident, two supervisors visited the area and were aware that much of the roof was as yet unsupported and could cave in. They said they told the miners to "work safe, don't take any shortcuts." They later told the investigators of the accident that they were not overly concerned about an accident because they had told the miners to stay away from the unsupported area through which they were dragging the buckets. But the slusher buckets began hanging up, and one miner jerked the cables to free the buckets. Eight tons of material came down on him. What is interesting here is the defense of the supervisors: We warned them to work safely. The Mine Safety and Health Administration investigators were presumably satisfied. They did not question why the company would have anyone working at all where there is a known danger from unsupported sides and roof.

A quarry mine accident in Georgia reminds us that while there is powerful machinery around, much of the work is done with shovels and sledgehammers, and these can be dangerous. A man driving a steel wedge into a granite block was killed when the wedge broke and a sliver of steel punctured a neck artery. "The accident was attributed to failure to properly maintain the wedge," presumably a failure to apply sonic testing after each blow! The MSHA investigators also found that if the man had been swinging the wedge from a standing position on top of the block, the "proper" position, rather than from the base of the block, "he might have been struck by the flying steel sliver in a less vital body area." Of such attributions of operator error are accident investigations constructed.

The next accident occurred in a Wyoming uranium mine. A worker walking with a probe over his shoulder was run down and crushed by a bulldozer that was backing up. The sound level in mines is frequently deafening when heavy equipment is working. The dozer operator could not see in back of him very well because the ripper blade was raised. The backup alarm was defective in the reverse gear that the driver was using. (No matter; the caverns reverberate with the beeps of backup alarms anyway.) But for the investigators, the "direct cause was the victim's failure to notice the bulldozer's movement before he started walking." Thus we have another certified "operator error," with a ritual admonition to the company to keep operating equipment in working order.

Another operator error accident occurred in a New York shale rock pit. The victim's job was to watch a bin and signal when it was filled by an inclined belt conveyor. There was an automatic bin level indicator, but it was considered less reliable than having someone observe the bin and signal. The conveyor stopped on the automatic signal instead of on the observer's signal, so someone went to see what happened to the observer. The article states:

> It was believed that the victim had stepped onto material which collapsed under him due to an unseen void or bridging, trapping his legs, and that he had then been covered by material still running into the top of the bin.
>
> Main cause of the accident, MSHA investigators found, was the observer's entry into the bin without a safety belt and tie-off line and without making

> sure a co-worker was on hand, as well as failure to deenergize and lock out the conveyor system. Contributing factors, investigators said, were the victim's failure to notify the loader of his intentions and management's failure to provide adequate job training and proper supervision.[25]

We do not know why the workman stepped into the bin, but it must have been to correct a problem. Whether the safety belt with a 5-foot nylon clip line would have helped him after he was covered by shale is dubious. No co-workers were around, the account makes clear, to call or to watch; and shutting down the conveyor for what was probably a minor problem would likely bring about censure. One cannot know; perhaps the workman simply did a very stupid thing. It happens with the best of us. But again, "operator error" is an easy classification to make. What really is at stake is an inherently dangerous working situation where production must keep moving and risk taking is the price of continued employment.

The article continues with several more accidents in this vein, selected to illustrate "the value of experience and training and the perils of ignoring the dictum, 'It can happen here.'" As I have tried to indicate, even this selection suggests that experience and training are perhaps less relevant than job pressures, careless management, and supervision, and above all, the inherent dangers of these enterprises. None of these are system accidents. The system is not very tightly coupled nor complexly interactive. We also examined extensive material, including unedited OSHA accident reports, that the public interest group INFORM obtained through a Freedom of Information action regarding fatal and serious accidents in the smelting industry. Smelting is quite loosely coupled and linear, resembling most manufacturing far more than, say, the chemical industry. Even though there are some resemblances to processes in the chemical industry, accounts of all fatal accidents for two years revealed none resembling system accidents. This is as we would expect.

Complexity and Coupling in Mines

Not all accidents are caused by the simple component failures described above. There are still elements of tight coupling and complexity in mines. They are being reduced through new regulations, fortunately (or were until the changed health and safety policies introduced by the administration of President Reagan in 1981). The U.S. Bureau of Mines now calls for multishaft ventilation and for segmented ventilation; this eliminates many common mode failures where the whole air supply is endangered by a collapse of one ventilation shaft or the failure of one set of fans. Single shaft mining was fairly common until the Bureau required multiple escape passages, in effect requiring multishaft mines. A jammed elevator alone could result in a disaster in a mine with only one escape passage. Another regulation required that unused sections of mines be sealed off. These sections have the capacity for creating toxic gases, or can be used for unauthorized activities (smoking, dumping chemicals); sealed off, the problem is lessened.

One obvious problem with mining is the lack of direct information about the system state, a characteristic of many systems with complex interactions. This is most apparent in roof falls, or "falls of ground" as they are usually called, a frequent and dangerous accident.[26] Gases and explosive dust constitute another serious uncertainty. Mine explosions are the biggest source of serious accidents. Methane gas escapes from underground rock, and is a particularly serious problem where coal seams exist, though methane explosions have also occurred where other mining or tunneling was going on. (A water supply tunnel under Lake Huron experienced a methane explosion in 1971, killing twenty-two miners.)

Methane or coal dust explosions are ancient history in mining. One in England in 1880 took 161 lives; one in Virginia in 1884 took 112 lives; a Hanna Mining Company explosion in Wyoming in 1903 took 169 lives, and despite public outcry, of the limited sort that you could raise in a company town in Wyoming in 1903, the mine reopened without any improvements in safety, and had another explosion in 1906, killing fifty-nine miners. A mine explosion in West Virginia in 1908 took 154; a fire in an Illinois mine in 1909 killed 259 miners and rescuers; an explosion in Alabama took 128 in 1911; 263 in New Mexico perished in a 1913 explosion; that year also saw a relatively minor disaster in a Colorado mine, which then had another one four years later, talking 129 lives.[27] The biggest of all, in the French mine of Courrieres in 1906, left 1,100 dead and required martial law and 25,000 soldiers to restore order.

Safety techniques have improved. A cursory search indicates the number of people killed has dropped since those years. From 1960 to 1978, for example, one source lists eight explosions with sizeable numbers of deaths, but the average number of fatalities for these accidents was only thirty-two. There was only one large mine accident that I know of between 1971 and 1978. Even the massive explosion in 1979 in the Belle Isle Salt Mine in Louisiana killed only five miners. The latter comes close to being a system accident, but the failures following the initial explosion were not independent of it.

This accident illustrates some of the complexity and tight coupling that can occasionally be revealed in these systems. Two miners set off a routine, scheduled blast from a distant blasting board. The blast apparently released about 4 million cubic feet of methane gas, but the cloud did not have sufficient air mixed in it to sustain an explosion. It drifted away from the face where the blast had occurred. The methane mixed with air, causing the bubble to expand in size. Ten minutes after the explosion, it had mixed with sufficient air to become explosive, and had drifted into another area of the mine where there were electric sparks, arcs, or frictional sparking. The resulting explosion created near hurricane-force winds, which blew huge trucks about, smashed doors, fans, and equipment, and created a blistering flash that reached temperatures up to 900° F. in a number of areas in the upper levels of the mine. One man groped his way to an escape tunnel to escape the winds and the swirling dust, debris and salt, only to be hurled 30 feet along the tunnel. The air movement had been reversed because the main fan had been destroyed. About 15,750 tons of salt were expelled from the mine by the explosion. There were only twenty-two miners in the mine at the time; seventeen survived, some with serious injuries.

The account in the mine safety journal notes, "Such outbursts of high-pressure gases and salt out of large vertical 'pressure pockets' have been documented in the Belle Isle and neighboring salt mines for many years. Sizeable outbursts have occurred when salt with high-pressure gases contained in rock structures has been penetrated by mine openings."[28] Yet it was only after the accident that the MSHA classified the mine and three others in the area as "gassy," thus requiring more frequent inspections and the like. It is not clear that the higher standards could have prevented this accident, however, or a future one if these pressure pockets are common. It would appear that there is an irreducible hazard in mining—an unpredictable environment for humans.

Losing a Lake

Our final accident in this chapter provides comic relief from the grim world of mining, and an example of a human-made modification of a natural system that produced a genuine system accident. It occurred in

1980 in Louisiana when the state lost an oil rig, a salt mine, and an entire lake. I will draw entirely on the very entertaining account of Michael Gold in the journal *Science 81*.[29]

It is an example of the proximity problem that briefly created an interactive and tightly coupled system out of two independent linear systems. We do not generally expect a Louisiana lake, an oil rig, and a salt mine to suddenly be subsystems of a larger system.

You may have guessed the interaction. Texaco was drilling for oil in Lake Peigneur in southern Louisiana. The drill was down 1,250 feet when it got stuck, and when pulled loose, the drill inexplicably jumped up and down in five- and ten-foot leaps. This is unusual for a device that weighs 40 tons. One hour later the men noticed that the rig was listing badly, and they abandoned it. Watching from the shore they were surprised to see it sink from sight in a part of the lake that was only 3 to 6 feet deep. Meanwhile, men working in a salt mine, part of which extended under the lake about 1,300 feet below it, noticed that their area of the mine was flooding, and they sounded the alarms. All fifty-one of them managed to escape as water rushed in to caverns that were 80-feet high and as wide as a four-lane highway. Meanwhile, on the surface, a whirlpool threatened some early morning fishermen, and eventually pulled in some barges and a tug associated with the drilling. The whirlpool grew in size until it pulled in 65 acres of the Rip Van Winkle Live Oak Gardens, a tourist attraction. A river (more like a canal) to the Gulf of Mexico, flowing out of the lake, reversed its course, eventually creating a 150-foot waterfall. An underground gas well ruptured and sent bubbles of natural gas to the surface, where they burned. After seven hours, the entire lake, once about one mile by two miles, had drained into the salt mine.

Diamond Crystal, the salt company, sued Texaco for ruining its very lucrative mine; Texaco sued the salt company, claiming they had not notified the company where their mine was wandering. The drilling company sued Texaco for the loss of equipment. The Rip Van Winkle Gardens sued both Diamond and Texaco for the loss of buildings and plants, including 30,000 poinsettias ready to bloom for Christmas. Eventually, seven lawsuits were filed. The damage ran into the hundreds of millions of dollars.

Everyone in danger, it turned out, was lucky. The fifty-one miners got out by the skin of their teeth, driving their trucks on the soggy salt underground highways. The two fishermen, noticing the lake dropping and the catfish jumping and swimming for deeper water, headed into the lake only to confront the whirlpool. They were blocked from behind by mud too deep to stand in, which would eventually go into the hole, and in front on one side by a row of barges cabled together, with the whirlpool to the other side. But two barges broke loose and were slurped up by the whirlpool, and this left room for the fishermen to motor at full speed for a bank, and to scramble out. The river, or canal, that began to flow backwards created a 20-knot current near the top of the waterfall, and boats moored along it were pulled in. A tug tried to lay a long barge across the river to stop the shrimp boats from being pulled in, but both the barge and the tug went over the waterfall, just after the crew jumped into the muck on one side of the canal. The barge and tug went into the whirlpool and down into the salt mine, where they still are. Seven other barges later popped up after the salt mine filled.

As the water poured into the mine, the air vented out through a ventilation shaft, keeping the emergency elevator bouncing around at its top. Then a 400-foot geyser erupted from the shaft, lasted for twenty minutes, and died. The next day the lake started to refill, after the mine had consumed an estimate 3.5 billion gallons of water. A few days later the owner of the Gardens had to hire a team of scuba divers to try to recover his wine cellar.

The salt mine was a big one, dug into the top of a salt dome that was about one mile across. Several shafts radiated out from the central shaft, the critical one about three-quarters of a mile out. Oil is often found near salt domes. When Texaco started drilling, they didn't bother to notify the Diamond Salt Company. Diamond learned of the drilling when they were asked by the U.S. Army Corps of Engineers if they had any objections to a dredging permit sought by Texaco. Texaco had charts showing where the mine was, but the charts conflicted; one put the mine under the drill hole, the others did not. Texaco never checked with its neighbor; Diamond never checked with Texaco. Texaco claimed that only Diamond knew where both the rig and the mine tunnels would be. Diamond claimed they had no information on how deep the rig would drill. A typical bureaucratic tale, one might say—lack of communication. But more to the point, it involved an unexpected interaction of two systems that were more proximate than anyone thought to notice. Had Texaco tried to hit a mine tunnel form available charts, even one as wide as a four-lane highway, one suspects they would have failed. System accidents are infrequent even in complex, tightly coupled systems. They are exceedingly rare, and truly freakish, when two large systems such as this, apparently independent, briefly interact. That is why no one bothered to check. Who would have thought that . . . ?

Conclusions

Some dams have considerable catastrophic potential, but while they are tightly coupled, so that recovery from failure is extremely limited, they are not subject to unexpected interactions—they are linear systems, in our terms. Preventing accidents then is largely a matter of preventing component failures, including proper design and construction. There is, however, a possibility of system accidents in the sense that unexpectedly a dam may become an active part of a larger system. Filling the dam may disturb the earth, causing earthquakes or the destressing that held prehistoric landslides in place. It might be some decades before such matters are properly understood, and in the meantime, we have the potential for unexpected interaction of failures, or the unexpected alteration of system states, since no "failure" in the ordinary sense is involved. But by and large, dam failures appear to be due to rather prosaic matters, in particular, ineptitude and deliberate risk taking. It was important for us to consider dam accidents because we needed an example of tight coupling without interactive complexity, and an example of catastrophic potential not related to system accidents. Not all high-risk systems are prone to system accidents.

Mining provided further indication of the usefulness of our scheme. While operating in a hostile environment, and subject to dramatic and catastrophic failures, again, system accidents did not seem to be present. Even the Belle Island explosion was caused by a component failure without any interaction for multiple failures, though tight coupling impeded recovery efforts. Instead, we had a glimpse of the prosaic failures in the DEPOSE system and of possible production pressures that plague all industrial activity. It also provided a useful glimpse into the subject of attributions of operator error—a kind of figured bass that runs throughout our analysis of high-risk systems.

Both dams and mining, and certainly the underground mining and surface drilling that produced the Lake Peigneur accident, have alerted us to the possibility of eco-system accidents—the unanticipated expansion of the system and thus the scope of failures. Systems not thought to be linked suddenly are. Most toxic spills, including the relatively trivial one of Church Rock, are not eco-system accidents in our terms, for they are planned, not accidental. It is perfectly clear that if toxic wastes are given shallow burial in impermanent containers (and for some wastes, there is no such thing as a permanent container) they will come to the surface, as at Love Canal. With dams in particular, however, eco-system accidents do occur because we are only beginning to understand the fragile character of our environment as our ability to move more and more tons of land and water is extended. As two geological experts testify: "Large dams and reservoirs create a complex new environment, and very little is known of the mutual interactions of the component forces, on a long-term basis." Some transformation processes such as nuclear fission and high-temperature, high-pressure chemical reactions, may never yield their secrets sufficiently to disclose all possible interactions. But in the case of eco-system accidents, I think we have a case where more time and motivation will give us the knowledge to understand multisystem interactions that create new, previously unanticipated systems. We need this knowledge "to keep the newly-created man-imposed systems and nature in a proper equilibrium."[30] In the case of dams, it may be possible to do so over the next two or three decades. But in the next chapter we shall encounter a far more fearsome potential for an eco-system accident where little attempt is being made to insure that human-created and human-imposed systems can remain in a proper equilibrium—the case of recombinant DNA research and production.

CHAPTER 8

Exotics: Space, Weapons, and DNA

This final data chapter deals with three very high-tech systems, and that is about all that links them together. One has almost no catastrophic potential (space missions); one has the ultimate catastrophic potential (nuclear weapons); and the third, recombinant DNA research and production, or DNA for short, has hardly begun, but could well develop in a direction that would be second in catastrophic potential only to nuclear war. Each system contributes to our argument in different ways, providing further evidence that system accidents are inevitable in complex, tightly coupled systems.

The first section, on space missions, will elaborate in a clear and dramatic fashion a point touched upon in the sea stories and the accounts of nuclear and chemical plant accidents: the attempt by the great designers of complex systems to do without the lowly operators. In the space program the operators were hardly lowly, since they were experienced, well-trained test pilots, the Waldo Peppers and the great Santinis of the space age. But they were treated as lowly operators until the designs came apart and their managers became befuddled, and only the pilot-turned-astronaut could save the missions. The space missions illustrate that even where the talent and the funds are ample, and errors are likely to be displayed before a huge television audience, system accidents cannot be avoided. I have argued throughout this book that we should give all risky systems more quality control and training than we do, but also that where complexity and coupling lie, it will not be enough. We gave the space missions everything we had, but the system accidents still occurred. This is not a system with catastrophic potential; the victims are first-party victims. Catastrophic potential resides in most, but not all, complex and tightly coupled systems.

The weapons systems have hardly been starved for funds or engineering talent either, but accidents abound. I think they are system accidents, though it is hard to tell from the limited public data. In any case, the catastrophic potential is infinite in the case of nuclear weapons. DNA involves genetic engineering, and we have had almost no experience with genetic engineering on a production basis (the most threatening part of that endeavor) and very little on a research basis, so we have no accidents to explore. But a scenario is imaginable, and if we are to be prudent as a society, scenarios should be explored.

We will start out on a relatively positive note by examining the space missions. The centerpiece of this part of the chapter will be an account of an extraordinary mission, Apollo 13, which commenced with a system accident, and ended with a recovery that dramatically illustrates the most exemplary attributes of both humans and their machines. The event will tell us something further about complexity and coupling: the recovery was possible because ground controllers were able to make the system more linear and more loosely coupled, and to put the operators back into the control loop that rarely included them.

No such encouraging lessons come from the section on nuclear weapons and early warning systems. We will not dwell on "the fate of the earth," that is, the destructive power of nuclear weapons, but on the limits of human capabilities and the even narrower limits of organizational capabilities. There is much to fear from accidents with nuclear weapons such as dropping them or an accidental launch, but with regard to firing them after a false warning we reach a surprising conclusion, one I was not prepared for: because of the safety systems involved in a launch-on-warning scenario, it is virtually impossible for well-intended actions to bring about an accidental attack (malevolence or derangement is something else). In one sense this is not all that comforting, since if there were a true warning that the Russian missiles were coming, it looks as if it would also be nearly impossible for there to be an *intended* launch, so complex and prone to failure is this system. It is an interesting case to reflect upon: at some point does the complexity of a system and its coupling become so enormous that a system no longer exists? Since our ballistic weapons system has never been called upon to perform (it cannot even be tested), we cannot be sure that it really constitutes a viable system. It just may collapse in confusion!

Finally, there is the recondite, novel, and thrilling field of recombinant DNA research. The potentials here for human benefit appear to be more extraordinary than all the other technologies put together. The potentials for human disaster are equally unprecedented, and rival that of nuclear holocaust—if there can be a question of rivals where extinction is involved. With breakneck speed we (that is, primarily the oil companies) are proceeding down a thoroughly unknown path, without brakes, without headlights, in search of undreamed amounts of private profits. The system accidents may, indeed, already have occurred, giving new meaning to Yeats's oft quoted line, "What rough beast, its hour come round at last, slouches towards Bethlehem to be born."

Space Missions

Industry Errs

The conquest, if it be that, of outer space illustrates a normal learning curve for technological activity. The rockets stopped blowing up and achieved remarkable reliability. But the history of space missions also illustrates that there remains an obdurate residual propensity for system accidents.

In Chapter 2, I argued that every industrial activity exhibits organizational failures, incompetence, greed, and some criminality. The space program certainly performed better in these respects than the nuclear industry. I know of no criminal activity. But even this vastly favored program exhibits enough incompetence and organizational failures to remind us once again that nothing is perfect. A major contractor, North American Aviation, was known as "Brand X" in the trade—a derogatory term indicating "cheap." The abundance of funds and cost-plus contracts may have encouraged such instances as a workman throwing the remains of his lunch into a costly automatic chart-recording instrument; a sealing device ruined by gouging it with a screwdriver while trying to force it into place; a booster rocket shutting itself off and leaving Walter Schirra and his *Gemini 6* on the pad because a dust cap had been left on a pressurizing line in one engine, even though a quality control inspector had approved the work as done.[1] No, McDonnell Douglas, the vendor in this case, was not duplicating its profit-maximizing DC-10 efforts; these examples come from facilities swarming with government inspectors in a program that, while profitable for contractors, would not generate the long-range profits that could come from aircraft sales.

The Mercury part of the space program, involving only orbital single-astronaut flights, was reviewed by NASA in a 440-page document that *New York Times* reporter John Finney called a "remarkably harsh indictment of American industry."[2] Among the problems: spare parts that were 50 percent defective; capsules with more than 500 defects; batteries with holes in them; vital electronic parts improperly soldered; valves improperly installed, leading to attitude-control problems on flights; dirty gas pressure regulators; and contaminated oxygen and water for breathing and drinking. The contractors denied they performed anything but superbly, but did not respond to the specific charges.

After the tragic fire on the launch pad that killed three astronauts, an Apollo inquiry board was appointed, and it in turn drew upon the research of twenty-one panels of experts—1,500 overall—when it prepared its report. Six of the eight board members were NASA employees, so there was some fear of a whitewash—NASA was investigating itself and its prime contractor, Rockwell International. However, the report, 3,000 pages long, was described in the press as fairly scathing. Among the findings: "Adequate safety precautions were neither established nor observed for this test." "The over-all communication system was unsatisfactory." "Deficiencies existed in command module design, workmanship and quality control, such as: . . . regulator failures, line failures and environmental control unit failures. . . . Coolant leakage at solder joints has been a chronic problem. . . . Deficiencies in design, manufacture, installation, rework and quality control existed in the electrical wiring."[3] A wrench and other "noncertified equipment items" were left in the command module at the time of the test; 113 "significant" engineering orders were not accomplished when the module was delivered and records did not match up. Rockwell protested most of the conclusions, and one of its officers speculated that astronaut Grissom might have kicked the bundle of wires thought to have caused the spark, thereby installing operator error into even this accident—which occurred while the three astronauts were patiently reclining in their seats waiting for the next part of the test.

The most probable cause was our old friend, a bundle of wires whose insulation had rubbed off. Unfortunately, in a pure oxygen environment, that meant a flash fire.

Hardly noticed in the wake of that accident was another one, four days later, in a module simulator where the effect of near-pure oxygen atmosphere on blood-forming organs in rabbits was being investigated. Two enlisted men were killed when a work-lamp wire shorted and ignited the atmosphere. Like the launch-pad tragedy, this one was most likely a component failure accident.

The Gemini flights, a prelude to the Apollo mission, were near perfect, though there were close calls. But another prelude, the Ranger flights, designed to survey the moon, had five out of nine failures. *Ranger 6* failed, incidently, for a reason that should be familiar to us—a safety device. In order to make sure the television cameras would come on to take pictures of the moon's surface, there were redundant power supplies and triggering circuits. According to a Babcock and Wilcox engineer, a short in a safety device (a testing circuit) depleted the power supplies by the time the Ranger reached the moon. The engineer notes that the more redundancy is used to promote safety, the more chance for spurious actuation; "redundancy is not always the correct design option to use."[4]

Most of the failures of the space program have not been death-dealing, and if they were, they were limited to first-party victims—the astronauts or technicians. However, in three cases of failures with plutonium power packs, the risks are potentially catastrophic, since plutonium is perhaps the most deadly substance known to humans. One of these power packs was retrieved successfully from the Santa Barbara Channel off California—barely off California; had it hit any place on land, whether near the city of Santa Barbara or not, the consequences would have been catastrophic. The second went into the Indian Ocean and is still there. The third was in a navigational satellite sent up in 1964 that failed to achieve orbit when its rocket engine failed. It reentered the atmosphere over the Indian Ocean and distributed 1 kilogram of plutonium-238 about the earth. By 1970 it was estimated that about 95 percent of it had settled on the ground or the earth's waters. The accident was estimated to produce a three-fold increase over the amount of plutonium contamination produced by all atmospheric nuclear weapons testing.[5] This received almost no publicity, in contrast to the breakup of a Soviet nuclear-powered satellite in 1978 and another one in 1983. The first public mention of it may have been in a 1967 item in the journal *Science.*[6]

The industrial problems continued right up to the present-day shuttle program, though in the shuttle's defense, it has not had the ample funding of the moon missions. The engines failed or exploded frequently, the pesky tiles kept falling off in shipment or merely from being bumped by a forklift truck, the cost overruns mounted, the expected cost of each flight more than doubled, and the scheduled first flight was delayed ten years. After two failures of spacesuits in a November 1982 flight, NASA convened another panel of experts, and the conclusion was not much different from the panel studying the launch pad fire: "Egregious oversights" by the prime contractor, United Technologies. In one case two small plastic pins were missing, creating a leak in the spacesuit; the inspection sheet indicated they were in place, and the employee's supervisor had signed it. Worse yet, a stray steel chip was found in an exhaust vent of the suit's oxygen supply system—it could have caused an explosion of the suit, which could have blown a hole in the side of the shuttle.[7] Tight coupling abounds in these systems. There can be little room for the all-too-human construction errors that will appear with the best of vendors, and certainly not for those we have been detailing.

But that is a minor point; more important is the appearance of system errors, and the means of recovery from failure. The space program provides a particularly insightful glimpse into something we have touched on before and will bring us closer to a major organizational issue: If there can be inexplicable interactions, who can best cope with them—the operators, or the design/management team? Some variants of organizational theory, reinforced by democratic values, generally assert that those closest to the disturbances in a system are in the best position to act upon them. Decentralization is the recommendation. But other variants and much of engineering logic assert that those close to disturbances cannot act fast enough, or with enough comprehension, to cope; therefore, designers should try to eliminate as many human tasks as possible and give them to machines, and managers should have the means to tell the system and/or its operators just what to do. Centralization is the recommendation. The conflict between these views runs through the space program. The regular appearance of mysterious interactions only escalated the conflict. We will consider this in the next section.

Twentieth-Century Taylorism

If everything in the DEPOSE system components could be perfect, we could drop the "O," the operator. The designer would be the operator; he or she would turn the system on and it would run, and the owner could turn it off when she wanted. The first suborbital and orbital flights were conceived of that way: the "astronaut" was not really needed, and went along as a test subject. (The term "pilot" was rejected; it implied control, or piloting. An astronaut was a "star voyager," a more passive term.) Indeed, the first suborbital flight and the first orbital flights had no one aboard; the second in each category had a chimpanzee aboard—Ham, and Enos. As with the astronauts, the chimpanzees had been extensively conditioned not to blow it. The chimps manipulated dials that had nothing to do with the ship; it was just to test reaction times. For the human passengers, it was almost the same, and at one time it was felt that they should be given tranquilizers of some sort to make sure they did not interfere. (I am drawing here on the immensely entertaining, and exceptionally perceptive book by Tom Wolfe, *The Right Stuff.* He will be our guide for this section.[8])

Think first of the "great designers," scientists and engineers who plan, or design, these spaceships. There are two basic sections, the rocket and the pod at its tip. The rocket has to go up, abort if anything is wrong, depositing the pod safely by parachute, or, if the launching goes well, shut off and detach itself from the pod at the proper moment, allowing the pod to hurtle like a cannonball into space. The "great designers" are also responsible for the ground control system that monitors the sensors in the rocket and the pod and intervenes if anything goes haywire. Within the rocket and pod they have put automatic systems that come on or shut off without anyone doing anything.

Next think of the ground controllers, or middle management—dozens of them sitting in front of keyboards and television screens with headsets on. They run the control and monitoring system that the designers have created. Finally, there is Ham, or Alan Shepard as the case may be, in the pod. Ham does not interfere in the system at all; Shepard is allowed to play with the thrusters that keep the pod from tumbling or turning, after the capsule is free of the rocket, or before that, to punch the abort button if there is an emergency. The abort button too could be automated. The hierarchy was clear: designers, controllers, and subjects. Project Mercury was supposed to be a scientific enterprise; astronauts were part of the test.

Actually, for the first suborbital flights, this was not inappropriate, since all that was involved was firing a capsule into the air like a mortar shell and having it come down in the right place. It was a bit more complex for the orbital flights, and immensely more complex for the moon missions. Still, the planet flybys—Jupiter, Saturn, Mars—were without on-board operators, and the middle managers at Mission Control did very well indeed without them. Did we need operators for the first orbital flights and the moon missions? They might have been nice to have there to handle emergencies, but they were expensive. Much of the complexity of the system that could create emergencies was the result of habitat and retrieval requirements once humans were aboard. Though the human subjects were not expensive in labor terms (they lived on their military salaries of around $200 a month and some perks), they were expensive to keep alive up there and to recover in one piece.

The question of their necessity is not easily answered. If we were only curious as to what happens to someone in a pod in space, we went to a lot of trouble to find out. If humans were, as one conference of scientists and engineers put it, "added to the system as a redundant component," they were still very expensive ESDs.[9] The unmanned probes of far space got along fine without them. If humans were to be significant nodes in the loop of design, equipment, supplies, and ground control, exercising some judgment, doing a bit of piloting in a "spacecraft" rather than a "capsule," there were no signs of that plan in the first flights. It is not clear that any policy or consensus existed on the matter. What Wolfe suggests is that in selling the space program to the public, the public response caused that decision to be made. The first seven astronauts were instant heroes after they were selected. They were to ride on top of rockets that were always blowing up, and come down in a large can in the ocean, and to do that, they must be heroes. When the seven found they were instant heroes, they began trying to dictate or at least influence the terms of their node in the loop. As we shall see, they demanded some modification in the pod and some control over it.

The importance of the question goes to the organizational heart of all high-risk systems: If they are risky, to operators and other potential first-party victims, as well as to the immensely expensive investments, why not eliminate the operators? A package could be landed on the moon and ordered to send back pictures, perhaps even samples. Some argued this would be safer and cheaper. Operators are unreliable, as well as alive. Most people who think about these things, as we have seen at length, feel that operator error account for 50 to 90 percent of the failures in complex systems. If the environment is as hostile as outer space, and the error rate of operators so great, why risk the operator of the system?

But, what if the designer errs, or the environment is uncooperative, or the builder leaves out a valve ring or makes a faulty switch? Can the Middle Managers in control recover? The evidence suggests that the answer is no, because the goal is not to explore space by shooting sensors into it. If that were all, we could get by with fly-by missions or soft landings of suitcase laboratories. But if we were to occupy outer space, control it, police it, use it to spy on the Soviets and shoot thunderbolts at them, that would require humans. Furthermore, we might even mine space someday, or move there. Any of these possibilities requires more than a recording cannonball. The national pride of landing an American on the moon also played its part.

The military-political aspect was there from the start and probably overriding. After the assassination of President Kennedy, the success of the Apollo project to put a "man" on the moon by 1969 was resoundingly reaffirmed by the new president. "I don't want to go to sleep by a Communist moon," President Johnson said, and that may have been all that was needed to be said. Occupying the moon required more than the moon fly-bys; those basically required good aim and some mid-course corrections, and toggle switches that turned cameras on and off. Even the orbital flights and the moon probes involved frequent failures that only operators aboard the vessels could recover from. Designers, builders, and controllers were not enough, it became clear.

The first astronauts were selected from volunteers who were test pilots. According to Wolfe this was expediency, not foresight. NASA needed a pool of applicants that could be assembled immediately after the Russian *Sputnik* insolently beeped through our heavens in October 1957. Test pilots would already have security clearances, evidence of physical and engineering skills, and could be given immediate orders to report for duty. While radar observers were probably the best trained for the passive skills needed to observe and be observed, test pilots seemed to be a convenient, accessible, and not inappropriate pool. The officials got more than they bargained for, because test pilots were not likely to be passive observers.

Test pilots, especially ones in the military services, had to be extraordinary human beings. Wolfe describes in gripping detail what they went through in their test pilot training. They generally had engineering backgrounds from college, received extensive technical and engineering training in flight schools and in their testing work, were extraordinary pilots, often aces from the most recent war (Korean, in this case, though some went back to World War II), and were unequivocally fearless. They were testing rocket planes and their forerunners, trying to break through the sound barrier of 660 to 760 miles per hour (the figure depends upon altitude, temperature, and the like). They lived dangerously on the ground (more Navy pilots, whether test pilots or not, died in automobile crashes than air crashes) as well as in the air. Most important of all, test pilots were in charge of their craft. Wolfe describes it thus early in his book:

> To take off in an F-100 at dawn and cut in the afterburner and hurtle twenty-

> five thousand feet up into the sky so suddenly that you felt not like a bird but like a trajectory, yet with full control, full control of Five Tons of thrust, all of which flowed from your will and through your fingertips, with the huge engine right beneath you, so close that it was as if you were riding it bareback, until you leveled out and went supersonic, an event registered on earth by a tremendous cracking boom that shook windows, but up here only by the fact that you now felt utterly free of the earth—to describe it, even to wife, child, near ones and dear ones, seemed impossible.[10]

The complexity and coupling of the ships these people flew were intense. The engines were unproven, or were pushed beyond design limits; test pilots' job was to break through the envelope and create and feel out new envelopes, or to "punch holes in the sky," riding on top of a "candle" that eventually had three-quarters as much thrust as the Redstone missile that powered the first suborbital flights. All kinds of things went wrong, and recovery was often impossible. Even after bailing out (if being ejected upwards at 100 miles per hour by an explosive charge can be called that), there was room for the unexpected interaction. The ace Chuck Yaeger bailed out only to find himself on fire. His seat was ejected with him by rocket propellant. The rocket propellant was still burning after he separated from the seat. Both he and the burning seat fell through the air. It was close by, above him. When the parachute opened, and slowed up Yaeger, the seat passed inside the plastic parachute shrouds, threatening to burn and sever them. This did not happen, but it did collide with him; as his chute slowed, the burning seat crashed into his visor. The burning propellant stuck inside the helmet and set one side of his face on fire. It filled the helmet with smoke and more fire as the seal melted in the oxygen hose, sending pure oxygen into the mask to fuel the flame. Ripping the visor off, he set a glove on fire and burned the flesh inside. He landed near a highway, and a passing youth lent him his knife to cut away the smouldering glove. He survived.[11]

Instantaneous decisions had to be made in rocket planes when they flamed out at 30,000 feet and tumbled in free falls like a brick, or lost the thrusters that stabilized them, or malfunctioned on takeoff. Eventually test pilots would fly to over 50 miles altitude at speeds up to Mach 7, or seven times the speed of sound, encountering and solving new problems at every step. The military's rocket plane program was cancelled in 1963 when the race to the moon was in full tilt, indeed, on the day that Yaeger was burned by his seat after ejecting from a F-104 fighter with a huge rocket bolted to it. But rocket planes surfaced again in the reusable space shuttle that first flew in 1982.

The military test pilots that were accepted for the first space probes were in for a rude shock. They were to be passive blotters absorbing the new conditions, monitored with electrodes and rectal thermometers, without even a proper window to look out of, only two small portholes on each side. The first American in space (the Russians had long been there, and were to stay ahead in everything for years), Al Shepard, did virtually nothing except watch the panel and be observed through a camera and wires and thermometers. In fact, much of the training of the astronauts consisted of conditioning—getting used to sensations in simulators that created high-gravity forces, learning to keep their hands off the switches, getting used to the sounds of the rocket motors. People with test pilot experience were not likely to easily accept this role. When they got an unexpected chance to act, they took it.

The "lab rats," as they called themselves during the long period of testing and conditioning at medical and psychological labs, decided to protest and insist on having some control over the flight. Their chances of succeeding were not good; humans are fallible, but it was presumed that designers are not. Even if builders goofed, the designers had redundant components to take care of the problem. The middle managers at ground control would have the most complete and accurate information, and have control over much of the spacecraft, either directly or through the automatic devices the designers had put in. But by now the seven astronauts selected for the Mercury missions were national heroes, and this gave them their opportunity. Their status in the project was rising. They asked for a window to look out of, and got it. They asked for an emergency hatch to get out after splashdown, and got it. (It was installed for the second suborbital flight; before that, other than climbing up the slim neck of the capsule, they had to wait until mechanics on the recovery ship unbolted the hatch from the outside.) They demanded control over the rocket itself, should it malfunction. They did not get this. They demanded complete control of the reentry procedure, over the hydrogen-peroxide thrusters that would control the attitude, pitch, and yaw of the pod in space. They wanted to establish the angle of attack manually during reentry, and fire the retro-rockets themselves, without the use of an automatic system. This one they won partially; they were given a manual override system, but the automatic one stayed.

The Operators Err

The initial flights suggest that pilot control by these grossly overqualified subjects was a mixed blessing as far as the designers and controllers were concerned. As Wolfe put it, not only did our rockets keep blowing up, but our boys kept botching it. Alan Shepard had little to do on the first short suborbital flight, but before lift-off he had his periscope out, and used the filter because of the sun. He forgot to remove the filter before it retracted (automatically, of course) for lift-off. He would not be able to remove it later, so he tried to remove it while it was retracting, but accidently brushed the Abort button—fortunately not enough to actuate it. He stopped trying to remove the filter, and thus was unable to report sighting various constellations and stars, or observing the orange band at the edge of the earth (events we know would take place because balloons had been up there already), since the filter made it all black and white. This may have contributed to his delay in starting the reentry countdown. Late in starting the countdown, he gave up trying to manually control the attitude of the craft (a task the astronauts had won control over) and put it on automatic. But he forgot to turn off the manual system, and thus precious fuel dissipated from the capsule. In this short flight it did not matter, but it would in a long one.

It mattered soon enough. Scott Crossfield, in the second orbital flight, played around so much with the thrusters on his flight that he ran out of time and power to properly align his module for reentry—a dangerous business, since one could skip back into space forever, or plunge into the atmosphere so sharply as to burn up; a few degrees off and that would be it. Furthermore, the automatic control system would no longer hold the capsule in position for reentry. (It had malfunctioned once before, with an ape, and would malfunction once again with John Glenn). He switched to automatic but forgot to change another switch, and for ten minutes was eating up fuel from both systems. When the time came to fire the retro-rockets, which would slow up the ship, he was behind, and the capsule was not positioned correctly; in addition he was late in hitting the firing switch. He was off by nine degrees in his angle. He had to release his parachute early, and by hand, since the automatic system was out of fuel. He overshot the target by 250 miles, and for forty minutes the impression created on television was that he was dead. Walter Cronkite was crying.[12]

These fabulously trained test pilots, asked to do a few maneuvers and make sure they set some switches properly, were not perfect. The most serious failure of all was in the second suborbital flight, by Gus Grissom. He lost the capsule. It qualifies as a system accident in our terminology, with multiple operator and procedure errors, and has the added interest that a safety feature started it.

Gus Grissom was the second man in space, flying a suborbital lob into the Atlantic near Bermuda just as Shepard had done. This time the "module," as it was called (it was later to be the "capsule" and then, finally, the "spacecraft"), had a significant modification demanded by the former test pilots—an emergency escape hatch with explosive bolts to be fired if the module did not right itself in the water and began to leak. There was no need to use it; the flight was perfect and Grissom busied himself with various housekeeping chores after landing, telling the helicopter to wait a few minutes. He may have been nervous; his pulse rate was significantly higher than that of Shepard's throughout the flight. After finishing his chores, he radioed for the helicopter to hook on, saying, as it had been practiced many times, "Okay, latch on, then give me a call and I'll power down and blow the hatch, okay?" The helicopter said yes, "will give you a call when we're ready for you to blow." But as the helicopter swung low with its long shepherd's crook to catch a lifting cable, the hatch blew. Out popped Grissom, since the capsule was filling with water, and he started swimming fast. The helicopter pilot was surprised but not worried; the astronauts had practiced for just such an emergency, and the pressure suit was more buoyant than any life preserver. The astronauts had even seemed to enjoy playing around in the water in what amounted to a large, form-fitted boat. So the helicopter pilot quickly sought to recover the capsule, which was taking on water and sinking.

Grissom swore ever after that he had no explanation for the hatch blowing—it just blew. Subsequently, extensive testing under all kinds of conditions in a sister capsule could not make it blow, and explosive bolts had been used for years in fighter aircraft without any mysterious actuations. (Still, there were all those contractor errors we cited early in this chapter.) After Grissom had armed the explosive charges, as required, the button could have been pushed in error or inadvertently while he was moving about in the capsule.

Grissom exited into the ocean and started to swim. But it just so happened that since he had landed safely and was about to be picked up and carried to the carrier deck by the helicopter, he had unplugged the oxygen inlet to his suit. Thus, his form-fitted boat had a large hole in it, and it was some time before he realized that instead of bobbing merrily, he was sinking. He was less buoyant than normal anyway, because the pocket of his suit was filled with souvenirs, such as two rolls of dimes and models of the capsule. He managed to reach down and close the inlet valve, but took a lot of water into his lungs in the process. In the choppy sea, the spacesuit with its water pulled him under as waves came by. Still, there was nothing to worry about because there was a second helicopter up there in case there was any trouble. It was watching him; in fact, he could see someone taking moving pictures of him. He waved frantically, and they waved back, assuming he was jubilant about being the second man in space. It had all been practiced so many times it was familiar. The helicopter hung there, as Gus swallowed more and more water and struggled to keep his head up, because the pilot was keeping watch over the other helicopter that was being dragged into the ocean by the sinking capsule. Aboard that other helicopter the warning lights were lighting up the control panel as the ship tried to carry 25 percent more than its rated lifting capacity, and finally the pilot cut the capsule loose; the helicopter's wheels were under water and the engine might blow. Then the back-up helicopter turned to Gus and he was hauled, incoherent, aboard. He remained incoherent for some time, furiously grabbing at life preservers in the cabin of the helicopter.[13]

The failures were trivial—accidental blowing of the safety hatch; open oxygen inlet valve; the preoccupation of the first helicopter with the capsule; the misreading of signals from Gus to the second helicopter; perhaps an overweight suit; a choppy sea. For any single eventuality there is an Emergency Safety Device, a redundancy, a back-up, a planned bit of overcapacity. But when they came together, we almost lost the astronaut. Had he not been there, of course, all would have gone fine; in fact, it would have been fine if the "safety" hatch had not been installed.

The Designers Err

But operator error is only part of the picture. There is also design or equipment error. As with any new, complicated system, even those subject to the most rigorous testing possible, there were bound to be failures. In the first suborbital flight with a living being aboard—the chimp named Ham—the rocket climbed at a slightly higher pitch than planned, and it came down at the same angle and missed the target by 132 miles. Ham almost died, since water seeped in during the two hours of bobbing in the ocean.[14] In the first orbital flight with a chimp, the electrical system malfunctioned and the flight had to be cut short. The automatic control system could not keep the capsule positioned properly, and used so much fuel that control feared they could not orient it properly for reentry (a pilot could have, it seems).[15] They brought the capsule down off the California coast, rather than the Florida one. The designers were doing no better than the hot test pilots.

The automatic control system also malfunctioned in John Glenn's first flight, and the indicators of the angle of the capsule with respect to the earth were off. He took over control, and eventually made a dramatic entry using visual sighting to position the capsule for the most critical part of the trip.[16] There also was a spurious warning that control picked up that a "landing pack" attached to the capsule might have been lost. To Glenn's outrage, they kept this from him for hours, asking him only to check this and that. They would not tell him, the pilot now, and not a blotter for sensations, what was wrong, only what to do. Though he had to override the automatic system for positioning, reenter by visually lining up the horizon, and had virtually no fuel left for corrections, he was still not treated as the pilot. Ground control was in charge, and would remain in charge throughout the space program.

Rationalization of a system replaced operators; it is the province of designers and managers. This process has a long history. But rationalization allows more complexly interactive systems to be built, and the failure manager, the operator, comes back in. This too has a long history, but the few years of the space program illustrate both tendencies. Initially, radar operators who would act as redundant sensors were all that was thought to be required. The first shots made it clear that more skill was needed, and in addition, the lionization of the astronauts gave them something akin to union power. The Gemini and Apollo programs made increasing use of the operator; we went from hollow cannonballs fired into orbit to spacecraft, still with manager control, but more reliance upon operators.

With the space shuttle, the operator has still more control. An automatic landing system there, already in place, must be monitored, and the need for override is expected to be frequent. Indeed, the operator's role is likely to be larger in the shuttle because the designers can't really figure it all out. The shuttle is designed with more redundancy than any previous spaceship, according to one journalist.[17] But this very fact, he notes, makes it more complicated than any other, and makes the spaceship's workings more mysterious and unpredictable, even to those who designed it. The director of flight operations is quoted as saying that the magnificent "architecture" of the craft makes it all the harder to learn to use the system. The numerous bugs in the first space shuttle flights testified to that.

We should not overemphasize the role of the operator. The rocket and spacecraft were full of automatic controls that in no conceivable way could be handled by a pilot, just as in nuclear and chemical plants and aboard aircraft. The operator must come last, the designer first. Managers generally have more information than operators and a wider purview. But despite their admitted contribution to error, operators make a contribution to recovery from other's errors or unexpected environmental conditions. The more linear the system, the less important this is, and the distant probes are more linear systems that moon- or earth-orbiting missions with landings. The more complex the system, the more important this vulnerable but exceedingly flexible component, operator, is. We will see that dramatically in the case of *Apollo 13.*

Apollo 13

In April 1970, the fruition of ten years and 10 billion dollars was at hand: *Apollo 13,* defying superstition by its very designation, was launched into a trajectory to the moon, equipped with a landing module for a moon landing. Two days into the flight, as they neared the point of no return, halfway to the moon, trouble developed. For the next four days, with full coverage from Houston Mission Control, the specter of three daring astronauts possibly becoming stranded in space and slowly asphyxiating gripped this and other nations. The recovery from the accident was truly heroic, and a technological marvel. The accident's origins, of course, were mundane. Our source here is Henry S. F. Cooper's intense and dramatic book *Thirteen: The Flight That Failed*—technological writing at its peak.[18]

Two weeks before the lift-off, a routine prelaunch test was made of the craft, including a test of the two oxygen tanks. These are large nickel-steel alloy spheres designed to withstand 900 pounds of pressure per square inch. They carried liquid oxygen, which is kept at an extremely low temperature, −297° F. Liquid oxygen is unstable, and the tanks have a capped dome with pipes and electrical wires going into it to run fans for stirring the viscous liquid, heaters to expand it and force it out if it doesn't flow freely, and all the associated gauges and switches. The two tanks, along with the hydrogen tanks, contain the elements for fuel cells that generate the electricity to drive virtually all the key systems in the spacecraft—guidance propulsion, power for communication, power for removing carbon dioxide, air to breath, and so on. At the end of the test the engineers had difficulty getting the liquid oxygen out of the tank. So they turned on the heaters and fans in the tanks, and left them on. There is, of course, a thermostat for the heater, designed to turn it off once the temperature reached 80° F. But the thermostat is designed to work on a 28-volt power supply utilized in the spacecraft; a 65-volt power supply was being used in the test. This would make no difference for most of the tests. But the thermostatic switch was designed to operate at low temperatures, as well as on low power. The high voltage probably prevented it from shutting the heater off at 80°, and the subsequent heat (estimated to have approached 1,000° F.) not only fused the switch, but burned all the insulation from the wires after they entered the tank.[19] The system was unplugged after a few hours' time without notice of the high heat.

There would be no sign of the failure of the switch or the insulation, and no occasion to check it; the tank had already passed inspection.

There are three modules involved in the spacecraft: the service module, where the tanks and other equipment and supplies reside; the command module where the crew resides and where the controls are; and the lunar module, a small, disposable craft used only to go a short distance from the spaceship to the moon's surface and back again. When the spacecraft was part way to the moon, John Swigert, the command module pilot, had trouble with the two hydrogen tanks; a caution light went on in the command module and, of course, at ground control. He also had difficulty with the quantity gauge for one of the oxygen tanks (the one that had been damaged during the testing). It is possible that the trouble might have been unrelated to the damage; the tanks are tricky, and that is why there are fans inside to stir up the liquid and heaters to build up pressure to force liquid oxygen out. Mission Control ordered the fans turned on so that the quantity, or level, gauge would function correctly. An electrical arc occurred because of the uncovered wires, and this created heat.

A short like this is conceivable, so there was a caution light warning of a short on the instrument panel (along with 250 other gauges). Since the hydrogen and oxygen tanks both empty out into the same place, the fuel cells, the system was designed in such a way that the caution light for the hydrogen system preempted that of the oxygen system. One does not need both, so to speak. But since the hydrogen one came on first, that is what they focused upon, and there may have been a problem there—but again, probably not, since false alarms are common in these highly interactive and tightly coupled systems.

In time the arc heated the oxygen tank sufficiently to blow off the cap. Other materials in the service module caught fire from the arc and, finally, one of the six bay covers (hatches) on the module blew off. The ship had lost its major source of oxygen — and support of life.

Though the ship was two days from earth, the explosion was seen on the earth. Next door to ground control an experiment was going on which involved tracking the spacecraft with a radio telescope. The image of the craft—just a dot by now even on the powerful telescope—was displayed on a television screen. The experiment was ending; the ship was too far away by now, and there were a lot of blips and other interference that registered on the screen. Then they noticed a bright spot on the monitor which grew to the size of a dime and then disappeared. It was the venting oxygen reflecting light from space. They attributed it to a problem with the monitor, and in any case, they had no operational connection with the ground controllers next door. Had the explosion occurred several hours earlier, it is conceivable they might have interpreted it as a malfunction of the craft and inquired, but probably not.

The explosion was also felt aboard the craft. There was a small jolt, and a distinct bang—not the sort of thing you expect in a voiceless void. Swigert attributed it to one of the other astronauts bumping into the hatch as he came in from the lunar module, but he wasn't sure. Mission Control got a brief interruption of telemetry, but that was not uncommon. Besides, ground control was now preoccupied with another problem; in addition to the hydrogen caution light, a warning light appeared, signalling trouble with the electrical system. The astronauts saw it too, of course, and we know they were worried because their pulses, routinely monitored by management, went from around 70 to 130. Management's pulses are not monitored, so we do not know what they felt, but they made light of the problem and were slow to be convinced that anything was really wrong.

As noted, the hydrogen and oxygen tanks fed a reaction to three fuel cells; these provided electricity for running radios, controls, et cetera, and also provided water for the system—for cooling essential parts, and for drinking. Finally, they provided oxygen for breathing. Only one of the two oxygen tanks exploded, but the second one was leaking; a pipe had been damaged by the explosion. In addition, there is a common-mode connection at the point of the fuel cells, such that the oxygen in the undamaged tank could find its way out through this path. Vital life-support systems were draining away; the redundancy of two tanks was defeated by design.

Aside from losing the source of air and the means of removing carbon dioxide, the loss of electrical power meant that the astronauts could not control the attitude of the spacecraft; that attitude was now upset by the oxygen venting from the service module. The spacecraft had to rotate every twenty minutes to even out the burning temperature on the sun side, and the supercooled temperature on the shade side. It wasn't doing this. In addition, the erratic motion threatened to put the spacecraft's inertial measuring unit used for navigation in a particular attitude or position known as "gimbel lock," from which it could not be moved, and which would render the guidance system ineffective. It would be like losing your compass. Finally, the motion interrupted communication with control—it was difficult for control to know which of the four antennas to tune to, as the spacecraft rotated erratically. For part of the time there was simply no communication.

For seventeen minutes no one could conceive of the source of the increasing number of failures that were occurring. One cannot see the service module, or inspect it. There is no gauge that reads, "Oxygen tank explosion," only gauges of pressure and temperature and quantity. The warning light for the oxygen tank was not even on, since the hydrogen one was on and preempted it. As with Three Mile Island, there was at least one gauge that might have pointed to the problem—the oxygen quantity guage—but there was no gauge to point to its significance. The astronauts noted the gauge, but disregarded it. The warning signals pointed to the hydrogen tanks and the electrical system, and they knew that the radios kept going out, the roll could not be controlled, two of the three fuel cells were losing power but not the third, and so on. None of these signals made any cumulative sense, but it seemed the problem was clearly electrical.

The search for the problem was conducted with the well-worn assumption we have been exploring in this book: since the system is safe, or I wouldn't be here, it must be a minor problem, or the lesser of two possible evils. Just as reactor cores had never been uncovered before, or that ship would never turn this way, or they would never set my course to hit a mountain, the idea that the heart of the spaceship might be broken was inconceivable, particularly for the managers and designers in Houston, all gathered together for the historic flight. Cooper puts it well:

> . . .they felt secure in the knowledge that the spacecraft was as safe a machine for flying to the moon as it was possible to devise. Obviously, men would not be sent into space in anything less, and inasmuch as men *were* being sent into space, the pressure around NASA to have confidence in the spacecraft was enormous. Everyone placed particular faith in the spacecraft's redundancy: there were two or more of almost everything.[20]

As with Three Mile Island, the extensive training was for straightforward problems of component failures, and not for inexplicable interactions of three failures (the problem with the hydrogen tanks, occurring just when there was an indication that the oxygen needed stirring, and a design that masked the subsequent failure of the oxygen tank). During the course of pre-flight simulations the ground controllers, who themselves rehearsed the flight many times since they would essentially control it, grew impatient with the sometimes bizarre and implausible accident scenarios that were programmed into the simulator. The previous Apollo missions had rather minor and straightforward single component failures, while the scenarios they were now being given sometimes seemed ridiculous. By the time of *Apollo 13* they had won their point, and the astronauts had joined them in the effort to restrict trials to "one-point" or "two-point" failures. "Four-point failures" or "way-out disasters" (what we would call system accidents) had been dropped from the simulation runs.[21] It may not have made any difference; the idea that a short would occur in an oxygen tank and not be signalled was probably inconceivable even to those who designed the bizarre simulations. The chief pilot, Swigert, said afterward that "nobody thought the spacecraft would lose two fuel cells and two oxygen tanks. It couldn't happen. If somebody had thrown that at us in the simulator, we'd have said, 'Come on, you're not being realistic.' "[22]

Faced with the increasing number of malfunctions, the managers set about following standard operating procedures. The most likely explanation was a telemetry failure. False readings from the gauges aboard the spacecraft were to be expected, and had occurred in the past. They proceeded to check instruments, or have the astronauts check them. Then the next most likely event would be a minor failure or problem with one or more parts of units. This was carefully checked out. Even after it was clear (first to the operators, note, then finally to the managers) that there had been an explosion, the managers hoped that the mission might still go ahead and there could be a moon landing. It was not until about forty-seven minutes after the accident, twenty-eight minutes after they knew something quite serious was wrong, that ground control completely gave up on the notion that a lunar landing might somehow still be possible. This hope, or production pressure, had operational significance: the managers were reluctant to take any action to contain the increasing damage, or to test the system to see what damage had been done, if that action would mean an end to the possibility of a lunar landing. It might be called, after Tom Wolfe, "Houston Right Stuff."

Diagnosis

The actual discovery of the failures exhibits further parallels with the system accidents we have been covering in this book, giving us deeper insight into the mental models that operators (astronauts or controllers) must construct. When the warning light came on indicating electrical system problems, the managers found that one of the two main electrical buses fed by the fuel cells was suffering a significant loss of power. (A *bus* is a kind of junction point where electricity gets distributed; it can be shut off, breaking the circuits.) The equipment powered by this fuel cell—approximately half of the command module's equipment—was also beginning to fail. But the warning lights in the capsule and on the ground went off, signifying normal operation. Both the astronauts and the controllers concluded a minor malfunction had cleared up; the astronauts' report of a jolt and bang receded from consciousness. Then another anomaly appeared in the bandwidth of the radio waves carrying information from the capsule. The ground controllers surmised that if there had been a temporary minor electrical problem in the bus that serviced the antenna, this must be a related problem. Now that the electrical problem seemed to be solving itself, communications would clear up momentarily as well. In fact, the radio disturbance was caused by the panel from the service module flying off into space and striking the antenna, but no one dreamed of such an explanation at this point.

Still within a minute of the accident, the astronauts had noticed that the quantity gauge for the oxygen tank had gone off the scale on the high side, but no one was drawn to this source as the cause of the problem. After all, they had been having trouble with the oxygen gauges right along; that was why the oxygen tank had been ordered stirred in the first place.

But now the other main bus was beginning to fail; ground control was beginning to get confused. Half the warning lights on the relevant control monitor were lit up. It was still hard to know where to begin to try to isolate the electrical problem. Two of the three fuel cells were also discovered now to be without any power whatsoever. Since the fuel cell that was still functioning drew its oxygen from the same tanks as the dead fuel cells, ground control saw no reason to suspect the oxygen tanks. To preserve the functioning of the equipment aboard the command module, reserve storage batteries that were designed to provide power during re-entry were hooked up. Eventually, as the scope of the disaster dawned on everyone, these batteries were disconnected to preserve power for re-entry to earth, but at this early stage there was still hope of going forward with the mission.

Finally, ground control ordered the astronauts to begin reading out loud all the gauges in the command module that had anything to do with the electrical system. When they got to the gauge for the pressure in the oxygen tanks, they discovered that the pressure gauge was reading zero, though the quantity gauge had gone off scale on the high side. One of the astronauts in desperation went to a window and saw a cloud of vapor venting from the service module. The astronauts now understood why they had been having trouble controlling the attitude of the spacecraft, and with the help of the pressure gauge reading, they finally had a grasp of the problem. Even though the astronauts reported their observations immediately, the managers went through one more check to determine whether all of the malfunctions including, presumably, the visual sighting of vapor, might still not stem from an instrumentation problem. The astronauts understood the dimensions of their problem thirteen minutes into the accident. It was another four minutes before ground control finally agreed and ordered the crew to start turning equipment off in the command module to ease pressure on what remained of the electrical system. Even at this late hour, however, ground control was still trying to keep its options open for completing a lunar landing. They carefully avoided turning anything off that would definitely end all possibility of completing the mission.

This difference between the analysis of the operators and management is revealing. The complexity of the system was more baffling to the controllers, even though they possessed more information than the astronauts, because their monitoring of it was all somewhat indirect. The astronauts *felt* a jolt and this remained an important part of their analysis throughout the process of trying to track down the problem. They were much more inclined to take the problem seriously, less inclined to look for an instrumentation failure. The astronauts finally looked out the window and *saw* gas venting from the service module. This left no doubt in their minds, but ground control went through one last attempt to find instrumentation failure that would mean production would continue. The inability to see and feel the system directly was a severe handicap for ground controllers despite the great sophistication of their electronic monitoring system. The only ones on the ground to actually see the failure, through a radio telescope hooked to a crude television monitor, did not know what they saw, and were not monitoring the operations.

The accident allows us to review some typical behavior associated with system accidents: (1) initial incomprehension about what was indeed failing; (2) failures are hidden and even masked; (3) a search for a *de minimus* explanation, since a *de maximus* one is inconceivable; (4) an attempt to maintain production if at all possible; (5) mistrust of instruments, since they are known to fail; (6) overconfidence in ESDs and redundancies, based upon normal experience of smooth operation in the past; (7) ambiguous information is interpreted in a manner to confirm initial (*de minimus*) hypotheses; (8) tremendous time constraints, in this case involving not only the propagation of failures, but the expending of vital consumables; and (9) invariant sequences, such as the decision to turn off a subsystem that could not be restarted. All this did not just take place with a few high-school graduates with some drilling in reactor procedures, or a crusty old sea captain isolated in his absolute authority, but happened with three brilliant and extremely well-trained test pilots and a gaggle of managers (scientists and engineers all) backed up by the "Great Designers" themselves, all working shifts in Houston and wired to the spacecraft. As great as all these folks were—and we shall see shortly how truly great they were—the complexity and the coupling of the system defeated the mission, and nearly led to the "final accident," the loss of the ship in space.

Recovery

The resources at hand in NASA for recovery did make a great deal of difference. Though the damaged equipment and dwindling supplies could not be replaced, nor the severing of subsystems undone even if they wanted to do so, and even though the subsystem failures were final, the system itself was partially intact, and an unexpected redundancy, one not designed-in but fortuitously available, existed—the little, disposable lunar module.

NASA had four complete teams, each with dozens of experts, available to staff its ground control system on a 24-hour basis. They were all trained in the system, familiar with it, and indeed some had helped design it. Such resources are not available to any other high-risk system in our society. NASA was able to free up about forty experts to concentrate on working out solutions to get the astronauts home safely; they were free of routine flight management duties. In addition, almost every step they devised could be quickly and realistically tested in a very sophisticated simulator before it was tried out in the capsule—something rarely available to high-risk systems. With these resources, they were able to work out completely new and heretofore unimagined procedures for recovery.

NASA had a completely dead command module with no regular oxygen supply and no regular electrical system halfway to the moon. An emergency oxygen supply of limited capacity was still available, along with batteries for reentry maneuvers, for powering up the command capsule for the last hours of reentry. But by now the ship had passed the point of no return and would have to spend days traveling out to the moon and using the moon's gravity to swing it around and head back to earth. The three astronauts would have to spend four days living in the lunar module, designed for two persons and for just a brief jump from the command module to the moon and back. The lunar module would have to establish trajectories and maintain attitude control for the whole spacecraft, while keeping the operators alive. Then, just before reentry, the lunar module would be discarded, since it had no heat shield and would burn up upon entering the atmosphere. The operators would have to return to the command module, bring it back to life after four days of dormancy, and have enough fuel and oxygen and electrical power to manage the difficult reentry problem.

All the designed-in redundancies of the spacecraft now were ridiculously limited. The limited oxygen supply in the lunar module would now have to last three people almost four days. Water would have to be rigidly conserved (and dehydration eventually led to reentry errors). The battery-generated electrical power was severely limited (that is one reason they transferred to the lunar module; it would use less). The lunar module did not have enough carbon dioxide filter canisters to cleanse the air for so long a period, but the command module canisters did not fit the lunar equipment (and who would have thought that this would ever be needed). Eventually, plastic bags for waste and electrical tape made a reasonable connection possible. The electrical system was designed so that the command module would support the lunar module; now it had to be reversed somehow so that the lunar module could power up the command module. No one knew if it would be possible. These were just some of the inconceivable problems facing the designers and managers and operators.

The operators of a spaceship work from checklists, just as ordinary humans do when they go Christmas shopping, move their households, pack for a ski trip, or study for a final exam. The checklists utilized by the operators of spaceships, however, take a team of engineers about three months to prepare, and could take three or more hours just to read off to the operators. It is not a matter of simply looking to see if switches are in the correct position; it involves entering values in the computers, running tests and subroutines, setting and resetting valves and switches. It configures the craft through a series of sequences. Faced with an unprecedented situation, teams of engineers had to create, in hours, new checklists—piloting directions for the operators. For example, all the checklists for powering up the lunar module were based on the assumption that it would be receiving power from the command module. Finally, a checklist was found that involved starting up the lunar module from its own batteries, but it was long and complex and would require two hours for the astronauts to complete. There was only fifteen minutes of power left in the command module, not two hours. A checklist was improvised on the spot based on the designers' and managers' intimate knowledge of the detailed workings of the system.

In a sense, what was involved was the design of a substantially new system based upon the components of an old and no longer functioning system. And along with creating a new system, a whole new set of procedures for operating it also had to be devised. Because of the uncertainties of such a project, in one sense the new system would be more complex than the original one. While it would involve interactions that had never been contemplated before and were therefore not well understood, it would also require a drastic simplification of the system. Complexity was reduced as well as safety devices. The moon-launching part of the mission, scientific experiments, television coverage, on board comforts and even automated devices were discarded, leaving them with a more single purpose, linear system. Some of this loosened the coupling of the system, but the complexity was greatly increased by the lack of redundancies and slack. Everything had to be conserved. To save fuel they continued the trajectory to the moon so that its gravitational force could be used to reverse the direction of the craft. This also gave controllers time to drastically simplify re-entry procedures and check lists, because the astronauts became error-prone with the lack of sleep, water and heat. At each point, controllers chose the option that gave them the most opportunities for intervening by themselves and for the operators to make corrections at subsequent stages. Thus, at each point decisions were made that would loosen the coupling of the system. As the flight approached the stage of reentry, some of the pressure on the consumables began to ease as reserves being held back for various contingencies were no longer necessary as crisis points were passed. So, as the flight got closer to home, some redundancy in the consumables reappeared, and the system, in effect, became less tightly coupled than it had been at earlier points in the recovery process.

The new system created by the accident also suddenly had the potential for a catastrophic accident that the old system did not have. Inside the lunar module there was a container of radioactive fuel, including some plutonium, for powering an experiment that was to be left on the moon. Now it was coming back home in the little lunar module. A similar canister in an earlier space shot had burned up in the atmosphere and scattered plutonium in the upper atmosphere. This canister had been designed to remain intact through reentry, but it might burst upon hitting the ground. In calculating where the command module would eventually be recovered, the controllers also had to calculate where this radioactive fuel canister would hit after the lunar module burned up in reentry. One early trajectory for the recovery of the command module would have brought the radioactive fuel down in a heavily populated area of Madagascar. The representatives of the Atomic Energy Commission at NASA lobbied against that course. (It is not clear what they would have said if it had been a lightly populated area of Madagascar.) In the end, the canister appeared headed for deep water off the coast of New Zealand. But the astronauts made a last minute error in attitude control; it would have been difficult and dangerous to try to correct. Fortunately, the new location for the casket was the Indian Ocean, and the AEC representatives were satisfied.

As it turned out, the lunar module was made into a much better "life-raft" than anyone had expected it to be, and it dragged the dead command module around the moon, back to earth, and successfully powered up the command module for a smooth reentry. It was a job no one ever imagined the lunar module would have to perform. It is an example of indigenous safety devices unplanned by designers, but just there, that are common in linear, loosely coupled systems but not in complexly interactive, tightly coupled ones.

The recovery from the failure was brilliant. Cooper captures it just as brilliantly in his long, detailed, and detached narrative of this phase of the mission. It was a technological triumph, but almost an inverted one. Cooper plays with this idea in the following:

> The supreme achievement of American technology had broken down utterly. All that was left was a spacecraft whose very complexity made it harder to handle, plus a group of flight controllers and three astronauts who were themselves products of the vast bureaucratic machine that had produced the malfunctioning spacecraft. On the face of it, this might appear to have made it all the more difficult for them to get outside the situation and impose their will on the wayward spacecraft. However, the accident had also demolished most of the technological appurtenances, such as checklists and flight plans, which substitute a sort of delayed time for immediacy, and also much of the automatic equipment aboard the spacecraft which performed tasks that earlier mariners would have performed for themselves. Now the flight controllers and the astronauts were no different from any other sailors facing disaster at sea. They would do a lot better by themselves than their elaborate paraphernalia had done by them.[23]

He is correct when he links the controllers and the astronauts together as mariners; in the recovery the distinction between manager and operator, indeed, between designer, manager, and operator tended to disappear. With *Apollo 13,* more than any other flight, the astronauts returned to their old test pilot status—professionals in a cooperative, extraordinary endeavor that made little of rank and control. The conditioned subject, the chimp, the redundant device, the man who had only to sit on top of a piece of fireworks with a thermometer up his rear to achieve hero status, disappeared forever. But so, hopefully, has the notion of the "Omniscent Designer," and the "Omnipotent Manager."

Early Warning Systems

Since Hiroshima and Nagasaki the ultimate in catastrophic accidents has been accidental nuclear war, an accidental determination of the fate of the earth, a holocaust. Headlines tell us of steady technological advances in missiles and the defense against them. The defense is largely more missiles or faster ones or quicker counterstrikes. Missile gaps continue to be debated; the 1960 presidential election was thought by many to be decided on the basis of a missile gap when missiles were few in number and overwhelmingly in our hands. As the gap has narrowed, the number of missiles has increased to thousands. (There are between twenty and thirty thousand U.S. nuclear weapons in the United States and abroad.) Complicating the picture is the increasing number of armed conflicts around the globe, their increasing ferocity and durability, and the increasing number of combatants with access to nuclear weapons. The nightmare of *Dr. Strangelove* has not gone away; in fact, that scenario, bizarre as it was in 1964, is pastoral in its loose coupling and linearity compared to scenarios that can be imagined today.

Dr. Strangelove was a hilarious but chilling film by Stanley Kubrick concerning our Strategic Air Command and national defense posture. In terms of system accidents it runs like this: General Jack D. Ripper (the humor is broad, to say the least) is a crazed SAC bomber commander who launches his squadron of B-52s with their nuclear bombs against the Russians. (Psychopathic behavior has been left out of all the accident scenarios in this book, but we know it exists. We will include it this one time, as failure #1.) Only General Ripper knows the secret code that will recall the bombers, and he will not reveal it. (Failure #2; a procedure error, or procedure risk. It resembles the rumored ability of submarine commanders to fire nuclear missiles without presidential orders under some circumstances, such as, of course, a loss of communication with our president.) The army is forced to attack his base, but (failure #3) the secret recall code dies with General Ripper in the attack (a common-mode failure). But there are still recovery possibilities, because, in those halcyon days, it takes a few hours for the B-52s with their missiles to reach Russia. (It is ten minutes now for submarine missiles, less than eight minutes for Pershing II missiles about to be based in Europe; none of our medium or long-range missiles can be recalled, or even destroyed in flight if fired in error. The system is immensely more tightly coupled today.) A lucky guess and the combined efforts of the Russians and the Americans make it possible to recall or shoot down all but one of the

SAC bombers But one is left, giving us failure #4. (One is reminded of the reconnaissance airplane that did not get the message during the Cuban missile crisis to keep clear of Soviet territory; it strayed over the border when we were "eyeball to eyeball" with the Russians, considering an air strike against Cuba, and commencing a blockade. It complicated the negotiations no end.[24])

The slack in the system had removed all but one potential source of catastrophe, it was thought, but it just so happened that the Russians had installed a "Doomsday" device that would destroy the world automatically if even a single nuclear bomb went off. The "Doomsday" machine was installed to deter nuclear attacks once and for all, but it would only work if we knew of it. Failure #5 was that the Russians had armed it but were waiting for an opportune moment to announce its existence to the world. Once operational, there was no way to disarm it, for its effectiveness depended upon neither the U.S. nor the Soviets being able to defeat it. (The closest counterpart today is "launch-on-warning," a posture threatened by both sides that would send irretrievable missiles to the enemy if it looked as if the enemy were launching an attack. That presumption of an attack will preoccupy us in this section.) At this point in the film, with appropriate patriotic music, no further human intervention is possible. The last remaining B-52 from the squadron, rejecting radio messages to return because the messages lacked the proper code, reaches its target, and the world goes radioactive. In the final scene U.S. officials are left discussing the possibility of preserving at least the U.S. version of civilization underground in abandoned mine shafts. The air force chief of staff warns the president of a possible "mine-shaft gap."

Today, the ridiculous accident scenario is not as ridiculous as it once seemed. With Soviet satellites watching for signs of missile launches, we have had a dropped wrench setting off a Titan missile (fortunately, it went only a few hundred yards), and a rumor of an accidental launching below the Canadian border that landed a missile in Canada. But there is so little information about accidental missile launchings or accidents that came close to causing launchings that I cannot even speculate on the matter.

This is also true of the matter of "broken arrows"—accidents with nuclear weapons, such as dropping them during ground transport or loading, dropping other objects on them, or dropping them accidently from airplanes as in the Palomares, Spain incident in 1966 (the cost of cleanup operations was well over $50 million; 5,000 barrels of the plutonium-contaminated soil are buried in lovely South Carolina).[25] There have been between "over twenty-seven" (official count) and one hundred twenty-five (International Peace Research Institute count) of these. In the official list, fourteen of the twenty-seven accidents involved the explosion of the detonating device (which in one case created a crater in Texas 35 feet across and 6 feet deep). But the detonating device must go off with extreme precision to explode the warhead; it consists of several explosive charges surrounding the nuclear core. It is thought that it is extremely unlikely that an accident could produce just the right pattern of explosions (indeed, we are not sure an intentional detonation can do it very often). However, we have probably had a few tries at it. In one case, five of the six interlocking safety devices on a bomb failed, and according to Dr. Ralph Lapp, head of the nuclear physics branch of the Office of Naval Research, only one switch prevented a 24-megaton bomb from destroying much of beautiful North Carolina.

One may easily imagine a scenario of commonplace industrial failures and military mishaps leading to such accidents; one need not even invoke the concepts of complexity and coupling. Even though a nuclear explosion is extremely unlikely, a broken arrow can spread deadly plutonium about, and the earth has been scraped in Spain, Greenland, North Carolina, Indiana, and probably elsewhere to remove the deadly contamination. While we no longer keep B-52 bombers aloft with armed nuclear weapons on a continuous basis, they do fly about and the weapons—unarmed—are transported by air routinely in cargo aircraft. (A helicopter carrying nuclear bombs once made a forced landing at Coney Island.) Their plutonium is available for distribution in an accident. While the destructive potential to third- and fourth-party victims is not as great as that entailed by an accidental detonation of one of the thousands of existing warheads, the probability of spreading plutonium about in, say, a 2-mile wide swath 25 miles long, contaminating earth and people alike, is much higher. The most destruction, of course, would come from an accidental launch of nuclear weapons, but that has the lowest probability of all. Let us see why.

We will focus upon the most credible example of accidental attack—false alarms. The matters of terrorism and insane commanders will not be considered.

The Warning

In a huge series of caves carved out of rock underneath Cheyenne Mountain in Colorado, the NORAD (North American Aerospace Defense Command) early warning command center waits for signals that the Russians are coming. It is not, though one might imagine it to be, an uneventful life. While the Russians are reasonably inactive, except for some test firings of missiles and space shots, everything else seems to be going off all the time. When something does go off, a "missile display conference" is called. In 1979, according to a Senate report, there were 1,544 missile display conferences. In just the first six months of 1980 there were 2,159 of them—over ten a day.[26] The enlisted men and women who staff the monitors, thus, are reasonably occupied.

The conferences are not much, though they must keep the managers busy there; they are telephone conferences, calling the duty officers at three other command centers to verify the accuracy of warnings. A "missile threat assessment conference" is more serious, though the telephone is again the medium. In this, persons higher than the duty officers at four separate command posts are called. There were seventy-eight threat assessment conferences in 1979, about one every five days, and presumably two or three times as many in the first six months of 1980. After a missile display conference and then the missile assessment conference we come to the third level of missile conference, one that has never been called, the "missile attack conference." The president of the United States is in on that one.

What keeps these outposts of our defense busy are primarily atmospheric disturbances that produce an infrared "signature" somewhat similar to that produced by an actual missile launch that is detected by our satellites. There are also flocks of birds to contend with, and space shots and testing, as well as miscellaneous anomalies in the atmosphere or the equipment. With so many warnings, the system probably has its responses worked out quite well. This was true on November 9, 1979, when the monitors indicated a massive Soviet attack. It was clear that it was not an anomaly; it showed both land-based missiles and submarine-based missiles were involved. In fact, it closely conformed to what the Pentagon anticipated the Russians would do if they attacked. It was a quiet day, internationally, which may or may not have contributed to the skepticism. The attack showed up simultaneously on monitors in the Colorado headquarters of NORAD, the National Command Center in the Pentagon, Pacific Headquarters in Honolulu, and "elsewhere."[27] A thousand Minuteman Intercontinental Ballistic Missiles (ICBMs), capable of hitting targets in Russia, were placed on low-level alert. Ten tactical fighters took off, but what they would have done is not clear to me; they cannot shoot down incoming missiles. But after six minutes the alarm was certified as false. That might seem speedy, but Russian submarine-based missiles will allow us only eight to ten minutes to mobilize our counterstrike, and in addition to arming and preparing to fire the Minuteman missiles the president has to be called, since only he or she can order a missile launch on such a warning. Over half of this time was used up before the alert was cancelled. But actually, it was suspected to be false within two minutes.

The realistic character of the false alarm was understandable: a training tape that simulated an expected Russian attack had been loaded on an auxiliary computer for routine purposes. Somehow the signal found its way into the active, on-line alert system. (One can visualize the enlisted personnel that monitor the screens saying, "Boy, that looks just like the real thing; just like what they trained us for," and pushing all the buttons in sight. Equally likely, however, would be this reaction: "That's *unreal;* nothing ever happens the way the training program says it will.") How did the managers know it wasn't real? They checked the reports from two independent sources, the satellites and the early warning radar systems—a matter we shall come to shortly.

Less than a year later, at 2:26 A.M. on June 3, 1980, the Strategic Air Command received an indication that two submarine-launched missiles were on the way; it came from NORAD headquarters. SAC called to confirm, and NORAD was unable to confirm, even though the message came from their computers. The SAC duty officer took the precaution of ordering all B-52 alert crews to board their airplanes and start their engines. It is essential that these bombers, armed with nuclear bombs, be aloft if there is a strike, or the Russian missiles will presumably have wiped out all their bases. It would take the B-52s hours to reach Russia, but they would be above the holocaust in North America. Shortly thereafter the SAC display indicated no missiles, and no other parts of the warning system indicated any either, so the B-52 crews were ordered to shut down their engines.

A few minutes later the SAC display monitor, receiving messages from NORAD headquarters, indicated Soviet land-based missiles on their way to targets in the United States, and a little later the monitors in the Pentagon showed submarine missiles coming in. The monitors do not "picture" these events, as a radar screen does, but only display numbers in the appropriate categories, indicating number of missiles, direction, et cetera. The duty officer at the Pentagon convened a missile display conference, the lowest level of alert involving a conference telephone call, and then went on to the next level, a missile threat assessment conference bringing in more officers. The commander of NORAD said that there was in fact no threat, and one minute later the SAC alert was terminated. In the meantime, the Pacific Command airborne command post actually sent bombers into the air from its base in Honolulu. The whole episode lasted three minutes. Since the Soviets have the capacity of monitoring the activity of our airbases and may have the capacity of monitoring telecommunication activity, if not the actual content of the messages, one wonders if they went on alert with their own missile crews and bombers.

The cause of the false alarm could not be determined. But NORAD knew it was false because neither the satellites nor the radar picked up signals of land or submarine-based missiles being launched or penetrating our perimeters. NORAD kept the system in the same configuration for the next few days to see if it would reappear, or so they testified in Congress.[28] Three days later the identical alarm recurred. SAC crews again started their engines, and again the alarm was determined to be false in less than three minutes. NORAD switched to a backup computer, and began the search for a possible malfunction. Eventually they found it. It was a defective, tiny silicon computer chip (cost, 46 cents), not in the computer itself that was set aside, but in the "multiplexer" that routes messages to the various command posts. This multiplexer sends a message to the command posts on a continuous basis to confirm that the channel is open and usable. However, this message was in the same format as that used to indicate a real attack. Why it was the same format is not clear. It seems unlikely that this was an inadvertent similarity, though that is possible; more likely, the form of the message was similar to insure that that form could be sent. The message has a space to indicate the number of missiles, the number given is zero in the routine, test message—we have no missiles today. Due to the malfunction of the chip, this zero changed to a "2" and then apparently to other numbers; it sent this data out to some, but not all of the command posts. The available testimony does not indicate what information other than "2s" for the number of missiles was also sent.

NORAD changed the routine message format so that it no longer resembled an indication of an actual attack. It also corrected another oversight, resembling the PORV warning light at the Three Mile Island plant. NORAD had been aware of what it told its equipment to send out, of course, but it had no monitors at its headquarters to show what actually was sent out. (At TMI, the operators were aware of what the automatic system had told the valve to do, but had no way of knowing what it actually did.) NORAD operators had no way of knowing that, without intending it, the signal was saying, "We have two missiles for you today." NORAD installed monitors, and presumably three new shifts of enlisted men and women now watch these and compare them with the signal that is supposed to go out. Now that this little anomaly has been discovered and corrected, one wonders what the next one will be that no one thought of, but was so obvious once it was discovered. The opportunities seem endless since NORAD is such a vast and complicated command, control, and communication system resting on radar and satellite stations that apparently transmit ten pieces of garbage—false warnings—a day.

Nevertheless, false information is not dangerous if it is not credible. Our early warning system has two major checks upon the credibility of computer errors of this type, and one on the credibility of sightings of missiles and, fortunately, these are reasonably decoupled. The instances we have reviewed were caused by false information generated at NORAD headquarters and sent to the command posts. But each post, and NORAD itself, has its own monitors showing what the radar and the satellite systems are seeing. Checking these, and seeing no indications in either one that bore any relation to the quantitative information from NORAD, the officers knew that it was a false alarm. That is why the president was never called. The system must work very fast, and in these three cases it did—within three minutes or less it was known or strongly suspected that the signals were false.

The two major sensors also detect missiles through different methods, and are not linked to each other. (Satellites pick up the launch, radar the incoming flight, so they are actually not two measures of the same event.) Land-based Soviet ICBMs would be detected first by the Ballistic Missile Early Warning System (BMEWS) strung across northern Canada and Alaska and, closer to impact, by the Perimeter Acquisition Radar Attack Characterization System (PARCS). (The elaborate and uncommunicative names are not reassuring; for me, the names and acronyms only reinforce a notion of unmanageable complexity.) PARCS is near Grand Forks, North Dakota, and is reported to be capable of a very precise identification of exactly what kind of missiles are involved and what their "multiple independent reentry vehicles" are aimed at. Of course, the Soviets have not allowed us to test this capability. For submarine-launched missiles off either the Atlantic or Pacific coast, the satellites are the first line of detection, followed by a special radar system called Pave Paws (I will not burden you with what that stands for). For those coming from the Gulf of Mexico, we have an older radar system.

The command centers can check all these. Now it is possible that two or three of these sensors could malfunction simultaneously, or nearly so, even though they are independent of each other and use different detection methods. But it is highly unlikely that the malfunction of the satellite would produce the signature that would just happen to match the malfunction of the BMEWS radar, and that there would also be a very similar malfunction of the PARCS radar. Not only would they all have to malfunction, but they would have to produce compatible malfunctions. With multiple and independent sources of information, the more detailed the information, the more unlikely an error. Information in industrial plants is generally crude (the valve is open or closed) or singular (the temperature is X). An independent source can mistakenly confirm such values because they are simple. If, however, the information has many parameters—number of missiles, trajectory, speed, and size—then an incorrect confirmation is much less likely.

But we do not know how similar the information must be to be credible. If our anti-submarine warfare satellite spotted a foreign submarine in the Gulf of Mexico, and if a satellite suggested three missiles were fired in the Gulf during a thunderstorm, but it was an ambiguous signal because of the weather, and it just so happened that the radar scanning the Gulf malfunctioned in such a way as to suggest two or five missiles were coming (the malfunction might have been associated with the storm), and if it seemed the missiles would arrive too quickly to wait for PARCS in Grand Forks to verify their arrival, and one of the four command centers had all this information at hand, the duty officer might just send the airplanes up, arm the Minutemen, and call for a second- and third-level alert conference. The three missiles suggested by the satellite might well "fit" with the two or five suggested by the faulty radar, especially since, in an attack, missiles would be fired off in rapid sequence and the number of missiles in the air would change rapidly as seconds passed.

Suppose the Soviet satellites also picked up the highly suggestive atmospheric signal from the Gulf and within three minutes put Cuba on the alert for a nuclear attack from the United States. We would notice that immediately, and it would reinforce the suspicion that Soviet submarines were attacking. We could even suspect that PARCS would not work that well when the chips were down, and thus was not tracking the missiles from the Gulf, and start trying to communicate with our submarines. The Russians would note that, and it would reinforce the partial information they had. They might even see our hardened silo covers roll back in the Midwest. There would be no time for a hot-line call, and we would not expect the Russians to tell the truth anyway, nor they us.

I am sure that military experts could point out many flaws in this hypothetical scenario, and I truly hope they can. But without a security clearance and a year to study the system I cannot be sure that this scenario, or some other, is completely out of the question.

It is also conceivable, though very unlikely, that a malfunction could reverse the path of communication between the sensors and NORAD.

We have seen that NORAD can generate false information on its own. Given the complexity of the system, it is not inconceivable that the failure that produces this might also produce a reverse flow, wherein the false information ended up on the satellite and radar screens through some devious route. After all, a training tape played to a packed house in the on-line NORAD system.

To summarize, the detection system—NORAD and the sensing systems—is moderately complexly interactive. Linearity is introduced because the subsystems are independent of each other, not proximate spatially, nor subject to many common-mode components (though there are some). Some loose coupling is present because recovery is possible from some low-level failures—the B-52s need not take off, or they can be recalled—and final action depends upon finding the president in time and convincing her or him. Some loose coupling is built into the system inadvertently (and inadvertent recovery aids are particularly valuable because they can cover contingencies designers did not think of). The PARCS system was originally designed as part of an anti-ballistic missile system to shoot down incoming missiles aimed at our ICBMs. That turned out to be impossible, but the system is now available as a check on the other sensors. The satellites went up to provide earlier warning than radar, but since the satellites are independent of the radar systems, they provide some check.

But note a final consideration. Each side attempts to complicate the job of the other side. Thus, incoming missiles can fire decoys that might just confuse PARCS and Pave Paws and BMEWS; aware of this possibility, these folks might mistake a confusion of geese and sun spots for the confusion the Russians were trying to spread. Then there is the disturbance produced by nuclear explosions in near space, which would disrupt most of our communications; could a blackout over an interlocked power grid extending from Colorado to the capital be misinterpreted as an electromagnetic pulse from a Russian weapon fired from one of their satellites? The defense system is one that grows on complexity. We are not merely adding safety devices to buffer a failure in the DEPOSE components; the "failures" we must guard against are those that an enemy (supremely clever and resourceful, it is assumed) is actively promoting. Each side mindlessly makes it less possible for the other to rest assured that it does not seek its total destruction. Thus, we confront another source of error that no other system we have studied has to cope with. Ironically, this new source of error is found in that system which could decide the fate of the earth.

The Response

The response system, once the credibility of a warning is thought to have been established, is apparently so complex as to provide considerable safeguard against an accidental war; it is hard to imagine that the system would actually work. Of course, this means that it provides little defense against an actual attack. Since there is really no "defense" any more, only retaliation and a possible mine-shaft gap, this may be a blessing. The World Wide Military Command System, of which NORAD is a part, operates so poorly and intermittently, despite billions of dollars, that we may suspect that its strategic weapons system is as failure-prone as its detection system. An admiral recently remarked that the strategic weapons response system is so complex, because of all its safeguards such as two-person control, multiple-command requirements, and conferences, that he wondered whether we could ever manage to launch our nuclear weapons even in the face of a clear threat.[29] The details of the presumed complexity of the system are not available for analysis; one may comfortably assume, however, that it is not a very linear system.

The response system is more tightly coupled than the detection system. A ballistic missile cannot be recalled or destroyed. A submarine commander and his chief officers presumably can fire their missiles if they believe that their lack of communication with their managers is because there has been a Soviet attack. It is probably the case that there is no signal to tell them that the lack of a signal is not a signal—a problem with tightly coupled systems. There is not likely to be any recovery from the consequences of an accidental strike; the Doomsday machine is, in effect, in place. In fact, it is worse; launch on mere warning has been threatened by both sides. It is something the generous Stanley Kubrick did not anticipate.

Conclusion

In conclusion, the early warning system appears to be moderately complex and coupled, but not disastrously so. Since its catastrophic potential is extremely high, we should well wish for further independence among the subsystems and more corroborating possibilities. In particular, we could hope that one aspect of complex systems, the presence of invariant sequences, could be reduced, such as by the ability to disarm the warhead and destroy our own missile in flight. But we should note that in this system, failure to deliver may be high, and since it is certain that false alarms are far more prevalent than true ones, this ineffectiveness may be a virtue. Indeed, one wonders if it is ever worth risking the high probability that an alarm is false by launching missiles at Russia. The output of such a response is mutual assured destruction, or MAD.

The response system is more complexly interactive and more tightly coupled than the warning system (though it is hard to separate the two). Missiles cannot be recalled; submarine commanders may be out of touch but able to act on their own; missiles may go off accidentally. The complexity of the system is such that it may limit its ability to share in the destruction of the earth, but it also means that an inadvertent first strike is possible. Our missiles could go off, not because of a false warning based upon the faulty interpretation of signals from the environment, but because internally-generated signals may go off by themselves.

With weapons system we confront new complexities. First there is the peculiar matter of failures of failures equalling success. If the detection system fails and falsely indicates an attack, but the response system fails to respond, we have success—that is, we have not started a war. One does not like to hope for failures to insure success. But the probability of each of these failures, separately or jointly (greatly reducing the joint probability of both failing) remains higher, I believe, than the probability that the Soviets would make a first strike.We are more likely to fail to respond to a false alarm than the Soviets are likely to intentionally attack.

However, we have a truly interactive system here. We do not know the probability that the Soviets would either accidentally launch an attack, or launch on a false warning. Presumably they are confronted with the same complexity and coupling problems that we have (though there is some belief that their missile system is, as their conventional weapons systems is thought be be, less technologically advanced—that is, less complex and tightly coupled—than ours). The permutations of risk analyses in this area could be an endless hall of mirrors.

Finally, we have, for the first time in this book, a system where the environment is intentional and self-activating. The enemy can intervene in the operation of the DEPOSE components through decoys, deception, disruption of communications, theft of designs, even perhaps the corruption of operators. We need safety devices to guard not only against the failure of a part or unit, but also against the ability to mask that failure. This adds another dimension of complexity to something already complex enough, and greatly limits the possibility of recovering from failures before the system itself fails. There does not seem to be any end to this increase in complexity and coupling. The judgment that, as of 1983, the early warning system is only moderately complex and coupled, may be overtaken by 1990 because of what we think the Russians are doing to intervene in our system—and what they think we are doing with respect to theirs. As noted before, it is ironic that the most fearsome system on earth should have this added burden of complexity and coupling.

Recombinant DNA Technology

With recombinant DNA processes, we will have to engage in an anticipatory analysis. The series of industries that focus on this technology are only just beginning to come into existence. There are no published descriptions of the production system in commercial DNA labs, and there are certainly no accounts of accidents in these labs as yet. However, we have a fair idea of the nature of the production system, and some people have written at length about the dangers of an accident. The system appears to be complex in its interactions and tightly coupled, but I caution the reader that I know even less about it than I do about nuclear weapon systems.

Recombinant DNA technology, or gene-splicing, is a microbiological technique that allows scientists to graft the genetic information from one organism into the cell nucleus of another. This enables the biologist to design new life forms for accomplishing specific tasks. Among the early successes of DNA have been the creation of bacteria that can produce complicated biological molecules used in medicine, which had been difficult and expensive to obtain. Human insulin, growth hormone, and interferon are among the first of these synthetic biological products that have commercial application. Much more is coming; indeed, whole sectors of the economy may be transformed. In the chemical industry, work is underway to create micro-organisms that can serve as catalysts for chemical reactions, thereby doing away with some conventional transformative processes requiring high temperatures and pressures.[30] Forms of bacteria might be created to consume oil spills or transform waste products into energy sources. Some biologists are speculating that agricultural production as we know it may well be transformed through the creation of new and exotic plant forms, and meat production could be transformed through the substitution of "single-cell protein cultures."[31] Finally, this technology holds out the potential for the application of bioengineering techniques on defective human genes—a technology for nothing less than designing or redesigning human beings. It is, in short, a process of remarkable potentials that could transform not only our economy but our sense of what is human.

The commercial potential of DNA is apparent in the venture capital flowing into the industry, and the increasing centralization of the industry taking place through cash-rich oil companies. The four original small gene-splicing companies that began with a handful of PhDs and venture capital in the mid- and late 1970s grew to a combined value of over $225 million by 1979.[32] By 1982 it was estimated that there were 350 companies involved in the recombinant procedures.[33] The larger pharmaceutical companies moved into the industry quite early, of course, and were quickly followed by chemical industries and, especially, the large oil companies. Large sums of capital have been committed to the industrial application of gene-splicing by such chemical companies as Dupont, Pfizer, and Monsanto, and among the oil companies, Arco, Standard Oil of Indiana, and Occidental Petroleum. Commercial success in the agricultural market may not come before the 1990s, but may come sooner for the medical field.

Unfortunately, with all these fantastic potentials go some fantastic risks. These industries will produce new, *living* technologies; life forms that are unique, unprecedented, and in some respects very poorly understood. In many of the proposed applications, new organisms will be introduced into the environment in massive quantities. Such quantities may produce totally unexpected interactions; there is simply nothing in our experience to go by. Once introduced, there seems to be very little and quite possibly no chance for intervention should unexpected and untoward interactions take place. As Pamela Lippe of the Friends of the Earth warned Congress in 1977:

> DNA is probably the most unforgiving technique we have yet developed. Radiation decays. We can stop making toxic chemicals. But a novel organism has a life of its own; once it has escaped or been released, once it has established an ecological niche, it is out and replicating, perhaps beyond our ability to control or clean up.[34]

The catastrophic potential of DNA is different from all other systems we have described in that it does not loose toxic or explosive substance upon the environment, but rather creates interactions between systems that were not previously linked at all, and perhaps could not be foreseen to have been linked. Once the linkage is made, it cannot be controlled by the operators. A catastrophe of literally epidemic proportions may be set in motion.

We know something about the interactions of toxic chemicals with the environment, and while these are not new organisms with unpredictable behavior, we can get some idea of the magnitude of the problem by briefly reviewing this relatively simple interaction between the system and its environment. Perhaps the most widely known example is DDT. It was the publication of Rachel Carson's *Silent Spring* in 1962 that first acquainted the public and many scientists with the chain of unexpected effects stemming from the application of DDT, endrin, dieldrin, and other related pesticides. The danger she uncovered was not that of direct poisoning, which was observable and well understood. Rather, Carson pointed out that these poisons "magnify" in living tissues. As they move up a food chain, from plants through small herbivores to a succession of larger and larger carnivores, the poisons become more concentrated in living tissues. The unanticipated interaction here was not a direct poisoning of people but rather that the food chain on which plants, small animals, and man are linked became impaired. The ramifications are still working their way through the planet in, for example, inhibition of photosynthesis in phytoplankton and the crippling of reproductive systems in birds.

The scenarios mentioned by scientists for possible accidents from the production of recombinant DNA are similar to the actual DDT accidents. They include an unexpected linkage between previously unlinked systems, limited prior knowledge of the interaction process, and indirect and delayed information on the consequences. These scenarios are *ecosystem* accidents that result from intended incursions on the ecological system.

Not all eco-accidents are system accidents in their origin. The oil spill in Santa Barbara, the leeching of toxic wastes at Love Canal, and the contamination of Seveso, Italy, by dioxin are component failure accidents. They are the result of poor design, operator failure, or equipment failure. Hooker Chemical Company knew of the danger of the toxic waste they buried at Love Canal. The drug firm, Hoffman-La Roche, in Switzerland was well aware of the danger of dioxin contamination in their plant in Seveso, Italy, and indeed plant officials were instructed to readily reimburse their neighbors for dead farm animals that continued to appear. Knowing that dioxin was a by-product of the pesticide the plant produced, they would not allow production to take place in clean little Switzerland, where their headquarters were, but instead had it produced in dirty northern Italy. When the chemical reactor exploded one weekend when no one was attending it, the safety device protected the plant by allowing the poison to blow up into the air through a stack, from where it drifted over the neighboring community. Plant officials avoided a panic by simply not informing the community. Components fail in such accidents as these, and the risk of an oil drilling accident, of leaking drums of highly toxic waste, and of exploding chemical reactors are presumably not only anticipated, but carefully calculated risks as well.

With eco-system accidents the risk cannot be calculated in advance and the initial event—which usually is not even seen as a component failure at all—becomes linked with other systems from which it was believed to be independent. The other systems are not part of any expected production sequence. The linkage is not only unexpected but once it has occurred it is not even well understood or easily traced back to its source. Knowledge of the behavior of the human-made material in its new ecological niche is extremely limited by its very novelty.

Eco-system accidents illustrate the tight coupling between human-made systems and natural systems. There are few or no deliberate buffers inserted between the two systems because the designers never expected them to be connected. At its roots, the eco-system accident is the result of a design error, namely the inadequate definition of system boundaries.

It is only since science has learned to replicate complex physical, chemical, and biological processes in the laboratory that its actions have been so consequential for the eco-system. The frequency of unintended interventions in the eco-system are likely to increase as the keys to more natural process are discovered. At least since Rachel Carson, concerned scientists and activists have worried about potential negative interactions. They have tried to anticipate the normally unexpected. This form of ecological consciousness was at its peak in the early seventies when molecular biologists first raised the possibility that an as yet untested technique, the "splicing" together of genetic materials from unrelated species and their implanting in either an intended or unintended host, might be potentially catastrophic. An analysis and discussion of the way in which this was handled by the scientific community will be instructive in its policy implications for other complex, tightly coupled systems. But before we get to that, a brief introduction to recombinant technology is in order.

Gene Splicing

The recombinant technology, popularly known as gene splicing, involves a set of techniques that allows scientists to use special enzymes to cut into pieces the long double strands of molecules that make up DNA and then to *recombine* the pieces with the DNA of a carrier, called the "vector." These recombined molecules are then inserted into a host where they will presumably propagate. By combining the foreign DNA into a vector that typically replicates in the host organism, scientists are able to induce the "expression" of the foreign genetic material in its new host. For example, if the foreign DNA carried the genetic information for human growth hormone, it might then be combined with a vector that replicated in some readily available host bacteria. The host bacteria would then begin manufacturing human growth hormone. This has actually been accomplished. Prior to this, all human growth hormone had to be extracted from human blood. As a result, it was rare and expensive. Its mass production in bacteria will be a tremendous asset to the treatment of childhood growth disorders.

But early in the development of these procedures controversies arose over the unrestrained application of these techniques to organisms that were clearly quite dangerous. A biochemist at Stanford, Paul Berg, was planning to put Simian Virus 40 (SV40), which causes tumors in monkeys, into *E. coli,* a bacteria that abounds in the human gut and is widely used for research. There was some evidence that SV40 could alter human cells to resemble tumor cells, and in this lay the potential danger. By slightly altering a virus that was potentially carcinogenic in humans, a new organism could be created that was even more lethal than the parent organism. But more important, by transplanting the offensive gene to a bacterial host such as *E. coli,* the far more disturbing prospect presented itself that experiments might create a previously unknown bacterial strand that, were it to escape, could conceivably cause a cancer epidemic. Robert Pollock, a microbiologist, learned about this planned experiment from Berg's student, who was visiting the Cold Spring Harbor laboratory for a summer seminar. Pollack was perhaps the first scientist to recognize the danger of creating a recombinant hybrid with unknown infectivity that might be capable of surviving in humans. His worried call to Berg touched off first anger and then concern in Berg.[35]

Over the course of the next six months Berg raised the issue with colleagues all over the country. As a result, he postponed the SV40 experiment. By June 1973 other scientists had expressed their concern to the National Academy of Science (NAS). The NAS asked Berg to organize a committee to discuss the recombinant DNA safety issue. The NAS group convened in April 1974 at M.I.T. Their decision was to call an international conference and to request a voluntary moratorium on specific forms of recombinant experiments thought by the group to be potentially hazardous. It is worthwhile to note that such a moratorium is probably unique in the annals of science. It was enforced through peer pressure, strengthened by the fact that the elite of the molecular biology world were its initiators.

The international conference came together seven months later, in February 1975, at the Asilomar Conference Center in Pacific Grove, California. The conferees began by trying to grade the risk of various experiments. Committees went on to assess specific controversial experiments. The outcome of the conference was a three-tier, graded series of containment measures to match the graded risk of various experiments. The intention was to put buffers between the recombined DNA molecules and the environment. Only a failure in containment could then result in an accident. Sixteen months later, after considerable debate and conflict in the research community, the rough plans laid out at Asilomar were turned into specific NIH guidelines.

In our terms, the scientists were reducing risks by decoupling a system that might potentially be tightly coupled with the environment. Their efforts were complicated 1) by the fact that all accident scenarios were hypothetical, based on limited knowledge, and 2) by the politics involved in restricting a burgeoning and highly valuable field of scientific research, one with tremendous commercial potential as well.

In setting up the NIH guidelines scientists had to grade the potential risk of a broad range of possible experiments and then to stipulate the conditions under which specific experiments might be done.[36] They began by trying to identify the highest risk experiments and simply prohibit them. Researchers were forbidden to work with a list of organisms classified by the Center for Disease Control and the National Cancer Institute as either pathogenic, causing disease, or oncogenic, causing tumors. Specific genes were also forbidden, among them genes that might code for toxins, plant pathogens, and drug resistance. Scientists were further forbidden from any deliberate release of a recombinant into the environment, such as use in agricultural research. Finally, all projects were to be small scale, limited to less than 10 liters of culture. By using lower-order, nonpathogenic organisms whose behavior in the lab was well understood, the writers of the guidelines hoped to restrict the realm of uncertainty.

But elimination of foreseeable risks was only a first step. More important, by far, were the guidelines for containment. Strategies for containment were of two kinds, biological and physical. Both strategies involved graded levels of containment that were to be matched with the expected risk of the experiment being performed. The principle behind biological containment is that each subsystem ought to be a transmission barrier.

That is, viral genes, vectors, and hosts would all be altered, insofar as possible, so that they accomplish their singular purpose without being capable of other interactions. Thus, plasmids (a vector) were to be chosen for their inability (or low-frequency ability) to exchange genes between cells. Hosts, such as *E. coli,* were to be sufficiently enfeebled so that they would not survive outside the lab.

Thus in biological containment the subsystems serve double duty as functional units and as buffers. The use of these buffers constrain the limits of permissible research. When the guidelines were issued, only a small number of vectors and hosts could be used in experimentation. These limits were being questioned even before the guidelines could be published. In fact, several cases arose where researchers were discovered to be working with prohibited viruses or vectors. In general, the greatest drawback to this form of containment is that it inhibits innovation and is then likely to be resisted or circumvented where the production system is part of a competitive economy.

The other form of containment, physical containment, is based on the more traditional procedures for separating potentially pathogenic organisms from the environment. The lowest level of physical containment is often compared to the usual lab procedures of a well-trained microbiologist. Most of these procedures are aimed at protecting the subject culture from contamination by wild strains, but the buffer effect works both ways. Higher levels of containment, known as P2 and P3, include sterilization, washing of hands, sealed windows, exhaust filters, and special safety cabinets if material might be sprayed into the air. The highest level, P4, requires the changing of all clothing, showers, and negative air pressure to keep all particles inside. At the time the guidelines were issued, several universities had begun the construction of P3 facilities. Only the government had a P4 facility, the former biological warfare center at Fort Detrick, Maryland.

Within a few short years, however, the scientific community did an abrupt about-face on the regulation of DNA technology.[37] Faced with the prospect of stringent federal guidelines restricting the kind of research that could be done, the biologists began to reassess the risks of DNA procedures in order to calm the surge of public fear. The most publicized experiment regarding the potential dangers was the study of Martin and Rowe. They asked, How dangerous could an escaped recombinant be if it had received a gene from a lethal donor organism? To answer this question, they grafted the genetic information from a cancer virus into *E. coli* and infected mice with the new bacterial strain. What they discovered was that the new bacteria were either noninfectious or far less infectious (by a factor of 10^{-9} or one in a billion) than the original cancer virus itself. Martin and Rowe felt this dispelled any fears that an escaped organism might pose some sort of health hazard. As Rowe explained it, the research demonstrated that there was nothing one could cut out of a smallpox virus which, inserted in *E. coli,* would be dangerous to work with on an open laboratory bench. "The same applies to all the tumor viruses and even the lethal Lassa fever virus," asserted Rowe.[38] The outcome of this research and the growing sense of familiarity with the technology led the biologists to successfully forestall the proposed federal legislation regarding restrictions on DNA research. By September of 1979 the Recombinant Advisory Committee (RAC) of the NIH had voted to exempt 80 to 85 percent of recombinant DNA research from all but the most standard of lab safety procedures. In fact there had been a rather radical change in the RAC's attitude toward risk in less than three years. The change has been characterized by Thomasson as going from protection against the worst possible case to protection against believable risk.[39]

The question must be asked, however, whether the new evidence actually warranted such a radical departure from the risk-conscious policies that had been implemented in the seventies. There are at least some informed scientists who contend that the Martin and Rowe experiments ought not to be taken as conclusive evidence that stringent safety regulations were unnecessary. Indeed, critics pointed out that the scientific community should not focus on the fact that the recombinant bacteria were less infectious than the tumor virus itself, but rather should note that these experiments demonstrated quite conclusively that the lethal trait could be transferred through DNA techniques. In the experiments conducted by Martin and Rowe, "about half of the mice that were injected with a bacteriophage containing a dimeric (two-copy) form of recombinant DNA" did contract the polyoma infection. To some observers this finding suggested that what was notable about the experiments was that they demonstrated how DNA research could in fact create a new vehicle for the transmission of such hazardous traits.[40]

There have also been other scenarios put forward suggesting that the peculiar and subtle complexities of recombinant organisms might lead to serious health hazards simply by interacting with biological systems in ways that are novel and consequently unbuffered. Perhaps the best example of a system accident scenario in this area is based on the autoimmune response in humans. Such a case requires the splicing of a protein from an animal donor into a host bacteria such as *E. coli.* Through a trivial accident, the *E. coli* could then establish itself in the intestinal tract of a laboratory worker. The recombinant would begin to produce the animal protein in the lab worker's system. The animal protein is similar in structure to human protein. The lab worker's immune system would be triggered and would attack the foreign protein. But because of the similarity in structure, the antibodies would not distinguish between foreign and indigenous protein. The immune subsystem would thus attack healthy tissue in the lab worker. This process is called autoimmune disease. While immunologists believe the probability of such a scenario is limited, their models cannot completely reject it.[41]

Whoever we choose to side with in this debate, however, one fact does stand out. The special irony of the regulatory history of DNA technology is that the serious and responsible concerns of the scientific community that were registered at Asimolar in 1975 have since rebounded in such a way that there is now a special aversion and stiff-necked resistance to the imposition of any nonvoluntary regulations on DNA research. Whether this aversion would be so unchallenged and widespread if the scientific community had not been forced into a legislative corner in the late seventies is a question about which we can only speculate. It does seem likely, however, that one important legacy of the experience of having to organize and implement an intensive lobbying effort in Washington has been the growth of a pointed reluctance among scientists involved in the DNA field to openly call into question the safety of these procedures. Such a reaction is not at all surprising, and the prevalance of such an attitude has been noted by a variety of commentators.[42] What is especially troubling about such a backlash, however, is the suggestion that certain institutional sanctions such as tenure and the availability of research funding are being used as an added incentive to discourage dissenting and specifically junior faculty from making waves. Whether we accept these allegations or not, it is notable that the current laissez-faire attitude toward DNA research clearly distinguishes the American research effort from the regulatory climate of other countries. Gene-splicing technology in Britain, for example, is constrained by a stringent set of regulations that require "medical monitoring, long-term epidemological studies, prior review of recombinant DNA experiments and periodic inspection of research facilities as well as containment levels that tend to be higher than those in the United States."[43] The Japanese, now technological entrepreneurs of the first order, have nevertheless implemented a strict set of national policies which are patterned after the original Asimolar-inspired National Institute of Health (NIH) guidelines.[44] The economic projection we noted at the outset, the great interest of private, for-profit firms, and the popularity of such U.S. firms as Genetech in our stock market may have something to do with this international difference.

When this discrepancy is brought to the attention of the American research community, their reaction has characteristically been to suggest that regulatory developments are unnecessarily restrictive and likely to hinder the advance of scientific knowledge. Noting that the Japanese have shackled themselves with stringent safety regulations, Dr. Peter Farley of the Cetus firm argues that as a result the Japanese have come to be less of a competitive challenge in the field. In 1980 Dr. Farley, estimating that the Japanese were three to five years behind the Americans, suggested that Japan:

> . . . on the other hand, is considerably less of a threat. Overall the Japanese scientists [sic] are absolutely superb microbiologists. However, they are unbelievably handicapped now because Japan has recently adopted the original very stringent NIH guidelines. This short-sighted action on the part of the Japanese regulatory agencies was received with amazement in our shop.[45]

But as the authors of the article quoted above quite astutely point out, Dr. Farley's concern with the potentially stifling effects of controls upon his shop were contradicted when, a year later, he expressed concern that the Japanese had made considerable headway in the field and were at that point only a year behind the Americans.

In his emphasis on the competitiveness of the research, Dr. Farley gives voice to what may well be the most important and perhaps the most disturbing change in the direction of the DNA field since its inception. The rush to develop the field may have less to do with the intellectual excitement of the subject matter than with the incredible economic incentives that are now apparent. It appears that research in the field is more and more coming to be perceived as an economic competition rather than a scientific one, and some universities are making the most of the economic potential.

Though I have not spent any time in either setting, it seems quite possible that the attention to potential risks is considerably higher in the university research laboratory than in the commercial lab. University technicians work under less commercial pressure, are probably better trained, work on a variety of experiments, giving them more generalized knowledge, and are closely supervised by graduate students and professors. University labs no doubt vary in these respects, but the careers of the graduate students (who do most of the work, and who will become Ph.D. researchers shortly) and the researchers (generally, but not always, professors), do not depend upon rapid, high-volume production or finding something that simply works. Their efforts must be carefully documented, repeated, written up for professional publication, and reviewed by their peers. The pressures to be "first" are great, but the system emphasizes care, documentation and, perhaps above all, scientific understanding or comprehension. I believe the risks of an accident are still considerable, but not as great as in a commercial lab.

In the commercial lab, economic pressures would appear to emphasize repeated trial and error to find results that work, without the probing inquiry that would lead to caution and reflection. Procedures and results are not as likely to be subjected to peer review, where errors can be detected; and there are pressures to move to commercial production as soon as possible, with an emphasis on production short-cuts and cost-cutting procedures. There are fewer buffers in this environment. There is no reason to believe that the commercial firms, under tremendous financial pressures to be the first to market the new organisms, will be any less casual or sloppy than the high-technology aerospace firms that left metal chips in spacesuits, caps off pipes, and contaminated the drinking water of the astronauts. It was easy to be aware of the implications of these failures. It will be extremely difficult to recognize the implications of failures in laboratory containment, or, once the products are marketed, in the more fearsome and unpredictable matter of unexpected interactions with a very complex and poorly understood natural environment.

Nicholas Wade argued the point in terms that will be familiar to us: "We can't even predict the behavior of totally man-made systems such as nuclear power plants, say, much less can we predict the behavior of very much more complicated systems such as *E. coli,* of which we only know about half the working parts."[46]

This comparison would be rejected by the biological community; in my experience they and chemical and aerospace engineers look on engineers in the nuclear power industry as some form of bungling drop-outs from a truly scientific world. But we have seen that complex systems go beyond the capacity of engineers and designers in all areas. A measure of humility, such as attended the Asilomar conference, would seem to be in order. In contrast, researchers in both the university and commercial laboratories seem to feel that we know all we need to know about the risks of the technology. In a personal communication, Sheldon Krimsky suggests that what has emerged is a type of orthodoxy that holds that "genetic materials will either do what they are supposed to do when they are displaced, or do nothing at all." From the vantage point of the systems examined in this book, this assertion of confidence seems quite unrealistic. In our rush for scientific fame or private profit we may be preparing the ultimate accident; indeed, it may already have occurred.

CHAPTER 9

Living with High-Risk Systems

A crucial question may have been at the back of your mind as you have read this book: What is to be done? After having looked at all these systems, what do I propose as a solution? I have a most modest proposal, but even though modest and, I think, realistic, it is not likely to be followed. I propose using our analysis to partition the high-risk systems into three categories. The first would be systems that are hopeless and should be abandoned because the inevitable risks outweigh any reasonable benefits (nuclear weapons and nuclear power); the second, systems that we are either unlikely to be able to do without but which could be made less risky with considerable effort (some marine transport), or where the expected benefits are so substantial that some risks should be run, but not as many as we are now running (DNA research and production). Finally, the third group includes those systems which, while hardly self-correcting in all respects, are self-correcting to some degree and could be further improved with quite modest efforts (chemical plants, airliners and air traffic control, and a number of systems we have not examined carefully but should mention here, such as mining, fossil fuel power plants, highway and automobile safety). The basis of these recommendations rests not only on the system accident potential for catastrophic accidents, but also on the potential for component failure accidents. I think the recommendations are consistent with public opinions and public values.

But though we have come a long way from the broken coffee pot that started this book, before we can conclude with recommendations we must still confront three substantial objections to any recommendations that involve abandoning systems or making drastic and costly modifications. (1) My recommendations must be judged wrong if the science of risk assessment as currently practiced is correct. Current risk assessment theory suggests that what I worry about most (nuclear power and weapons) has done almost no harm to people, while what I would leave to minor corrections (such as fossil fuel plants, auto safety, and mining) has done a great deal of harm.[1] Risk assessment is a new science, but it has clearly occupied the high ground in government and many intellectual circles and will expand in the decades to come. Thus, this science deserves scrutiny.

(2) My recommendations could also be wrong if it can be shown that they are contrary to public opinions and values, or, if they are not contrary, that those public opinions and values are ill informed and should be corrected, rather than respected. It turns out that there is a fair bit of quite interesting work in cognitive psychology suggesting that the public is ill informed, and ill equipped, because of the way it reasons, to properly make important decisions about very complex matters. I think this work is in error, and will briefly outline a critique of it, though a full critique would require more detail, space, and arcane discussion than we can afford here.

(3) A third objection to my recommendations is more basic to the theory of the book. It says there is a way to run these systems safely; it simply requires authoritarian, rigidly disciplined, error-free organizations, such as, for example, the naval nuclear submarines appear to have. There is an organizational solution, this view argues; it is just that we have not been willing to put such organization in place. We will have quite a bit to say about this objection, since our whole book has been, in a sense, an organizational analysis.

We have four tasks, then: to examine the new field of risk assessment, since it counsels taking risks that I think are unacceptable and improperly evaluated; to examine the field of decision making, since it argues that the public is poorly equipped to play a role in decisions on risk; to examine organizational dilemmas inherent in high-risk systems; and, finally, to show how the analysis of these three problems, plus our analysis of system accidents and system characteristics, would lead to some modest recommendations for reducing the risks we have been told we must run. Above all, I will argue, sensible living with risky systems means keeping the controversies alive, listening to the public, and recognizing the essentially political nature of risk assessment. Ultimately, the issue is not risk, but power; the power to impose risks on the many for the benefit of the few.

Risk Assessment

Not surprisingly, the appearance of so many catastrophic man-made processes (the sexist pronoun is, for once, fitting) has occasioned public concern and, in turn, a response from the vendors of these processes and a number of social scientists. A whole new field of inquiry has resulted—risk-benefit analysis, or risk assessment. While not as dangerous as the systems it analyzes, risk assessment carries its own risks, and so we will look at both it and at the "ill-informed" public concern. As we have seen through our running commentary on the subject of "operator error," the realities are not always as the experts see them.

The activity of risk assessment is not new. People with power have always commissioned risk assessments. No important decision is likely to be made without some crude calculation of the probable benefits and the probable costs. Shamans, priests, court advisors, astrologers, lawyers, and so on have been the handmaidens of rulers and property owners throughout human history. But as power has expanded and become more centralized, so have the consequences of decisions grown, making assessment even more important. As the risks increasingly came from technological activities, scientists and engineers replaced the shamans as advisors.

Many of our potentially catastrophic systems today are not new, and already have been subject to a primitive form of risk assessment. Mining, chemical, and munitions disasters have been with us for two centuries; the failure of bridges and ships goes back to antiquity; even railroad and aircraft disasters are not all that ancient as the pace of change mounts geometrically. But third- and fourth-party victims were not present in catastrophic numbers for the older high-risk systems. Odysseus' vessel neither polluted the Mediterranean shoreline nor could destroy much of

Texas City; the World War II bombers could not crash into a building holding nuclear weapons, as happened at an unidentified overseas base in 1956;[2] chemical plants were not as large, as close to communities, or processing such explosive and toxic chemicals; airliners were not as big, numerous, or proximate to such large communities; and it is only recently that the risk of radiation from a nuclear plant accident has been visited upon almost every densely populated section of our country. The older systems now have more catastrophic potential because they are bigger and closer to us, and we have new systems that are inherently even worse. With nuclear power, nuclear weapons, and arguably, recombinant DNA, we have entirely new systems with catastrophic potential for third- and fourth-party victims—the innocent bystanders and future generations. There are few boundaries to these catastrophes, in space or in future time.

When societies confront a new or explosively growing evil, the number of risk assessors probably grows—whether they are shamans or scientists. I do not think it is an exaggeration to say that their function is not only to inform and advise the masters of these systems about the risks and benefits, but also, should the risk be taken, to legitimate it and reassure the subjects. With the increase in risk and public concern today, a new field of risk assessment has grown up, giving advice and (usually) legitimating the decisions of elites in the private and public sectors. At the behest of Congress, regulatory agencies have appeared in large numbers, and another function of risk assessors is to second-guess these agencies' awkward attempts to do a very difficult job. Risk assessors, interestingly enough, usually call for less regulation and are severe in their criticism of the agencies.[3]

The professionals in this field are generally engineers, scientists, and social scientists; they are based in universities, research organizations, government regulatory agencies, military establishments, and industry trade groups. Private, profit-making research or consulting groups, such as management consulting firms, undertake this profitable business for government and industry. Trade associations such as the Electric Power Research Institute conduct or sponsor risk-assessment studies. General Motors recently sponsored a conference on risk, and published the proceedings under the title, "How Safe Is Safe Enough."[4] The leading experts in the field were there, and a primary concern was not the risks of military or industrial activity, but the risks of regulation.

In addition to these groups, the large government and private foundations have sponsored risk-assessment studies. The National Science Foundation established programs and even a division devoted to the topic, funding the work of management consulting firms, the Brookings Institute in Washington D.C., the Rand Corporation, and university research teams. The National Academy of Sciences (a semi-governmental body) established a committee for risk assessment some years ago headed by a prestigious authority, Howard Raiffa. The Russell Sage Foundation, arguably one of the most cautious and conservative of the large private social science foundations, recently declared this a high priority for its funding. Finally, most of the large universities now have study centers or research programs dealing with risk assessment, drawing their funding from government and industry grants, and professional journals exist to market the research results. The need is great, and the response has appeared.

This is a very sophisticated field. Mathematical models predominate; extensive research is conducted; and the esoteric matters of Bayesian probabilities, ALARA principles (as low as reasonably achievable), "discounted future probabilities," and so on are debated in courtrooms as well as academic conferences. Some of the best scientific and social science minds are at work on the problem of "how safe is safe enough."

Yet it is a narrow field, cramped by the monetarization of social good. Everything can be bought; if it cannot be bought it does not enter the sophisticated calculations. A life is worth roughly $300,000, one study concluded;[5] less if you are over sixty, even less if you are otherwise enfeebled. After taking into account age and potential earning power, a life is a life.[6] Death by diabetes should have the equivalent impact on people as death by murder, is the implication of a study that deplores the public's unawareness that the former is a cause of many more deaths than the latter. "Unfortunately," they say, "there is evidence that peoples' perception of risks are subject to large, systematic biases. . . . Such biases may misdirect the actions of public interest groups and government agencies resulting in less than optimal control of risk."[7] This bias in reasoning is due largely, they imply, to sensationalism of the media.[8] But consider how a murder death affronts human values such as dignity, and the desire for security and predictability; the researchers themselves note it is not to be equated with a diabetes death, and public estimations of death rates reflect that, but the public is still held to be "biased." To take another case, for some economists and risk assessors (often the same people) there is no difference between the death of fifty unrelated people from many communities and the death of fifty from a community of one hundred. Social ties, family continuity, a distinctive culture, and valued human traditions are unquantified and unacknowledged. Fifty thousand highway deaths a year are equivalent to a single catastrophe with fifty thousand casualties for these experts, and they deplore the fact that the public protests nuclear plants and estimates highway deaths to be only half of what they are.[9]

The field acknowledges the difference between voluntary risks such as skiing and hang-gliding, and involuntary ones such as leaching of chemical wastes.[10] But it does not acknowledge the difference between the *imposition* of risks by profit-making firms who could reduce that risk, and the *acceptance* of risk by the public where private pleasures are involved (skiing) or some control can be exercised (driving). All are bundled up in a vague reference to market principles, as if we would not have heat and light without X number of dead miners or irradiated nuclear glow boys. The literature reflects a rational, calculative marketplace theory of cost-benefit analysis. The technical literature is fond of pointing out that we spend millions of dollars in safety devices to save one nuclear power worker, but refuse to spend $80,000 to save the driver of an automobile. (That is, the benefits from, say, an emergency core cooling system and an automatic seat belt are figured on the basis of how many lives one expects to save, and of course the cost of the ECCS and the seat belt are vastly different, as is the number of lives to be saved.)[11] It is thus irrational to spend that much money on the nuclear plant;[12] we should spend it on seat belts, highway guardrails or anti-smoking literature. It is as if there were a fixed budget category for safety, regardless of whether corporation profits or private needs are involved, and the budget, being fixed, cannot be enlarged when new risks come along.

Reading this discussion can lead one to imagine a scenario such as this: At the board meeting of a large corporation, the vice president for finance has received advice from the risk assessors. He announces that by not installing some safety devices we will kill one more worker per year. This will not affect the supply of candidates for this death, since workers will still take jobs at the plant because the labor market is depressed, and each worker decides there is a very good chance that someone else will be killed. On the benefit side, killing that worker will mean that the corporation will avoid a cost of $50 million for the safety devices. This avoids a price rise or a cut in the dividends or management bonuses. The vice president might estimate that they can avoid a one-dollar price rise on 20 million items, and avoid a cut in dividends of $30 million. By killing the worker, the public and the stockholders will obviously greatly benefit. What is a life worth? Well, he figures, 50 million is pretty high for a random, anonymous worker, so let's do it. The vice president for finance is correct; in risk analysis terms, it is a good bargain. Something similar took place at the Ford Motor Company when it decided not to buffer the fuel tank in the Pinto, and at the General Motors Company when it rejected warnings from engineers that the Corvair would flip over for the lack of a $15 stabilizing bar.[13]

Risk-benefit analysis, with its monetarization of cultural goods and values, has been succeeded by cost-benefit analysis, with its more open concern with the dollar as the ultimate solvent for all things social. Baruch Fischhoff, in a thoughtful examination of cost-benefit analysis (the article has the engaging title, "Cost-Benefit Analysis and the Art of Motorcycle Maintenance"), notes another consequence of the monetarization of social good by economists.[14] Cost-benefit analysis is "mute with regard to the distribution of wealth in society," he notes. "Therefore, a project designed solely to redistribute a society's resources would, if analyzed, be found to be all costs (those involved in the transfer) and no benefits (since total wealth remains unchanged)." Risks from risky technologies are not borne equally by the different social classes; risk assessments ignore the social class distribution of risk.

Cost-benefit analysis also relies heavily upon current market prices for evaluating costs and benefits. Yet these reflect current economic arrangements that many might question and wish to change. For example, people with low earning power can receive lower prices on their lives. The current market price for temporary nuclear workers is quite low, given a long recession; when calculated, the costs of replacing steam generator tubes reflects this.[15] This can mean that the cost of an accident is low only because the economic system places a low value on some people. Property values near a chemical plant are likely to be low because of odors, fumes, and fire and explosion risks. When an accident takes place, the damage to the environment is calculated in terms of values already depressed because of the accident potential, rather than what the land would be worth if an electronics plant were there, or a nice park.

Another consequence of the prevailing assumptions is the argument that new risks should not be any higher than existing ones we have "already accepted" (Have we really had much choice?), and the corollary that if other industries get much riskier, the safety levels of nuclear plants or chemical plants can be lowered. This argument recently influenced the decision of the Nuclear Regulatory Commission when it set safety goals for its nuclear plants. As a colleague, Kim Scheppele, points out, as society gets more dangerous, the NRC can allow plants to get more dangerous.

Another argument that we hear much of these days is that we must push ahead with risky endeavors or other companies or nations will beat us in the competitive race to the marketplace. This line of reasoning notes that our country was founded in risk and grew powerful by taking risks, and the social benefits have been enormous. We should, for example, push ahead with genetic engineering or the Japanese will beat us. But why push ahead if they can do it with more safeguards? If there are vast public benefits to come from genetic engineering, does it matter all that much if we buy these benefits from Japanese firms rather than from U.S. firms? It will certainly matter to the owners and investors in U.S. firms, because the oil companies and the pharmaceutical and chemical companies will lose private profits and dividends to stockholders. (Some jobs will be lost through such innovations, but those are not labor-intensive industries.) But if the Japanese will reduce the risk of a catastrophe by even a tiny amount with their better controls, the private profit losses and the indirect losses to our economy will be well worth it. What risks should we run to promote the private profit of a few? This cannot be factored into an economist's model. There, a dollar saved is a dollar saved, no matter who gets the dollar or who lives a riskier life to save it for them.

There are those who argue that we are losing our moral fiber because we no longer want to take risks with technologies.[16] But it is striking that those who feel we have abandoned risk in our search for security are speaking only of technological risks associated with large corporations and private profits, or aggressive military postures. The corporate and military risk-takers often turn out to be surprisingly risk-averse (to use the jargon of the field) when it comes to risky social experiments which might reduce poverty, dependency, and crime. Though the following list of proposals may strike one as fanciful, they all involve substantial risks that liberals and leftists have suggested, but the corporate and military risk-takers would not want to try because of the consequences for the class structure, their power, and their values. The proposals include guaranteed income maintenance plans (as found in Europe); truly progressive taxation; investment in poor and declining areas (U.S. banks pay one of the lowest income tax rates in the nation—about 3 or 4 percent—but refuse to risk investing in poor inner city areas); heroin maintenance programs to reduce crime; unilateral nuclear disarmament (a risky venture that could not only reduce the risk of nuclear accidents but promote economic prospects); withdrawal from Central America; and so on. The risks that made our country great were not industrial risks such as unsafe coal mines or chemical pollution, but social and political risks associated with democratic institutions, decentralized political structures, religious freedom and plurality, and universal suffrage.

Nor do the risk-assessment and risk-benefit studies distinguish between addiction and free choice in activities (the equivalent of our distinction between forced and unforced operator error). Along with highway fatalities, lung cancer from smoking is the favorite referent of the new body-counters. It is treated as a voluntary activity, like hang-gliding. But most of us who smoke today do so because we were barraged with advertisements and inducements that soon addicted us. In World War II every packet of field rations held its five cigarettes per meal, and the sale of cigarettes to the armed forces was heavily subsidized and untaxed. Airliners used to pass them out gratis, presumably to calm one's nerves after takeoff. No Hollywood hero was without them. The promotion was intense, and so were the private profits. So addicted was the whole economy that the government subsidies to tobacco farmers today far exceed all that is spent on warnings and research by the government. Young people see a large number of adults who have not been able to break the habit and all are still barraged by advertising and smoking TV stars. Ironically, the health of an important sector of our economy (tobacco growing, cigarette sales, and advertising) depends upon the illness of the victims; the costs of stopping smoking are not only individual (because of addiction) but corporate. This is not a matter of free marketplace decisions made by informed consumers, to be ridiculed and compared to these same people's "irrational" attacks on nuclear weapons, nuclear power, or Love Canals. Smoking is a government-supported program of addiction, for immense private profit. An individual's addiction to smoking should not be compared to the costs industry must be forced to incur to reduce brown lung disease or make safer Christmas toys.

We could say the same of alcoholism and other forms of drug abuse. These also are the referents of the risk assessors, cited to show that the public is incapable of making sensible choices between the cost of more airline safety and the cost of "substance abuse."[17] Liquor advertising is substantial; the willingness of physicians to prescribe tranquilizers and other drugs is well known, and their abuse is said to far outdistance the use of illegal substances. There is little free market choice at work here.

Finally, the risk-assessing field only infrequently distinguishes between those activities over which the person has some control, however illusory it may be, and those where she or he does not. Driving is a key one. We appear to accept risks more readily when we think our skill will play some part in avoiding the hazard. We fear and reject risks where we are passive recipients of harm. The plant, we feel, not unreasonably, should not blow up, the dam break, the air controller goof, the Ford executives fail to protect a gas tank from exploding; over these risks we have little control. But we are willing to take our risks with driving, skiing, and parachuting. Risk assessors treat the difference as one of voluntary or involuntary risk, but I think that misses a key point. Driving to work for many of us is about as involuntary an activity as there is, but at least we have some control over it. On the other hand, although we voluntarily fly in an airliner to a distant vacation, we have no control over the aircraft or the airways. We voluntarily attend large events at stadiums that sometimes burn or collapse, but we have no control over the architects or the construction firms, or the owners who always seem to lock the safety exits. Furthermore, "active risks," as we might call them, are generally not pursued for someone else's private profit; "passive risks" generally are.

For active risks, those that the individual performing the activity has some control over, the marketplace provides at least a rudimentary, though imperfect, way of addressing safety issues. Safer skis sell better; people stopped buying Corvairs and Pintos. Though there are exceptions, people make sensible choices when they have meaningful choices, and in time manufacturers respond. This is not nearly as true for those activities where we are passive recipients of risks in the control of organizational leaders. Though the airlines are interested in safety, and it is to their economic advantage to have safe travel or usage will decline, we still require a Federal Aviation Administration to encourage, require, and police them. For such activities as nuclear waste disposal, brown lung disease, toxic contamination of Times Beach, Missouri, or the Midland area of Michigan, or the Teton dam, we cannot count on any "market" to automatically incur the costs of more safety. These activities are beyond our control. For these, the government must step in.

In more and more areas of our life the government must step in. This is not the result of the cancerous growth of government, but rather is essential because our personal control over our environment and our activities is being steadily eroded by systems that we participate in, or are passively affected by. In some cases Congress recognizes this danger. There is a huge regulatory apparatus in place trying to control the generation of nuclear power, the NRC; but the regulation of chemicals and chemical plants is quite modest. It is even more modest in the case of mining, and almost absent in the case of DNA production. Nevertheless, one frequent refrain in the risk assessment literature is lamenting over-regulation. Economist Ron Howard of Stanford would do away with all regulation;[18] Starr and Whipple, pioneers in the field (from the pioneering trade association, the Electric Power Research Institute), rather than being concerned about the risks industry imposes needlessly, pity industry. The costs of regulation "stem from the litigation, misplaced investment, retrofit, and costly delays." And what causes these high costs? They "result from industry's inability to predict the acceptance of risk by the public.[19] Perhaps risk assessors might advise industry that it is easy to predict an aversion to such passive risks as mercury poison, dioxin, DES, asbestos, and brown lung disease.

Two dangers of "active risks" should be noted. Consumers will not always voluntarily pay for safer products, and often are not attentive to the risks even if they are well known. I assume it will always be thus. Second, active risks are attractive; we like to take some risks, if we feel we have personal control over them. This means that as the risk declines with better equipment, more people may be brought into the activity, those who now feel the risk is reduced to the level they will tolerate. The end result is that the accident level may not change with the new safety devices. For example, as better ski equipment appeared, and the slopes were better groomed and more safely designed, the ski resort industry began to advertise heavily to attract novice skiers. More novices meant more accidents, including more accidents for the advanced skiers they ran into. While the play was more safe, the risk was increased because of more inexperienced players. The safety record of any active risk activity must include the number of participants and the proportion of new, unskilled ones.

The risk assessors, then, have a narrow focus that all too frequently (but not always) conveniently supports the activities elites in the public and private sectors think we should engage in. For most, the focus is upon dollars and bodies, ignoring cultural and social criteria. The assessors do not distinguish risks taken for private profits from those taken for private pleasures or needs, though the one is imposed, the other to some degree chosen; they ignore the question of addiction, and the distinction between active risks, where one has some control, and passive risks; they argue for the importance of risk but limit their endorsement of the approved risks to the corporate and military ones, ignoring risks in social and political matters. As I indicated earlier, risk assessment is not as risky as the systems being assessed, but it has its unfortunate consequences for our society nevertheless.

One unfortunate implication of quantitative risk assessment is that the public should be excluded from discussions that affect them.[20] Few of the risk assessors call for this outright, most imply it; some state that the public must be involved, but only on the risk assessor's terms, and a few reject the implication and genuinely think the public has something to contribute (principally the Decision Research group, an important private company in Eugene, Oregon, that has done the best work in the field

I believe, though they hedge on the value of the contribution). Thus the range of opinion on this is broad. Most seem to take the middle position: bring the public in but control them. This is to be done by "closing the gap between the expert and the public" (that is, them and us); but the gap almost always is to be closed in one direction only—by bringing the public over to the experts' side through education.

Why the gap? Ignorance on the public's part is the main reason offered. "Public perceptions and reality dramatically differ," says Howard Raffia, a leading expert at Harvard, and the experts should be concerned even if public perceptions are based on "infirm and dubious facts"; the public should be "informed."[21] Or, as William Clark asserts, "Society's attitudes towards risks such as cancer and nuclear reactors are not readily distinguishable from its earlier fears of the evil eye."[22] By definition, the experts should know more than nonexperts. I am sure the gap exists, as the experts define it, but that could be remedied in time. There is another assumption made, however, and to that we will turn in the next section: Even if they are given the facts, the public is deficient in proper reasoning powers. Here the risk assessors have powerful support from the cognitive psychologists, and cite them.[23] Humans in general do not reason well (even experts can be found to make simple mistakes in probabilities and interpretation of evidence); heroic efforts would be needed to educate the general public in the skills needed to decide complex issues of risk. At the basis of this is a quarrel about forms of rationality in human affairs.

Three Rationalities

Why would the public puff away on cigarettes while voting against nuclear power or marching toward disarmament? One possible reason is emerging from some clever and striking work in the area of decision making and cognition. We do not reason well, the psychologists tell us: we minimize some dangers and maximize others, and do not calculate the odds as a statistician would recommend. Some of their data is convincing, but is of little relevance, I will argue; much of their data is probably spurious. While rationality in humans is certainly limited, or "bounded," as it is termed, it is possible that precisely when confronted with disorderly data and discordant goals, this limitation is its greatest strength.

It is convenient to think of three forms of rationality: absolute rationality, which is enjoyed primarily by economists and engineers; "bounded" or limited rationality, which a growing wing of risk assessors emphasize; and what I will call social and cultural rationality, which is what most of us live by, although without thinking that much about it. Absolute rationality we have encountered in the description of risk assessors' wherein calculations can be made about risks and benefits, clearly showing which activities we should prefer—such as nuclear power over coal-fired power plants. Even including deaths from the whole fuel cycle from mining to production, nuclear power is close to risk-free (the probabilities of a meltdown and breach of containment are minuscule), while coal power kills an estimated 10,000 people a year (through mining, transportation, and pollutants from old plants without scrubbers that remove particulates from the emissions that cause acid rain). The choice is obvious in terms of absolute rationality. Why then do we have from 20 to 40 percent of the people worried about nuclear power? The answer of some is that this is irrational thinking, and the implication—though it would be unwise to do more than imply it—is that the irrational public is incapable of participating in decisions about risk. The public is "hypercritical" about nuclear power, which is one of society's "worry beads."[24] If the public is unfamiliar with a suspected carcinogen, this "can lead to irrational concern and unnecessary alarm."[25] But if it is an irrationality, the public response is a burden the experts and the elites must bear. In their view, the social harm of such response is extensive; it includes protests, demonstrations, and a Congress that refuses to follow the experts.

Cognitive psychologists, those who study the process of thinking or cognition, have generally shared this view of absolute rationality, yet conducted hundreds of experiments that showed that people were considerably less than absolute in their rationality. Gradually, in trying to find out how people think, they began to abandon notions of ignorance and irrationality, and instead began to talk of the limits on rationality, or in Herbert Simon's terms, used in another context, bounded rationality. The limits on our ability to consistently and easily make fully rational decisions might be due in part to neurological limitation, to limits on memory and attention, to lack of education, and to lack of training in probabilities and statistics. But it also seems to be due to some very practical problems and concrete experiences in daily life. Hunches and rules of thumb and rough estimates and guesses appear to be patterned and widespread. Cognitive psychologists call these guesses "heuristics," from the word for discovery, and are now identifying several specific ones. For example, the "availability heuristic"[26] suggests that rather than examining all existing cases of some phenomenon, and then basing their judgment on all this experience, people tend to judge a situation in terms of the most readily available case, the one most easily remembered. If there has recently been an airline crash, we focus on that event and ignore all the successful flights when we think about the probability of a crash while deciding whether to take a flight or not. Or, if asked whether the letter *r* occurs more frequently as the first or third letter in thousands of common English words, we tend to say "the first" because words are more "available" to us as we recall more words beginning with *r* than containing *r* as the third letter.

It is universally granted that heuristics are useful, time-saving devices, even if they sometimes or even often get us into trouble. While some cognitive psychologists are busily giving names to rules of thumb that people use (and pointing out how fallacious they are, and how they impede rational decision making), a few have begun to search further.[27] Why might these studies be useful? The inquiry is only beginning, but I think we can make some useful observations. First, heuristics prevent a paralysis of decision making; they prevent agonizing over every possible contingency that might occur. Second, they drastically cut down on the "costs of search," the time and effort to examine all possible choices and then to try to rank them precisely in terms of their costs and benefits. Third, they undergo revision, perhaps slowly, as repeated trials led to corrections of hunches and rules of thumb, and do so without expensive conscious effort. Finally, I think, they facilitate social life by giving others a good estimate of what we are likely to do, since we appear to share these heuristics widely. We may do something an expert would disagree with, but at least joint action with other nonexperts (the vast majority of people, by definition) is possible, even if that action is not the one best line of action.

Heuristics appear to work because our world is really quite loosely coupled, and has a lot of slack and buffers in it that allow for approximations rather than complete accuracy. Because our social, daily life is like this, it takes special training to perform with precision in technological worlds that are tightly coupled. These violate our common sense experience that "things will work out," or that they are not precisely and tightly coupled. Furthermore, not everything we attend to is all that important (certainly this is true of the famous test of estimating the frequency of the letter *r* as the first and as the third letter of words—much of the work subjects are required to do in the experiments is of this doubtful relevance to real life). It is possible that for decisions we realize are crucial, we set aside the convenient but possibly faulty rules of thumb and make much more careful decisions. Unfortunately, the psychological literature is very weak on what people actually do in real life. Even the striking work on medical decision making by Eddy actually concerns the recommendations that medical text and reference books make with regard to diagnosis and treatment, rather than actual practices. The books turn out to ignore or misuse probabilities and underlying rates and thus the confident recommendations are dramatically faulty, and could have severely unfortunate medical consequences.[28] Laboratory experiments simply fail us here. They can tell us how unmotivated subjects (largely undergraduates in college psychology courses) take shortcuts to solve unrealistic and often extremely complicated problems, and while suggestive, the experiments may be misleading in ways we shall shortly consider.

One important and unintended conclusion that does come from this work is the overriding importance of the context into which the subject puts the problem. Recall our nuclear power operators, or the crew of the New Zealand DC-10 on its sightseeing trip, or the mariners interpreting ambiguous signals. The decisions made in these cases were perfectly rational; it was just that the operators were using the wrong contexts. Selecting a context ("this can happen only with a small pipe break in the secondary system") is a pre-decision act, made without reflection, almost effortlessly, as a part of a stream of experience and mental processing. We start "thinking" or making "decisions" based upon conscious, rational effort only after the context has become defined. And defining the context is a much more subtle, self-steering process, influenced by long experience with trials and errors (much as the automatic adjustments made in driving or walking on a busy street). If a situation is ambiguous, without thinking about it or deciding upon it, we sometimes pick what *seems* to be the most familiar context, and only then do we begin to consciously reason. This is what appears to happen in a great many of the psychological experiments. Without conscious thought (the kind that can be easily and fairly accurately recalled), the subject says, "This is like *x*: I will do what I usually do then." The results of these experiments strongly suggest that the context supplied by the subjects is not the context the experimenter expected them to supply. With an ill-defined context, the subject of the experiment may say, "Oh, this is like situation A in real life, and this is what I generally do," while the experimenter thinks that the subject is assuming it is like situation B in real life, and is surprised by what the subject does.

For example, take the supposedly widespread failure to take the "base rate" into account (all the past events, such as the number of flights where there were no accidents, or the proportion of single people in a community). If the problem is presented in one way, the subject will use the frequency and probabilistic reasoning to estimate a rate (since we do use them in many situations in life); if the problem is presented in another way, with some vague but misleading cues, the subject will rule out probabilistic reasoning. This would explain several well-known experiments, by Kahneman and Tversky, such as the one where subjects are to estimate the probability that "Jack" is an engineer or lawyer, where 70 percent of the population are lawyers and 30 percent are engineers. When no description of Jack is given, the subjects estimate the probability on the basis of the base rate. When a supposedly "irrelevant" and "useless" description is given, they ignore the base rate. But the subjects probably figure that it would be silly for the experimenters to give a description that has no meaning, so they search for the meaning, and find something. In fact, there *is* a cue in what Kahneman and Tversky assume is a "useless" description. Jack has no interest in political or social issues, and therefore is not likely to be a lawyer, but he has many hobbies such as carpentry and mathematical puzzles, which suggests that Jack is likely to be an engineer.[29] Given these cues, the subjects may ignore the base rate because they assume the cues are meant to be used as guides, in spite of the base rate.

Finally, heuristics are akin to intuitions. Indeed, they might be considered to be regularized, checked-out intuitions. An intuition is a reason, hidden from our consciousness, for certain apparently unrelated things to be connected in a causal way. Experts might be defined as people who abjure intuitions; it is their virtue to have flushed out the hidden causal connections and subjected them to scrutiny and testing, and thus discarded them or verified them. Intuitions, then, are especially unfortunate forms of heuristics, because they are not amenable to inspection. This is why they are so fiercely held even in the face of contrary evidence; the person insists the evidence is irrelevant to their "insight."

This happens in the psychological experiments; subjects insist, even after their error is explained to them, that if a fair coin has come up heads twenty times in a row, it is slightly more likely to come up tails on the twenty-first toss. Subjects insist tails is the best bet because in the long run the proportion of heads can only be .5, or 50 percent, so this unusual run of solid heads has to end. They are wrong, the experts insist, because each toss is independent of every other toss; the coin has no memory and on each toss the odds are .5 that it will be tails, and .5 it will be heads. I must admit that I am a victim of this "gambler's fallacy" too. I reason that in 1,000 tosses there will be about 50 percent heads. Twenty-one tosses is more likely to reproduce this pattern of 50 percent heads in 1,000 tosses than, say, three or four tosses, so the odds have to be made slightly better than .5 after 20 heads, and this encourages me to choose tails.

The implication of the psychologists' work is that the public is not qualified to participate in decisions about the risks they will have to endure. The public pursues an informal, probably messy "logic" that the experts do not share. At least one cognitive psychologist has recently defended intuitions. Baruch Fischhoff wonders if our intuitive judgments, or logic, might not be utilized even if it is denigrated by the experts. "It is worth asking," he writes, "whether there is not a method in people's apparent madness. Are there not decision-making criteria overlooked by formal analysis yet essential for human welfare or psychological well-being?"[30] As we shall shortly see, there are such criteria, if we examine the work of Slovic, Fischhoff, and Lichtenstein carefully.

The difference between the absolute rationalists and the cognitive psychologists who emphasize bounded rationality can be illustrated by examining the public reaction to the accident at Three Mile Island. For the rationalists, TMI was merely the occurrence of a rare event, that would be expected to occur, say, once in three hundred years for that reactor. That it occurred a few months after start up in 1979 rather than in the year 2079 is insignificant. Estimates indicate it is likely to occur sometimes, but rarely; this was simply that rare time. For the bounded rationality group this is true, but not the point. This was an unfamiliar problem, and appropriate heuristics have not been developed for it. A significant event such as this is an indication to the public of what is possible; it is a signal that these plants can have serious troubles even though the experts say they will do so only very rarely.[31] Since some experts appeared to think it could almost *never* happen, expert predictions might reasonably be questioned by the public. If experts are wrong, then this may not have been the one time in three hundred years, but the first one of many, many times over three hundred years. (Note the bounded rationality theorist is not siding with the public on the risks of nuclear power, only saying that it is not irrational that they could come to their conclusion.)

Furthermore, the public might say, if there is even a remote chance of a catastrophe, why risk it? For the bounded rationality theorist, a quite efficient and understandable logic is at work here, even though it is technically faulty. It is an efficient logic, given the fact that experts, like everyone else, are fallible and have been proven wrong in the past. It is efficient to question them. Such logic is also efficient because it motivates the public to demand: "Remove that threat; I don't want to live with a lot of threats; you are not counting in my psychic costs; find another energy source."

Finally, bounded rationality is efficient because it avoids an extensive amount of effort. For the citizen, think of the work that could be required to decide just what the TMI accident signified. In the experts' view, the public should make the effort to answer the following questions, and if they can't, should accept the experts' answers. Did the accident fit in the technical fault tree analysis the experts constructed in WASH-1400, the Rasmussen Report (a matter of several volumes of technical writing)? How many times in the past have we come close to an accident of this type? Can we correct the system and thus learn from the accident? Was it accurately reported? Do the experts agree on what happened? Did it fit the base rate of events that led to the prediction that it was a very rare event? And so on. The experts do not have answers to some of these questions, so the public, even were they to devote some months of study to the problem, could not be assured that the answer could be known.

We should note that the public is far from hysterical about nuclear power. Even after TMI from 60 to 70 percent of the public indicate more nuclear plants should be built.[32] But this does not mean that those favoring nuclear power make this judgment on the basis of a rational calculation using extensive knowledge any more than does the small minority—from 15 to 20 percent—who favor shutting down all plants. The lay proponents of nuclear power may not be using the correct decision procedures recommended by the experts any more than the lay opponents. But note that even the more enlightened of the bounded rationality proponents, while defending the efficiency of the public's reasoning and understanding the "natural" bases of its errors, might still conclude that public opposition to nuclear power is wrong. The public's fears must be treated with respect, and a way found to bring them into policy considerations, the bounded rationalists would say; but the gap is still to be closed by bringing the public over to the experts side.

Social Rationality

The third view of rationality, social and cultural rationality (or *social rationality* for short), departs from the absolute rationality of the risk assessors and economists even more significantly than bounded rationality does. It recognizes the cognitive limits on rational choice, but holds that such limits are less consequential in accounting for poor choices than cognitive psychologists believe and are, in fact, quite beneficial in other respects. Our cognitive limits may make us human in ways we treasure.

There are at least two general reasons why we might be thankful for our limited cognitive abilities. People vary in their cognitive abilities in absolute terms, but they also seem to vary with respect to different thinking abilities for different tasks. You and I may be equally intelligent, when measured over a number of areas, but you are good at counting while I (as I tell my quantitative colleagues) don't count. Yet I have learned how to visualize, or model, things in three-dimensional space, or perhaps have an innate capacity for it. Because of my limitation in counting, I need you, and vice versa. Our limitations bring about social bonding. Bonding by diversity in skills (which is related to limitations in cognition, incidentally) is more stable and perhaps more satisfying than bonding by addition of equal talents. That is, the standard illustration of two people moving a rock that neither could move alone as the basis for social life is a very minimal one; any partner would do, and once the rock is moved, we can part. But bonding because sometimes we need to count and sometimes we need to visualize, so we had better have each other around when these tasks appear, is a strong basis for social life. If everyone were equally rational, we would not need economists. Since we are not, we need both economists who try to see where rational, quantitative solutions will work and sociologists who try to see how social bonding can be utilized and maximized.

A second cheer for our limitations stems from your propensity, for example, to see all problems as one of measurement and counting, and my propensity to see all problems as one of social interactions. If we have a common problem and it seems to have a lot of numbers, rates, proportions, and so on in it, you are likely to move quickly to a mathematical solution. Your "heuristics" are better than mine if numbers are included. But because of your expertise, you are very likely to end up deciding that the problem should be seen in a manner that allows a quantitative analysis. Your "framing" of the problem prejudges the problem and prejudices the answer. So does mine. For you, the choice of nuclear or coal can be measured by toting up the deaths per megawatt of power produced to date by each activity. The risks of DNA research can be measured by seeing how many experiments have gone on without any accidents. But I might define the power generation problem in terms of potential deaths in a rare but conceivable catastrophe, the fact that the deaths would involve related people (communities), and potential contamination of large land areas for generations to come.

A working definition of an expert is a person who can solve a problem faster or better than others, but who runs a higher risk than others of posing the wrong problem. By virtue of his or her expert methods, the problem is redefined to suit the methods. Because you can count, and because we have data on deaths, the choice is defined as a problem of toting up known figures. Because I look for social relations, symbolic values, and human progeny, I define the problem as one of potential consequences, not observed ones. Your concern with power lawnmowers or automobiles, where we have good accident statistics, may save more people from injury and death, than my concern with nuclear war, which happened only once and has a very low chance of happening again. We both have bounded rationalities, but our world is immeasurably enriched because our limits are not identical, that is, because we emphasize different skills or cognitions. (The enrichment will be reduced if each of us refuses to recognize the advantages of the alternative view.) The rationalist would have all be completely rational; that does not seem possible. The bounded rationality theorist would point to the unfortunate limitations on most or even all people's cognitive ability. The social rationalist would say the limits are far from unfortunate; they necessitate interdependence, and differences promote new perspectives and solutions that no one person is likely to have. Two cheers, then, for bounded rationality; it promotes the human qualities of interdependence, and it allows legitimately different values to come into play and conflict. The third cheer, not relevant here, is that the limits on every human make domination of the many by the few more difficult.

For those holding the third perspective, emphasizing social rationality, the fears of the citizens of Middletown, Pennsylvania, are a part of the cost of nuclear power. A technology that raises even unreasonable, mistaken fears is to be avoided because unreasonable fears are nevertheless real fears. A technology that produces confusion, deception, uncertainty, and incomprehensible events (as the crisis over the several days did) is to be avoided. The technology affects social bonding and social interaction, and individual psyches and peace of mind. A worker's death is not the only measure of dread; the absence of death is not the only criterion of social benefit.

Thus we can think in terms of three types of rationality: economic or absolute rationality, which requires narrow, quantitative and precise goals; bounded rationality, which emphasizes the limits on our thinking capacities and our inability to often achieve or even seek absolute rationality; and social and cultural rationality, which emphasizes diversity and social bonding. Cost-benefit analysis emphasizes the first; risk assessors are increasingly moving from the first to the second; our discussion has emphasized the last. Next we will try to show that the public also emphasizes social rationality. The public is uninformed in many respects, and certainly can make errors in reasoning, but for matters of catastrophic risk these errors seem less disabling than the alternative of neglecting the rationality embedded in social and cultural values. The bounded rationality viewpoint has had a tremendous impact upon organizational theory. It was tentatively set forth in an influential book by James March and Herbert Simon in 1958, and then developed by March and his colleagues and students and now challenges all mainstream organizational theory.[33] This theory holds that organizational problems and solutions are tossed haphazardly into a "can" and people pick and choose as opportunities arise and move off as interest flags or new cans appear. Problems get intertwined, solutions are looking for problems to give them life, and participation is unpredictable. I think that the notion of social rationality, sketched above, is an extension of garbage can theory, indicating some reasons why the apparent inefficiency of organizations, when judged from the rational model, may prove to be quite efficient from a social rationality point of view.

The Discovery of Dread

Actually, some careful public opinion polling by Decision Research and members of a Clark University group supports the social rationality view.[34] The researchers were exploring the basis of the presumably irrational view of the public about some technologies, such as nuclear power, and compared the views of experts in various fields with the views of some members of the public, in this case, college students, members of a local business and professional association, and members of the League of Women Voters.

The experts and the lay members of the public agreed on the riskiness of several of thirty activities. Both groups rated as highly risky motor vehicles, handguns, smoking, drinking, and motorcycles. Rated as low in risk were vaccinations, power mowers, food coloring, and home appliances. But the experts and the public disagreed on others, especially nuclear power. Where 1 equals most risky and 30 the least, nuclear power was ranked as 1 by both students and League of Women Voters members. The business and professional club members ranked it 8, but the experts ranked it a very low 20 out of the 30 activities. Why the discrepancies?

The researchers had noted in other studies that the public's estimates of fatalities were greatly "biased," in part by sensational newspaper coverage and the ease of remembering recent events (the "availability heuristic"), so they checked on this. They asked the groups about the number of people likely to die next year in the United States from each of the activities, if it were a normal year as deaths go. (If you think this might be a difficult exercise you are correct; what do you think the figure for scuba diving might be, for example, or power mowers?) The experts, being experts, gave figures closely resembling the actual figures for past years. The lay people's estimates were not very accurate; they might greatly overestimate the number of, say, power mower deaths. But more important, even though they might think that many people were killed this way, when asked to judge the riskiness of the activity by ranking the 30 activities, they might judge it as low in risk. Another activity that they felt (correctly or incorrectly) killed fewer people might be judged a more risky activity. Here we would seem to have more evidence of the inability of the public to judge risk: the public's estimates of risk do not conform with their own (often inaccurate) estimates of harm.

Fortunately, interested in this discrepancy, the Oregon researchers had included another question: What if it were a particularly disastrous year? For most of the activities the estimates of deaths did not change much, for some it changed substantially, and for nuclear power the estimate of fatalities in a bad year shot way up for the lay persons. Thus, disaster potential apparently explained the discrepancy between the perceived risk and the actual annual fatalities for nuclear power and a few other hazards.

That made sense. The public judged some risks on the possibility of a disaster, not the historical record; that would explain why the public would worry about nuclear power regardless of the number of people killed on the highways. But there were still many discrepancies between the public and the experts. Probing further, the researchers then asked the respondents to rate each of the 30 activities on the following dimensions: the degree to which the activity's risks were voluntary, controllable, known to science, known to those exposed, familiar, dreaded, certain to be fatal, catastrophic, and immediately manifested. Now the study began to pay off. Here the difference between the experts and the public all but disappeared: all of the groups gave similar ratings to each of the activities on each of the dimensions. Most strikingly, nuclear power scored at or near the extreme on all of the undesirable characteristics, for both experts and the public. "Its risks were seen as involuntary, delayed, unknown, uncontrollable, unfamiliar, catastrophic, dreaded and fatal."[35]

Note that the experts agreed with the lay people in this characterization of nuclear power, but in the same questionnaire still ranked it as only 20 in riskiness out of the 30 activities, while the three lay groups gave it ranks of 1, 1, and 8. Dread and the unknown, uncontrollable aspects were recognized by the experts, but not thought relevant in judging riskiness. But not so for the public. In fact, for the lay groups, one could predict almost exactly their assessment of risk, based upon their assessment of how much dread was involved in the activity and the likelihood of a mishap being fatal. The ratings of dread and lethality also closely predicted their estimates of the number of fatalities that could be expected in a bad year. But this was not true of the experts. The degree to which a risk was a "dread risk" and likely to be fatal did not influence their judgment of the overall risk. Apparently, to the experts a death is a death, whether from scuba diving or irradiation.

The researchers then conducted a larger study with ninety hazards and eighteen instead of nine risk characteristics. The results were consistent with the earlier study, but more elaborate. They could be represented by three "factors" (clusters of interrelated judgments, where the three clusters are fairly independent of each other). The most important factor, which they labeled "dread risk," was associated with:

- lack of control over the activity;
- fatal consequences if there were a mishap of some sort;
- high catastrophic potential;
- reactions of dread;
- inequitable distribution of risks and benefits (including the transfer of risks to future generations), and
- the belief that risks are increasing and not easily reducible.

High on this "dread risk" factor were nuclear weapons, nuclear power, warfare, nerve gas, terrorism, crime, and "national defense." Note that nuclear weapons, nuclear power, and military activities in general are systems we classified as complex and tightly coupled—cell 2 in our I/C chart (see Figure 9.1).

It is striking that there is a parallel between the "dread risk" factor and interactively complex and tightly coupled systems. The correlation is not perfect, but still, two independent and totally different systems of classification converge, one based upon a theory of system characteristics that is independent of catastrophic potential, and another that includes catastrophic potential but much else besides, such as lack of perceived control, inequitable distribution of risks, and the belief that risks are increasing and not easily reducible. Clearly, the public's classification scheme is broader than ours in this book; it combines an intuition about systems with a perception of social characteristics and consequences that go beyond catastrophic consequences.

Factor 2, labeled "unknown risk," included risks that are

- unknown,
- unobservable,
- new, and
- delayed in their manifestation.

FIGURE 9.1
Interaction/Coupling Chart

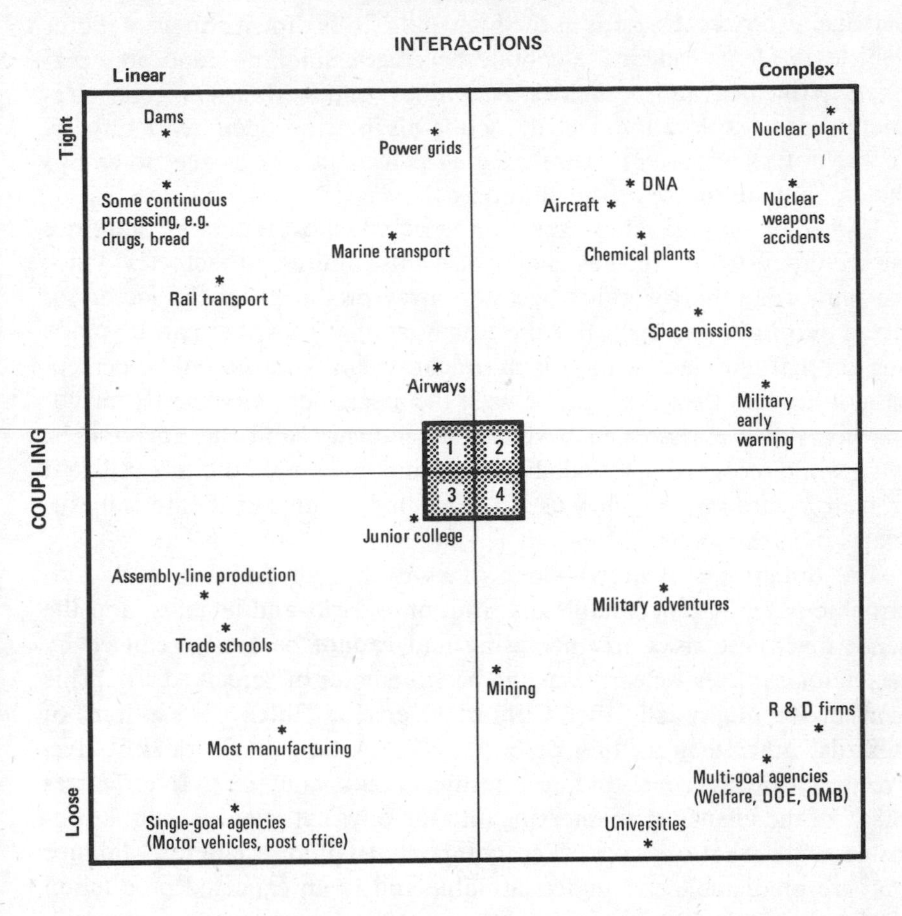

High on this factor were solar electric power, DNA research, earth orbiting satellites, space exploration, food irradiation, lasers, and nuclear power. Nuclear power is both high on dread and fairly high on the unknown dimension.

The factor of unknown risk is not conceptually related to our I/C chart as well as the dread factor. Some are processes rather than systems (solar cells, food irradiation, and lasers) and cannot be placed on our chart. There is some association of "unknown risk" with interactive complexity and tight coupling, but not for all the activities. DNA research, earth orbiting satellites and space exploration, and nuclear power are systems that we placed in the high complexity and tight coupling cell and the respondents associated with unknown risks.

Less information is given about the third cluster, "societal and personal exposure." This involved the number of people exposed and the rater's personal exposure. Hazards at the high end of this dimension were motor vehicle accidents, caffeine, alcoholic beverages, smoking, food preservatives, herbicides, and pesticides. At the low end were lasers, solar electricity, space exploration, laetrile, scuba diving, and open heart surgery. It was not as important in influencing perceived risk as the other two factors, but still made a contribution.

Unfortunately, the rated risks do not include most industrial activities we are interested in, such as chemical plants, mining, or factories. Thus, we cannot lay the I/C chart over the array produced by the factors of dread and unknown risk. But the rough comparisons that can be made suggest that our classification is more consistent with the public perception of hazards than it would be with the assessment by experts, including many risk assessors in government, industry, and the universities, who, while recognizing dreadedness and unknown risk, did not utilize it in their hazard ratings. They used simple body counts or theoretical estimates of such counts.

The dimension of dread—lack of control, high fatalities and catastrophic potential, inequitable distribution of risks and benefits, and the sense that these risks are increasing and cannot be easily reduced by technological fixes—clearly was the best predictor of perceived risk. This is what we might call, after Clifford Geertz, a "thick description" of hazards rather than a "thin description."[36] A thin one is quantitative, precise, logically consistent, economical, and value-free. It embraces many of the virtues of engineering and the physical sciences, and is consistent with what we have called component failure accidents—failures that are predictable and understandable and in an expected production sequence. A thick description recognizes subjective dimensions and cultural values and, in this example, shows a skepticism about human-made systems and institutions, and emphasizes social bonding and the tentative, ambiguous nature of experience. A thick description reflects the nature of system accidents, where unanticipated, unrecognizable interactions of failures occur, and the system does not allow for recovery.

Running Risks, or Trials Without Errors

For most of the ills that come with advanced industrialization, an increased measure of surveillance, penalties, and prohibitions would suffice to limit them. Smoking can be taxed more heavily, the subsidies to the tobacco industry eliminated, and smoking in all public places and places of work forbidden. Mine safety can be improved; smokestack scrubbers used; fluid sandbed furnaces required (they are more efficient and drastically reduce pollutants). Toxic chemicals can be much more closely regulated and illegal disposal eliminated (though organized crime might have to be eliminated to do so, if newspaper stories of their involvement in illegal disposal are correct).[37] Automobiles and highways could be made safer, and so on. We have made some of these improvements already, and could do more—if elites were willing to run the risk of alienating the offending industries. (One enormous risk which the industrialized nations may be facing is not considered in this book on normal accidents; eliminating this ill would require much more drastic measures than any of the above: This is the problem of carbon dioxide produced from deforestation primarily, but also from burning fossil fuels such as coal, oil, and wood. This threatens to create a greenhouse effect, warming the temperature of the planet, melting the ice caps, and probably causing an incredible number of other changes, most of them disastrous. If it is significant—the experts do not agree—we may have a few decades to handle this; but it may be too late. It is one of the strongest cards the nuclear addicts can play, though the enormity of the problem, by some accounts, would dwarf the capacities of nuclear industry. We would have to divert our energy and natural resources from much of industry and use it to build nuclear plants for the next generation to meet some estimates. Battalions of scientists, engineers, and operators would have to be recruited and trained, and so on.)

If the problem is not that great, it is possible that a massive and nearly impossible shift to conservation and solar power could make the difference, but recall that solar cells and collectors also require significant amounts of energy to build and distribute, though hardly as much as a thousand nuclear plants in the U.S. alone. Producing the safe systems cannot be done without further pollution, for several decades, by the major culprits, the industrialized nations. The international planning and cooperation required would be way beyond any precedents.

But it is fair to ask whether we have progressed enough as a species to handle the more immediate, short-term problems of DNA, chemical plants, nuclear plants, and nuclear weapons. Recall the major thesis of this book: systems that transform potentially explosive or toxic raw materials or that exist in hostile environments appear to require designs that entail a great many interactions which are not visible and in expected production sequence. Since nothing is perfect—neither designs, equipment, operating procedures, operators, materials, and supplies, nor the environment—there will be failures. If the complex interactions defeat designed-in safety devices or go around them, there will be failures that are unexpected and incomprehensible. If the system is also tightly coupled, leaving little time for recovery from failure, little slack in resources or fortuitous safety devices, then the failure cannot be limited to parts or units, but will bring down subsystems or systems. These accidents then are caused initially by component failures, but become accidents rather than incidents because of the nature of the system itself; they are system accidents, and are inevitable, or "normal" for these systems.

Much can be done to make these systems somewhat safer, but accidents cannot be entirely avoided. Quality control, operator training, design experience, and environmental controls will help, but will not be sufficient. These are benign steps to reduce the frequency of system accidents. But there is one other solution that may not be so benign: instituting highly centralized, authoritarian organizational structures.

The discussion we had of thinking, or cognition, is particularly relevant to the inescapable fact that our risky enterprises are organizational enterprises. Just as some hope our risk analysis can be made more rational and scientific, so do some envision more tightly controlled, authoritarian organizations to run our risky enterprises. In their view, we simply have to eliminate "operator error." We have been dealing with organizations throughout the book. It is now time to confront explicitly this context for catastrophe. What do we know about organizations and does what we know reassure us that risks can be managed? We know quite a bit, intuitively, by simply living in the world. We also learn a lot by reading books on almost any important topic, since most important topics have organizations at least in the background. Finally, those of us who are familiar with decades of research on organizations may know an additional bit more that is useful. The topic of organizations is quite relevant to high-risk systems, at least as important as the possibilities of technological fixes.

This book started with a commonplace system accident without catastrophic potential. Better organization of our affairs might have made a difference in the coffee pot and job interview story, but not much; better luck would have been more to the point, so that one of the several safety devices was not defeated. We then moved on to the nuclear industry, and I argued that we should not expect nuclear power to be above the industry average in organizational competence or honesty; the lack of both did not bring down TMI, though they contributed to the problem. Well-run plants have system accidents too. So do the much better-run (I suspect) chemical plants. Our review of some aircraft manufacturers suggested undue indifference to safety, but by and large, both the airways and the aircraft are not plagued with organizational problems. The marine industry is; it was analyzed as an error-inducing system, and all the variables used to make that analysis were organizational ones, including the suggestion that rather than give more technology to support the authoritarian decision-making system aboard ship, we should reorganize the decision-making system. The space missions dramatically support the notion that the operator must be left in the loop because the designers' loops are far from seamless, and faulty information and reality construction by managers, along with pressures for production, can lead to problems. Building dams, in our two case studies of the Teton failure and the radioactive tailings dam, suggested run-of-the-mill organizational failures similar to the nuclear power industry, and mining also reflects poor organizational behavior, especially in the attribution of operator error and the failure to take elementary protection in the case of potential explosive mixtures. With DNA we were especially critical; elementary precautions may be absent in the rush to publish and profit.

But we must come now to a more systematic limit on some organizations, a kind of Pushmepullyou out of the Doctor Dolittle stories (a beast with heads at both ends that wanted to go in both directions at once). The organizations at risk are the complexly interactive, tightly coupled ones in cell 2 of our Interactive/Coupling chart (see Figure 9.1). It should be considered with Figure 9.2 in order to highlight the dilemma. I will argue that complex but loosely coupled systems (cell 4, such as universities) are best decentralized; linear and tightly coupled systems (cell 1, such as pharmaceutical plants) are best centralized; linear and loosely coupled systems (cell 3, most manufacturing) can be either; but complex and tightly coupled systems (cell 2, including nuclear power) can be neither—the requirements for handling failures in these systems are contradictory.

Systems with interactive complexity (cells 2 and 4) will produce unexpected interactions among multiple failures. But while these are troublesome and unwanted, they need not bring about accidents—that is, dam-

FIGURE 9.2
Centralization/Decentralization of Authority Relevant to Crises

COUPLING	INTERACTIONS: *Linear*	INTERACTIONS: *Complex*
Tight	CENTRALIZATION for tight coupling. CENTRALIZATION compatible with linear interactions (expected, visible). Dams, power grids, some continuous processing, rail and marine transport. 1	2 CENTRALIZATION to cope with tight coupling (unquestioned obedience, immediate response). DECENTRALIZATION to cope with unplanned interactions of failures (careful slow search by those closest to subsystems). Demands are incompatible. Nuclear plants, weapons; DNA, chemical plants, aircraft, space missions.
Loose	3 CENTRALIZATION or DECENTRALIZATION possible. Few complex interactions; component failure accidents can be handled from above or below. Tastes of elites and tradition determine structure. Most manufacturing, trade schools, single-goal agencies (motor vehicles, post office).	4 DECENTRALIZATION for complex interactions desirable. DECENTRALIZATION for loose coupling desirable (allows people to devise indigenous substitutions and alternative paths), since system accidents possible. Mining, R&D firms, multi-goal agencies (welfare, DOE, OMB), universities.

age to a subsystem or the system as a whole. Accidents will be avoided if the system is also loosely coupled, (cell 4, universities and R&D units) because loose coupling gives time, resources, and alternative paths to cope with the disturbance and limits its impact. But in order to make use of these advantages of loose coupling, those at the point of disturbance must be free to interpret the situation and take corrective action. Since the disturbances are generally (not always) likely to be experienced first by operators (which include first-line supervisors and other on-duty personnel such as technicians and maintenance), this means the system should be decentralized. These personnel have two tasks: analyzing the situation and acting so as to prevent the propagation of errors. Unexpected and incomprehensible interactions will not allow immediate analysis of the cause of the accident, but given the slack in loosely coupled systems, this is not essential. It is enough that personnel perceive an unwanted system state (even though it is also unexpected and its causes are mysterious), and do so before it interacts with other units and subsystems. To do this they must be able to "move about," and peek and poke into the system, try out parts, reflect on past curious events, ask questions and check with others. In doing this diagnostic work ("Is something amiss? What might happen next if it is?"), personnel must have the discretion to stop their ordinary work and to cross department lines and make changes that would normally need authorization. This is a part of decentralization. This delay and experimentation is possible because of loose coupling, but required by the interactive complexity.

Once an untoward state is identified, and some notion of possible consequences formed, recovery actions must take place to prevent the failure from spreading. In a loosely coupled system there is sufficient slack, resources, alternative paths, fortuitous substitutes, and safety devices to effect recovery. But the persons best able to bring these into play are those at the point of the disturbance. Thus, where systems are both complexly interactive and loosely coupled, decentralization is efficient both for diagnosing and recovering from errors. (It is also appropriate for these systems for other reasons than dealing with failures, since these systems can benefit from unexpected interactions of nonfailure events, called synergy, flexibly cope with unstandardized "raw materials," and can utilize professionals that have been socialized through professional training and thus can be left unsupervised because they have absorbed the appropriate norms.)

At the other extreme (cell 1, linear and tightly coupled), consider a continuous processing operation with a well-established technology, standardized raw materials, and linear production system. Here tight coupling is required for efficiency, and can be tolerated because the technology is well understood and the materials well controlled. When failures occur, as they inevitably will, they will not interact in unexpected and incomprehensible ways, but in expected and visible ways. The system programs responses for these infrequent but expected failures; the responses are determined at the top or in the design, and employees at all levels are expected to carry them out without question. They must be carried out immediately and precisely because of the tight coupling present, otherwise the failure could expand from part to unit to subsystem. This does not represent a problem; employees expect it, and desire it. There is no time for rumination, no fortuitous substitutions or safety devices available to bring into play, no alternative sequences to use other than those programmed in. Repeated drills will insure fast and appropriate reactions, but the reactions are prescribed ahead of time, by the central authority. The system operates much the same way in the production mode as it does in the failure mode, so the centralization is appropriate.*

For organizations that are both linear and loosely coupled (cell 3; most manufacturing, and single goal agencies such as a state liquor authority, vehicle registration, or licensing bureau) centralization is feasible because of the linearity, but decentralization is feasible because of the loose coupling. Thus, these organizations have a choice, insofar as organizational structure affects recovery from inevitable failures. The fact that most have opted for centralized structures says a good bit about the norms of elites who design these systems, and perhaps subtle matters such as the "reproduction of the class system" (keeping people in their place). Organizational theorists generally find such organizations overcentralized and recommend various forms of decentralization for both productivity and social rationality reasons.

For the interactively complex and tightly coupled system (cell 2, including nuclear plants, nuclear weapons systems, chemical plants, space missions, and DNA) the demands are inconsistent. Because of the complexity, they are best decentralized; because of the tight coupling, they are best centralized. While some mix might be possible, and is sometimes tried (handle small duties on your own, but execute orders from on high for serious matters), this appears to be difficult for systems that are reasonably complex and tightly coupled, and perhaps impossible for those that are highly complex and tightly coupled. We saw the space missions move from a highly centralized mode in the first missions to a more decentralized one in the moon shots, and the somewhat less complex and tightly coupled space shuttle may allow still more decentralization. But I predict that the tensions between the two modes will remain, and consume a good deal of organizational energy. Chemical plants are also not extremely high on complexity and coupling, but we saw safety engineers ruminating about the problem of bringing in operators more as the process got more complex, and we saw operators overriding the centralized, automatic system to decouple parts of the plant and to manually shut down affected parts on their own. I do not know enough about this industry to say that the problem of being both centralized and decentralized is expensive and continuous, but I would expect that it is.

In the case of nuclear power I am confident that it is. I once participated in some discussions with Nuclear Regulatory Commission and

*Organizational theorists, at least since the pioneering work of Burns and Stalker, 1961 and Joan Woodward, 1965 and others in what came to be called the contingency school, have recognized that centralization is appropriate for organizations with routine tasks, and decentralization for those with nonroutine tasks. For an early statement see Perrow, 1967, and Lawrence and Lorsch, 1967. The present formulation suggests an extension of that view.

nuclear industry executives about the optimal organizational structure for nuclear power plants. We cycled endlessly through the problem of insuring rapid, unquestioning response to orders from on high (or orders in the procedures manual), and at the same time allowing discretion to operators. Regarding discretion, the operators would have the latitude to make unique diagnoses of the problem and disregard the manual, and be free of orders from remote authorities who did not have hands-on daily experience with the system. We could recognize the need for both; we could not find a way to have both. Predictably, the NRC and industry personnel, faced with the incompatibility, chose the centralized model.

The centralized design for nuclear plants recommended by most (but not all) regulatory and industry personnel goes beyond that of most manufacturing or continuous processing plants because of the catastrophic potential present. It thus has implications for society. It calls for a wartime, military model in a peacetime civilian operation. A military model reflects strict discipline, unquestioning obedience, intense socialization, and isolation from normal civilian life styles. (I am drawing upon an idealized model here; no suggestion is made that our current military organizations resemble this.) Designed to "defend a free society," it must violate that freedom. Because the military is concerned with protection from violent attacks upon our existence, it is given a latitude no other group has. Its structure is not something we generally endorse, but rather is something society tolerates because of the special problems and circumstances. Efforts to extend this model to industry in the nineteenth and early twentieth centuries failed; it was too incompatible with American social values and culture. The question arises, then, of whether we are not designing more systems that are at once both complex and tightly coupled, and that will require the extension of a military model of organization to more and more activities in society—in the name of progress, defense, competition, or whatever.

The issue came up in the discussions with industry representatives I mentioned above, but it came up far more forcefully in the inquiry into the accident at Three Mile Island.

These Ordinary Men

Immediately after the accident at Three Mile Island, President Carter appointed a prestigious commission to inquire into it, the Kemeny Commission as it came to be called. After the commission had taken testimony for some months, they met in closed sessions to draft their final report. The sessions were transcribed, and the following summary and quotes from just one meeting (September 15, 1979; the transcriptions are available in the NRC's Three Mile Island Reading Room), illustrate some key questions of this book: Do high-risk systems have to operate any differently than low-risk systems? What is the price of trying to manage both complexity and tight coupling at the same time? These questions were intensely discussed; these were immensely practical issues for a commission that would grab headlines and could influence the president and the Congress.

First, the commission had to confront the reality of how nuclear plants actually ran when they were running normally. They were surprised. Taken on tour by the utility, Metropolitan Edison, of the undamaged Unit 1 plant, they found an alarming lack of basic housekeeping and indifference to that fact. Stalagtites and stalagmites 3 feet long were growing out of leaky valves; pools of radioactive water lay about; and piles of radioactive tools, materials, and protective clothing were scattered around with slips of paper on them laconically saying "hot." Loose wiring hung from the ceilings. On another occasion one of the consultant groups working with the commission was taken on a tour of Unit 1. It reported to the commissioners at the September meeting that the engineers from the utility conducting the tour appeared not to understand the basic design of the system or the importance of such problems as loose wires hanging about.

One of the commissioners had vast industrial experience—Patrick E. Haggerty, Honorary Chairman and General Director of Texas Instruments, and executive committee member of the extremely influential Tri-Lateral Commission (which is decidedly pro-nuclear, advising the key, advanced industrial nations in the West and Japan to go nuclear). He said that there were always loose wires hanging about in a Texas Instruments plant and engineers often don't know how the whole system works. (That may be fine for TI, someone on the commission might have said, but not for a nuclear plant.) He and some other commissioners also argued that a nuclear plant can't shut down for every leaky PORV, because that creates other problems. "We all know," he said, "that the best thing you can do is to keep something running. . . ." And if there are leaky valves or plant shutdowns, the plant should not tell the local public about the problems they are having. The sociologist on the commission, Professor Cora Marrett, said "it would be ridiculous" for the utility to provide information on everything that happened that might be counter to safety, thus defending the utility's right to secrecy. The influential Washington lawyer and reputed power-broker in the Lyndon Johnson administration, Harry C. McPherson, echoed the sentiment. They should do no more than General Motors plants; there is no difference.

Commissioner Peterson, President of the National Audubon Society and former research director of duPont, was not convinced. "They've got some damn dangerous stuff out there in that building" he mused. "There is a little sign on the desk as we were in the plant," he continued, "saying right there, 'Nuclear Power Is Safe.'" Commissioner Pigford (professor of nuclear engineering at the University of California, Berkeley, and former employee of a nuclear vendor) insisted that "they thought it was safe." (One wonders why they needed the sign to remind them if that were the case.) Peterson retorted, "They knew damn well it was dangerous," and went on to say, "It's easy to work safely when you're making a suit of clothes, but not nuclear energy." Later he brought up the difference again: "There's one big helluva difference" between nuclear plants and garment factories or General Motors plants, "and the community knows that now." Commissioner Pigford stubbornly disagreed with Peterson about the difference between industries, and thus disagreed with the thesis of this book. "I don't know that; I'll examine each one," he said.

The controversy was important. I have argued all along that nuclear plants are, unfortunately, run no different from most of industry, and that we cannot expect much out of most industrial activity in the way of high reliability and safety. But they should be different because of the complexity, coupling, and catastrophic potential. The pro-industry commissioners were presented with dramatic evidence that TMI was, indeed, run like Texas Instruments and General Motors, so they argued it did not matter. I think they lost their point, though they were the majority of the commission, because McPhearson finally conceded there was a difference. But then he proposed how the difference should be handled: "There is a model of a nuclear system different from the one we have in our country, in our commercial power system," he said. "That's the naval reactor program, run with an iron fist, every decision made at the top, nobody budging down below, intense training, intense discipline on the operators." It merits our consideration, he went on, because so many of the current plants have undisciplined, untrained and, unmotivated people. "We ought seriously to consider the question of nationalization."

The discussion that followed generally approved of McPhearson's iron fist criteria, but did not like the idea of nationalization. Perhaps, in private hands, some thought, the iron fist might work. Then Commissioner Lewis, Dean of Journalism at Columbia University, broke in and finally focused the discussion squarely on risk assessment:

> I really don't believe what I hear. We are going to say in order to continue with

> nuclear power, we are up-front going to say, a certain number of lives we're willing to risk? We are going to centralize and militarize this very dangerous source of power? I mean, I don't like the Rickover thing applied to a peacetime operation. I just think in social terms it bothers me. And we're deciding to do all of these, for what? We haven't explored whether there are alternatives to getting ourselves electricity. I mean, we're really saying we're willing to risk health and safety, a serious potential accident, to centralize a source of power with a military type of group?

Acrimonious discussion followed, as one might imagine, for some time. It was not what those with ties to the industry wished to hear. Then, still frustrated, Lewis spoke again on the issue of a special kind of a system:

> You have seen the NRC and I guess the question I am asking is do you really want to trust your family's life to these people? This is the question I keep asking all through our hearings. I looked at Mr. Galena, and I looked at Mr. Denton, and I looked at Mr. Hendrie [top NRC officials active in the recovery attempt and frequently interrogated by the Commission] and the whole bunch of them; and I am saying, this is a terrible thing I am putting in the hands of these ordinary men.

But Professor Pigford would not let up. "There is no such thing as no accidents," he said. "So we have got to bite the bullet and realize that we are not going to be able to determine what is acceptable."[38]

This became the majority view. From the assertion that "plants are all the same, no special precautions are needed," they moved to a need for military management because they are special, to finally saying, "to hell with the risk assessments, let's forge ahead." The commission's report, while excoriating everyone, merely said, "Now let's all do a better job."[39]

Pigford was correct; there is no such thing as "no accidents." Normally, even space shots do not pose a catastrophic potential; but when we send up caskets of plutonium, we create one. The answer is to not send up plutonium, and we and the Soviets have stopped. But some things cannot be stopped. There have been catastrophic accidents in Admiral Rickover's nuclear naval program too, "run with an iron fist, every decision made at the top, nobody budging down below." So Professor Lewis is correct too; since there will be accidents, we should consider whether any peacetime activity is worth having if it requires that kind of a social order. Even it cannot prevent accidents. Two nuclear submarines have gone to the bottom of the ocean with all their crew.

Organizational theorists have long since given up hope of finding perfect or even exceedingly well-run organizations, even where there is no catastrophic potential. It is an enduring limitation—if it is a limitation—

of our human condition. It means that humans do not exist to give their all to organizations run by someone else, and that organizations inevitably will be run, to some degree, contrary to their interests. This is why it is not a problem of "capitalism"; socialist countries, and even the ideal communist system, cannot escape the dilemmas of cooperative, organized effort on any substantial scale and with any substantial complexity and uncertainty. At some point the cost of extracting obedience exceeds the benefits of organized activity.

What Is to Be Done

The question of what is to be done about our high-risk systems depends upon what we think the problem is. By now, you might correctly assume that I would not say it was dumb operators, but there are three other candidates I would like to dismiss fairly summarily, namely technology, capitalism, and greed, before we discuss a better candidate, "externalities." The technology argument I am concerned with here is not the one that ran through the book, arguing that elites have decided on highly risky technologies that will inevitably have system accidents. The more conventional one I wish to dismiss states simply that we are in the grip of a technological imperative that threatens to wipe out cultural values, nature, and so on. We could cite the failure of collision avoidance systems to work, and the invention of multiple warheads, in support of this. Systems are too complex, and our reach has surpassed our ability to grasp. There is a bit of this philosophy in this book, particularly in the argument for social rationality. But, first, there is no imperative inherent in the social body of society that forces technologies upon us. People—elites—*decide* that certain technological possibilities are to be financed and put into place. Second, most of our technologies do not threaten our values, nature, or our lives. One can no more be antitechnological than they can be anticulture or antinature; the hoe and the wheel and cooking meat and baking bread are technologies.

The next candidate for possible blame is the group of people making decisions—in our society, these are reputed to be capitalists, or government agents of capitalists. Critics argue that organizing an economy around private profits leads to short-run concerns and neglects the longrun consequences. We have encountered the role of production pressures (for profit) repeatedly. Furthermore, capitalists can ignore the "externalities" or social costs of their activities (pollution, large accidents, enfeeblement of the working class) because they are borne by the public as a whole or by segments of it. The social costs to each capitalist or corporation are very small, since they are spread over society, or nonexistent if they are borne only by particular groups (workers, residents of Times Beach, Missouri, or Seveso, Italy). Government does attempt to redress the balance through regulation, but since government is overwhelmingly business oriented, or made up of businesspersons (including the lawyers who serve business interests), it does not succeed. Furthermore, government requires large-scale organizations to deal with the large-scale organizations capitalism finds it needs, and these are inherently limited and inefficient. Even if the regulatory groups were made up of impoverished environmentalists, there would be substantial inefficiencies and ineffectiveness. So runs the capitalist organization account of our problem with high-risk systems.

While I believe that capitalism accounts for a great deal of what goes on in the world, once the system was established the world was changed so substantially that invoking capitalism as a cause of these specific problems is pointless in the context of this book. Socialist countries (in part because they must compete with capitalist ones, and in part because of the limits of organized activity) behave in much the same way. They pollute, ignore the long-run costs, and in at least the Soviet sphere, enfeeble workers much more than capitalist societies do (because the workers cannot fight back). Production pressures appear to be as high or higher in some socialist countries, and vary substantially in capitalist countries (high in mines and on ships, low in air transport). Capitalism, per se, then, is not a useful or meaningful explanation.

What, then, of a less ambitious explanation than capitalism, namely greed (whether due to human nature or structural conditions)—or to put it more analytically, private gain versus the public good? This is useful for highlighting differences within capitalist and within socialist societies. Some social systems have higher ceilings on private gain than others, and some kinds of activity allow the more rampant expression of private gain than others. Presumably state bureaucrats pursue private gain as assiduously in socialist countries as they do in capitalist ones, though the private benefits are considerably limited (vacation homes and elite schools for children in socialist societies, in contrast to enormous fortunes and relatively unchecked power in capitalist ones). I don't have the impression that ceilings (such as taxation or some other form of redistribution of wealth) have much of an impact upon the creation and operation of high-risk systems, though I think it is essential for a just society. But within each society we can identify activities where private gain is easily realized (chemical plants) and where it is much harder to realize (space missions). It would be hard, however, to classify our systems by only this criterion; some systems with the most catastrophic potential are government activities (dams, nuclear weapons) and systems with the least catastrophic potential may be private (air transport). Private gain, then, does not seem to be the overriding problem.

The question of what the problem really is has a lot to do with the role of "externalities." These are the social costs of an activity (pollution, injuries, anxieties) that are not reflected in the price of the activity. These social costs are often borne by those who do not even benefit from the activity, or, if they do, are unaware of the externalities. Externalities are important in the case of high-risk systems because of, for example, the costs of cleanup from toxic substances, or of rebuilding after a dam failure. The price of electricity from nuclear power plants does not reflect the very large government subsidies, nor the costs of the unsolved problem of long-term waste storage, nor even the unknown costs of dismantling reactors after their forty allotted years, if they run that long. Had all these been properly considered in the 1950s and included in the cost, this book would have not been written because no utility would have ordered a plant. The externalities of coal-fired power plants without proper scrubbers are enormous, and the externalities drift over several states and the Canadian border. We are now beginning to acknowledge this. Externalities are found in both profit-making and governmental activities. The Corps of Engineers is not a profit-making organization, but the possible externalities of dam failures are not included in the budget they request. The published figures on our weapons systems do not include money set aside for broken arrow accidents. Were externalities built into the price of the product, the consumer of electricity, defense, or motorcycles could make a better choice.

Outputs that are sold to the final consumer directly are able to hide externalities less easily than those that are indirect. Systems, whether public or private, that have identifiable and predictable victims are more likely to consider externalities than those where victims are random, anonymous, and probable rather than certain. Systems that have high-status, articulate, and resource-endowed operators are more likely to have the externalities brought to public attention (thus pressuring the system elites) than those with low-status, inarticulate, and impoverished operator groups. To protect their own interests, the operators in the first category can argue the public interest, since both operators and the public will suffer from externalities. The contrast here runs from airline pilots at the favorable extreme, through chemical plant operators, then nuclear plant operators, to miners—the weakest group.

This suggests that in designing or redesigning high-risk systems, we need to consider not only the technology but the role of a variety of groups, including types of victims, and the true long-run social and economic costs, or externalities. Private gain still has to figure heavily in the analysis, but "structural" variables are also important. Such an analysis would consider whether or not there were direct sales to the final consumer, identifiable victims, and the degree of political independency of the operators. This would be combined with probability of loss, structure of the insurance industry, and the possibilities of lawsuits (as opposed to the much more ineffective system of workman's compensation), the amount of federal presence, and the costs that are externalized. Rather than attempt this, we will conduct a much more primitive and highly impressionistic inquiry into what can be done.

Let us state the problem as follows: How risky, in terms of only the catastrophic potential, are the high-risk systems we have been considering, and how costly would be the alternative ways of producing the same outputs, if there are any alternatives? This is essentially a risk-benefit question, but it now includes a variety of concepts that are normally not even considered.

Catastrophic Potential

Our Interaction/Coupling Chart has served us well, but now is inadequate. It presents only the theoretical placement of systems in terms of their interactive complexity and coupling, and asserts that those in cell 2 are more likely to have system accidents than the others, assuming only that there will inevitably be failures in the DEPOSE components. Establishing that argument has been the major task of this book. It says that these failures are inevitable. But by itself, this argument is without practical application. It has nothing to say about the matter of risk for society.

Now we must venture into far less categorical statements and estimates. One of these estimates is of the catastrophic potential of these systems from either system or component failure accidents. This is independent of system accident potential; once the plutonium caskets are removed from space missions there seems to be almost no catastrophic potential, despite the probability of system accidents; dams do not have system accidents, but can kill 3,000 people at one blow. Furthermore, to complicate things, system accidents even in systems with catastrophic potential may be limited to subsystem failures with no serious damage to any humans, or even a total system failure (TMI) with no damage to humans. The marine transport system has system accidents, but catastrophic potential exists primarily where toxic and explosive cargoes are involved.

Estimating catastrophic potential from a large accident is difficult. I have ignored first-party victims entirely, assumed that for second-party victims (mostly passengers) that over 100 deaths would count as a catastrophe, but the following analysis would remain the same if the deaths totalled over 200. For third- and fourth-party victims the most catastrophic systems are estimated to be nuclear power plants, weapons systems, and DNA accidents; all of these could be very, very large indeed. Somewhat further behind are chemical plants (largely vapor cloud explosions and release of such toxins as chlorine gas) and marine accidents with toxic chemicals at sea or in port, or explosions in port. Both chemical plants and marine accidents would involve third- and fourth-party victims, but in the hundreds, normally, rather than the thousands or millions.

In column 1 of Table 9.1 I have very roughly ordered the systems of concern in this final chapter by their intrinsic potential for system accidents plus their catastrophic potential if such an accident occurred. This is the catastrophic potential from inherent system accident potential. Space missions have a fair propensity for system accidents, but almost no catastrophic potential, so they are at the bottom; so are dams, which have no system accident potential, and so no catastrophic potential from this form of accident. Nuclear power and DNA are high on both system accident and catastrophic accident potential. (They are not placed at the top, because there are other considerations. The numbers in the figure are just for ease of comparison; they have only relative significance, not absolute significance.)

But we have seen that the potential for a system accident can increase in a poorly-run organization. If there is poor regulation, poor quality control, or poor training, there is an increased chance of failures in the DEPOSE components, and these can make the unexpected interaction of failures more likely, because there are more failures to interact. This should be considered also, as it is in Column 2 of Table 9.1. This column constitutes not just the catastrophic potential from inherent system accident potential, but the catastrophic potential from this source plus what I take to be current rate of DEPOSE component failures that is in excess to what could normally be maintained. This is, how badly run is the facility? There will always be DEPOSE failures, but if there is no effective regulation (as in DNA research) there will be more than necessary. Column 1 is

TABLE 9.1
Catastrophic Potential (CP)

	1 *Inherent System Accident with CP*	*2* *Actual System Accident with CP*	*3* *Component Failure Accident with CP*		*4* *Net Catastrophic Potential (2+3)*	*5* *Cost of Alternatives*	
(High)							*(Low)*
10				20		Space	1
		DNA	Weap			Weap	
9		Weap/NucP		18	Weap	NucP/Mar	2
8	NucP			16	NucP		3
	DNA				DNA		
7	Weap		NucP	14			4
6	Fly	Fly	DNA	12		DNA	5
					Fly		
		Chem			Chem		
5	Chem	Mar	Mar/Fly/Chem	10	Mar	Dam	6
4	Mar			8			7
		Airw	Dam				
3	Airw			6			8
					Airw		
		Mine	Mine		Dam/Mine		
2	Mine		Airw	4		Chem/Mine	9
	Space	Space			Space		
1	Dam	Dam	Space	2		Air Transp	10
(Low)							(High)

the best we can expect if these industries make all the necessary efforts; Column 2 is what we are likely to expect, since they do not seem to be capable of making those efforts.

But accidents can occur simply from component failures. I think some systems are more likely to have catastrophic accidents from system failures than component failures (nuclear power plants), but with others (dams), it is the reverse. Column 3 presents some rough estimates of the potential for catastrophic accidents from component failures, given current experience.

We now have two sources of catastrophe: system accidents and component failure accidents. The simplest thing is to add them together. (There are problems with that—and indeed with all the columns—but we are conducting only a very crude analysis here.) This is done in Column 4, the net catastrophic potential of each system (or, in the case of marine transport, that part of the system that has catastrophic potential).

The result is not all that different from Column 1, but there are reorderings and some clumps have formed.

The first four columns in Table 9.1 indicate a number of conclusions. Dams have few or no system accidents, and even mismanagement is not likely to produce them. They do have component failure accidents, however, and catastrophic potential (the large ones and the small ones with radioactive wastes). If space missions have accidents, they are likely to be system accidents rather than component failure accidents; this is because it is a well-run program and the system has many redundancies. Space missions have little catastrophic potential, though. Mining has little system accident potential, and little catastrophic potential even from component failure accidents. Therefore, while it could be made much safer for first-party victims, its net catastrophic potential remains small. Marine transportation of toxic and explosive materials is inherently but modestly prone to system accidents; this potentiality is increased by poor management and regulation. We must add to that the moderate potential for component failure accidents, and the net result is a strong showing on risk.

The petrochemical industry is moderate on all counts; as with marine transport, there is room to improve the system, and reduce the risk. The airways (largely air traffic control) have managed to reduce coupling and complexity, and because of extensive backups and possibilities for decoupling, component failure accidents with catastrophic consequences are less frequent than system accidents. The airways system, then, is doing well. Flying, however, has substantial system accident potential, even though it is not increased by poor management. Component failure accidents are not common because of the extensive redundancy and safety-consciousness of aircraft builders and the skill of pilots. It is in the middle with regard to the net catastrophic potential, and probably always will have about that much potential because new technological innovations only seem to push the speeds higher and increase the flying in bad weather and crowded skies.

DNA, nuclear power, and nuclear weapons remain at the high risk end throughout. DNA's net potential, however, is substantially due to the lack of regulation, and lack of safeguards in production that we hypothesize exist. Genetic engineering could be made substantially safer. Note that it is not expected to have major component failure accidents, nor is the nuclear power industry. The reasons were different. With recombinant DNA, the failure of any step along the way will signal trouble and result in termination; it is the unexpected result of a nearly successful series of steps that is most likely to be dangerous. With nuclear power, though there are many component failures, there already are extensive backup devices and defense in depth, such as the possibility of scram. While DNA could be made safer, it is unlikely that nuclear power could be. I suspect that this is even more true of nuclear weapons accidents. Unfortunately, component failure accidents with nuclear weapons could be catastrophic. Unlike the military early warning system (not included in this analysis), which has some degree of loose coupling that protects it from component failures, a Titan missile can literally go off with the drop of a workman's wrench and possibly release plutonium. Though unverified, I was told of a slip in a maintenance step that sent an armed missile into Canada. As described, it was not a complicated accident at all, but rather, akin to what one might expect in industry in general. Thus, this system appears to have the ultimate catastrophic potential, with a high probability of both types of accidents and of the release of lethal substances.

Do we need these systems? Given the risks, what are the benefits? What are the costs of alternative means of getting the output of these systems? Here even more subjective estimates are made, presented in Column 5. I assume that air travel cannot be substantially reduced and high-speed ground travel is not practical over distances of roughly 300 miles. Enormous investments in rail transport would be needed to make it safe and sufficient to compete with air transport even for short distances. Yet in marine transport, there is no overwhelming need to transport deadly poisons or explosives such as LNG long distances through winter storms in unsafe and poorly designed ships operated for private profit. The world economy and the economy of individual countries would not suffer if the poisons and explosives were not transported at all, or in small quantities in super-safe containers aboard specially designed and run ships. The cost of transport might double, but that is a trivial matter compared to the cost to the ecosystem. Such reductions in risk are not feasible for the chemical and mining industries. Our economy and our lifestyle is built around these industries; while some substitutions are possible, not many are likely.

We could move housing out of the flood plains and use the plains for agricultural activities only, avoiding the need for dams. This could be done in many cases, but not all, and we have to consider the "sunk costs" already in place. Moving Los Angeles (a giant flood plain) would be impossible. And dams generate electricity as well as protect flood plains—another reason we are unlikely to give up dams.

The question of nuclear weapons cannot be sensibly discussed in a paragraph or even a chapter. I have simply followed those who call for extensive, unilateral disarmament of ballistic nuclear weapons and nuclear missiles. I do not see the Soviet threat as all the administrations have since World War II, and thus I think the benefits from abandoning strategic nuclear weapons (and tactical ones too, though that is not at issue here) would be enormous in terms of reducing international tension, improving the economy, and halting the drain of a quarter or even half of our scientific talent from the civilian economy.

The case for abandoning nuclear power strikes me as very strong. There are two pitfalls. First, the government has allowed utilities building nuclear power plants to count as profits during the building years the expected profits that will come after the plant opens. This is called Allowance for Funds Used During Construction, and shows up in the revenue side of the annual financial report. At times, this figure is as much as half or more of the profit in that year, and is paid out in the form of dividends to stockholders and bonuses to management. Were we to close down all construction and abandon these projects, not only would billions go down the drain, but some utilities might be literally and technically bankrupt. The effect upon those rich enough to invest in the stock-market, and the effect upon union and other pension funds would be extensive. The effect upon the stock market as a whole and thus our economy and the free world economy could be serious. If it occurred during a shaky time of defaulted loans to poor countries or of high unemployment, the interaction of failures could produce a subsystem or system accident in this and even other economies. Still, that might be better than a nuclear accident that contaminates a populated area of the earth.

The second problem is that of capacity and distribution. Nuclear energy produces only around 11 to 12 percent of our electrical energy. Were it to disappear, some industries and other heavy users of electricity could convert to nonelectrical energy (such as fluidized beds using coal and scrubbers), but most could not. The problem is transmission over long distances. Chicago, thc Carolinas, and parts of the Northeast would need electricity from the West, Southwest, and the northeastern Canadian provinces. Recent hydroelectric capacity in Canada and the exploitation of vast natural gas fields offshore would substantially reduce the shortage in some northeastern states, but would not eliminate it. Efficient transmission to Chicago from the Pacific Northwest is not in place, though it is technically possible. Though we have excess electrical generating capacity (one reason for the cancellation of nuclear plant construction), the excess is not always in the time and place that makes its use feasible. I am sure there are some who argue the conversion back to non-nuclear power can be done without much cost or effort, and others who deny this. I will just leave it as one of the two major problems with closing all nuclear plants. Should that possibility suddenly become attractive (if, for example, another nuclear plant has all the safety systems fail as the Salem plant did during 1983, and it is also running full-tilt and has a LOCA), I suspect we could cope with the capacity and transmission problem. We might begin planning for it now; I would expect a worse accident than TMI in ten years—one that will kill and contaminate.

Even considering those two problems, neither of which would have results as disastrous as a major release of radioactive materials, the case for shutting down all nuclear plants in the United States seems to be clear. There will be more system accidents; according to my analysis, there have to be. One or more will include a release of radioactive substances to the environment in quantities sufficient to kill many people, irradiate others, and poison some acres of land. There is no organizational structure that we would or should tolerate that could prevent it. None of our existing reactors has a design capable of preventing system accidents. Perhaps a safe one will be discovered—loosely coupled and linear—but I am doubtful. Were nuclear energy all that stood between us and starvation, that could be another matter. It is not.

I place DNA near the middle because, on the one hand, we have done well enough without it, but on the other, the potential benefits are said to be enormous. I will give it a very considerable benefit of the doubt by accepting some of the potential for good it is said to have. The benefits advertised exceed the benefits from air transport, nuclear power, and nuclear weapons. One hesitates to be optimistic, given the overly optimistic predictions about nuclear power (eliminate want from the face of the earth, for example), but the opportunities seem to be incredible for food production, for energy saving, for disease control, for repairing genetic defects, even for toxic waste disposal. The rapid abandonment of minimal safeguards probably did not occur because of these potentials for human life, but because of the prospect of immense profits and the excitement of research and knowledge accumulation. Were the prospect for human betterment the only driving force, we could go very slow and very carefully; on the scale of humankind a couple of generations is trivial. But the economic and scientific impulses are much more uncontrollable. Too many rewards for individual scientists and economic elites are at stake. We will not go slow.

Should DNA be combined with biological warfare, the immense dangers of the latter would be multiplied. There is a fair prospect that the

FIGURE 9.3
Policy Recommendations

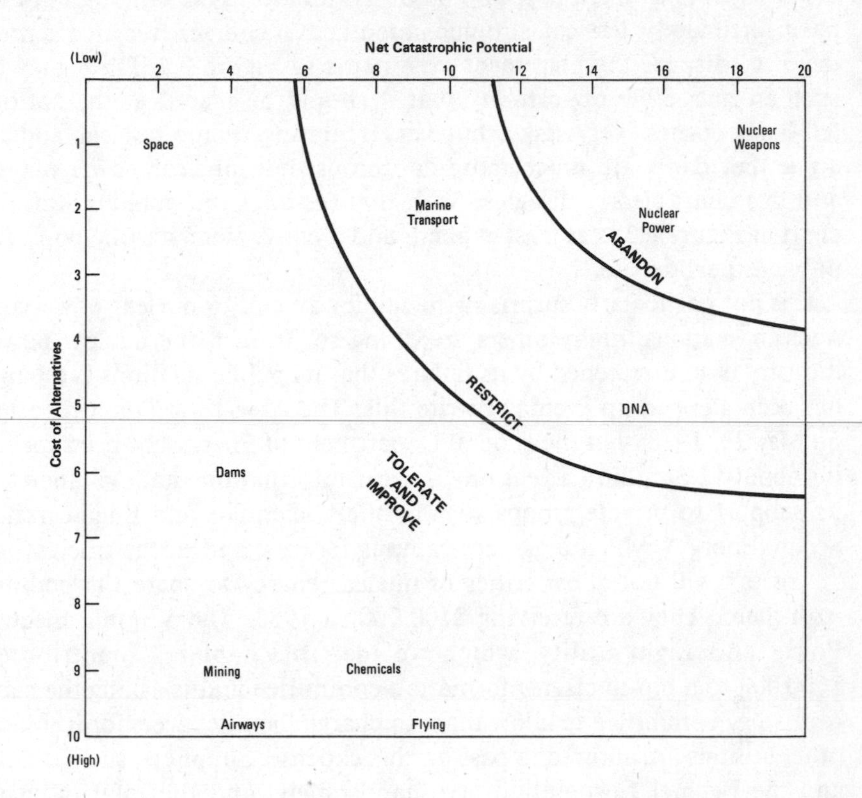

next world war will not be fought with nuclear weapons, but with biological ones. Both the USSR and the United States are known to be developing these weapons. I am ignorant of the matter, but were Genetech or Cetus (two leading DNA firms) to be found doing classified military research, I would fear that we have put the second of the two ultimate dangers to humankind into the same hands we have put the first—the military. On a long string of worry beads this one has to be near the top.

When we take Column 4, the net catastrophic potential, and combine it with Column 5, an estimate of the cost of alternatives for these systems, we produce the array given in Figure 9.3. This constitutes the policy recommendation generated by our efforts in this book. It indicates which systems are both highly risky and not essential, and thus could be abandoned, and which that we would find it hard to do without, but that have, fortunately, less catastrophic potential. Of course, everyone is privileged to dispute the placement of systems in Figure 9.3. DNA may be such an incredible opportunity that it should be placed at the bottom left-hand corner: very risky, but very tempting. Some nuclear addicts argue that dams are much more dangerous than nuclear power plants, and this chart clearly disagrees with that (as does the public). Still, the chart in Figure 9.3 is at least explicit, and a convenient starting point for public-expert debate.

It is not particularly surprising to call for an end to nuclear power and nuclear weapons; many others are doing so. In fact, the nuclear power industry is so threatened by its failures that its public relations campaign has been stepped up even as I write this. The *New York Times* reported on May 23, 1983, that the Federal Department of Energy has been spending about $2.5 million a year on "nuclear information" and will increase its support to private groups such as the "Scientists and Engineers for Secure Energy" which organizes campus forums, and in the interests of "fairness" will not allow critics of nuclear energy to share the podium with them. They are receiving $100,000 in 1983. The Virginia Electric Power and Light utility, which we met in Chapter 2, contributed $600,000 to a pro-nuclear information committee and is asking the state regulatory committee to allow them to charge the ratepayers for it. Many other utilities are allowed to pass on this expense. Suppliers, such as G.E. and the Bechtel Corporation give handsomely, and the total industry contributions will amount to $25 or $30 million. This suggests there is something to our argument in this book; were nuclear power as safe as the decal indicated at TMI Unit 1, this campaigning would all be unnecessary.

The nuclear freeze movement seems to be spreading at the time of this writing. However, in the field of genetic engineering, the biological laboratories on campuses and in private industry have kept a very low profile; we might hope this is because the extraordinary precautions that are needed are being taken, but I have my doubts. There is nothing at all that I know of going on with regard to improving or protesting the transportation of toxic and explosive materials by sea; indeed, the same problem on land, one that we did not consider here, appears to be growing in intensity.

Much can be done in all these systems to increase safety, but our record in each one has not been encouraging. I hope this book will remind some of that record; it has been, after all, a dismal and dismaying travelogue through the world of high-risk systems. At each turn, even in the best of the industries, we found rampant attribution of operator error to the neglect of errors by the Great Designers and the Centralized Managers. We found organizations that could not carry the burden of error-free operation, and sometimes seemed insensitive to the damage they did or could do. We may be thankful for the regulatory agencies, but too often they were shown to be ineffective, sometimes natteringly so, sometimes even criminally indifferent or co-conspirators.

But, important as these problems are, they were not the main point. The main point of the book is to see these human constructions as *systems,* not as collections of individuals or representatives of ideologies. From our opening accident with the coffeepot and job interview through the exotics of space, weapons, and microbiology, the theme has been that it is the way the parts fit together, interact, that is important. The dangerous accidents lie in the system, not in the components. The nature of the transformation processes elude the capacities of any human system we can tolerate in the case of nuclear power and weapons; the air transport system works well—diverse interests and technological changes support one another; we may worry much about the DNA system with its unregulated reward structure, less about chemical plants; and though the processes are less difficult and dangerous in mining and marine transport, we find the system of each is an unfortunate concatenation of diverse interests at cross-purposes.

These systems are human constructions, whether designed by engineers and corporate presidents, or the result of unplanned, unwitting, crescive, slowly evolving human attempts to cope. Either way they are very resistant to change. Private privileges and profits make the planned constructions resistant to change; layers upon layers of accommodations and bargains that go by the name of tradition make the unplanned ones unyielding. But they are human constructions, and humans can destruct them or reconstruct them.

The catastrophes send us warning signals. This book has attempted to decode these signals: abandon this, it is beyond your capabilities; redesign this, regardless of short-run costs; regulate this, regardless of the imperfections of regulation. But like the operators of TMI who could not conceive of the worst—and thus could not see the disasters facing them—we have misread these signals too often, reinterpreting them to fit our preconceptions. Better training alone will not solve the problem, or more gadgets, or promises that it won't happen gain. Worse yet, we may accept the preconception that military superiority and private profits are worth the risks. This book's decoding asserts that the problems are not with individual motives, individual errors, or even political ideologies. The signals come from systems, technological, and economic. They are systems that elites have constructed, and thus can be changed or abandoned.

LIST OF ACRONYMS

AFCS	automatic flight control system
ALARA	as low as reasonably achievable
ASD	automatic safety device
ASRS	Air Safety Reporting System
ATC	Air Traffic Control
BMEWS	Ballistic Missile Early Warning System
BWR	boiling water reactor
CAS	collision avoidance system
CDTI	cockpit display of traffic information
CPA	closest point of approach
CRT	cathode ray tube
DEPOSE	design, equipment, procedures, operators, supplies and materials, environment
ECCS	emergency core cooling system
ESD	emergency safety device
ESF	engineered safety device
ETA	expected time of arrival
FAA	Federal Aviation Administration
HPI	high-pressure injection
ICBM	Intercontinental Ballistic Missile
IMCO	Intergovernmental Maritime Consultative Organization
INS	inertial navigation system
LER	licensee event report
LNG	liquefied natural gas
LOCA	loss of coolant accident

LPG	liquified propane gas
MRIT	marine radar interrogation transponder
NAS	National Academy of Sciences
NIH	National Institute of Health
NORAD	North American Aerospace Defense Command
NRC	Nuclear Regulatory Commission
NTSB	National Transporation Safety Board
OSHA	Occupational Safety and Health Administration
PARCS	Perimeter Aquisition Radar Attack Characterization
PORV	pilot-operated relief valve
PRA	probabilistic risk analysis
PWR	pressurized water reactor
TCA	terminal control area
TMI	Three Mile Island
VEPCO	Virginia Electric Power Company
VLCC	very large crude carrier
VTS	Vessel Traffic Services

NOTES

Introduction

1. John Kemeny et al., *The Need for Change: The Legacy of TMI,* Report of the President's Commission on the Accident at Three Mile Island (Washington, D.C.: Government Printing Office, 1979), 2, 11, 113–16; and Charles Perrow, "The President's Commission and the Normal Accident," *The Accident at Three Mile Island: The Human Dimensions,* ed. David Sills, Charles Wolf, and Vivian Shelanski (Boulder, Colorado: Westview Press, 1981).

2. Essex Corporation, *Human Factors Evaluation of Control Room Design and Operator Performance at Three Mile Island-2* (NUREG/CR-1270, vol. 1, 1980).

3. Robert Jervis, *Perception and Misperception in International Politics* (Princeton, N.J.: Princeton University Press, 1976); and Karl Wieck, "Educational Organizations as Loosely Coupled Systems," *Administrative Science Quarterly* 21:1 (March, 1976): 1–19.

Chapter 1

1. David Bird and Frank Prial, *New York Times,* 1, 2, 9, and 25 November 1982, 7 December 1982, and 25 January 1983.

2. Matthew L. Wald, *New York Times,* 21 September 1981.

3. John Kemeny et al., *The Need for Change: The Legacy of TMI,* pp. 2, 11, 113–16. Report of the President's Commission on the Accident at Three Mile Island (Washington, D.C.: Government Printing Office, 1979).

4. Babcock and Wilcox, *Press Conference;* and *Science,* 19 October 1979.

5. Babcock and Wilcox, *Press Conference,* pp. 82–83, 90.

6. President's Commission on Three Mile Island, *Hearings,* 30 May 1979, 57.

7. Earl A. Gulbransen, "Not Safe Enough," *Bulletin of the Atomic Scientist* (June 1975): 5.

8. Nunzio J. Palladino, "Defends Zirconium," *Bulletin of the Atomic Scientist* (March 1976): 5.

9. President's Commission on Three Mile Island, *Hearings,* 30 May 1979, 57.

10. *Washington Post,* 29 February 1980.

11. Richard D. Lyons, "Crews at Reactor Criticize Cleanup," *New York Times,* 28 March 1983.

Chapter 2

1. Irvin C. Bupp and Jean-Claude Derian, *Light Water: How the Nuclear Dream Dissolved* (New York: Basic Books, 1978), 49.
2. Ibid., 74.
3. Ibid., 75.
4. Ibid., 155.
5. Ibid.
6. Matthew L. Wald, *New York Times,* 21 September 1981.
7. Ibid.
8. John B. Emshwiller, *Wall Street Journal,* 24 October 1979.
9. Ibid.
10. Ibid.
11. NBC, *Nightly News,* 24 July 1982.
12. David Perlman, *San Francisco Chronicle,* 6 November 1981.
13. Ibid.
14. Nunzio J. Palladino, quoted in Walker Turner, *New York Times,* 2 December 1981.
15. Victor Gilinsky, "Full Ahead for Nuclear Power?" *Technology Review* (February March 1982): 10.
16. Richard E. Webb, *The Accident Hazards of Nuclear Power Plants* (Amherst, Mass.: University of Massachusetts Press, 1976): 194–95.
17. Olson McKinley, *Unacceptable Risk: The Nuclear Power Controversy* (New York: Bantam Books, 1976): 22.
18. Union of Concerned Scientists, *The Risks of Nuclear Power Reactors,* 44–51.
19. *Washington Post,* 29 February 1980.
20. *New York Times,* 26 September 1981.
21. *New York Times,* 28, 29, 30 October; 12 December 1981.
22. W. R. Castro, "Safety-Related Occurrences Reported in October-November 1970," *Nuclear Safety* 12:2 (March-April, 1971): 145.
23. W. R. Castro, "Operating Experiences," *Nuclear Safety,* 12:3 (May-June 1971): 249.
24. Castro, "Safety-Related Occurrences," 145.
25. Castro, "Operating Experiences" (May-June 1972): 236.
26. Castro, "Operating Experiences" (March-April 1975): 233.
27. Webb, *The Accident Hazards,* 197–98.
28. Anna Gyorgy and friends, *No Nukes: Everyone's Guide to Nuclear Power* (Montreal: Black Rose Books, 1979), 111–12.
29. John G. Fuller, "We Almost Lost Detroit," in *The Silent Bomb,* ed. Peter Faulkner (New York: Random House, 1977), 46–59.
30. Ibid., 46.
31. Ibid., 49.
32. R. L. Scott, Jr., "Fuel Melting Incident at the Fermi Reactor on October 5, 1966," *Nuclear Safety,* 12:2 (March-April 1971): 123–34.
33. Ibid.
34. Ibid., 133.
35. D. C. Hunt, "Restricted Release of Plutonium—Part 1. Observational Data," *Nuclear Safety,* 12:2 (March-April 1971): 85–89.
36. Ibid., 88.
37. Robert L. Seale, "Consequences of Criticality in Accidents," in *Nuclear Criticality Safety,* ed. R. Douglas O'Dell (U.S. Atomic Energy Commission, Technical Information Center, 1974), 16–24.
38. U.S. Atomic Energy Commission, *WASH-1192: Operational Accidents, 1943–1970* (Washington, D.C., 1972).
39. Gyorgy, *No Nukes,* 60.
40. P. A. Morris and R. H. Engelken, "Safety Experiences in the Operation of Nuclear Power Plants," in *International Atomic Energy Agency Principles and Standards,* proceedings of a IAEA, 1973 symposium, Vienna, 429–46.

41. Ibid., 438.
42. Ibid., 440, 444.
43. U.S. Nuclear Regulatory Commission, *NRC Licensee Assessments* NUREG 0834 (Washington, D.C., *USNRC*, August 1981).
44. *New Indicator,* "You Have One Hour to Evacuate." University of California, San Diego, 7:6 2 December–4 January 1982, 1–2.

Chapter 3

1. Julie Graham and Don Shakow, "Risks and Rewards: Hazard Pay for Workers," *Environment,* 23:8 (October 1981), 14–45; and Christopher Cyr, Julie Graham, and Don Shakow, "Risk Compensation—in Theory and Practice," *Environment,* 25:1 (January-February 1983), 14–40.
2. Paul Bagne, "The Glow Boys," *Mother Jones* (November 1982): 24–27.
3. U.S. Nuclear Regulatory Commission, "Safety Goals for Nuclear Power Plants: A Discussion Paper," *NUREG* 0880 (Washington, D.C.: USNRC, February 1982), 15.
4. E. W. Hagen, "Common-Mode/Common Cause Failure: A Review," *Nuclear Safety,* 21:2 (March-April 1980): 184–92.
5. See a similar conclusion concerning the Ranger moon mission in Weaver, "Pitfalls in Current Design Requirements," *Nuclear Safety,* 23:3 (May-June 1981), 328–29.
6. National Transportation Safety Board (NTSB)—Marine Accident Reports, MAR-75-5, 6 July 1978.
7. Larry Hirschhorn, "The Soul of a New Worker," *Working Papers* (January-February 1982), 45.
8. Eliot Marshall, "NRC Takes a Second Look at Reactor Design," *Science,* 207 (28 March 1980): 1445–48.
9. For a discussion of the problems of human factors engineering, see Charles Perrow, *The Organizational Context of Human Factors,* Technical Report, (DTIC number ADA 123435), U.S. Navy, Office of Naval Research, Washington, D.C., November 1982; and "The Organizational Context of Human Factors Engineering," *Administrative Science Quarterly* (December 1983).
10. Wilson and Zarakas, "Anatomy of a Blackout," *Spectrum* (15 February 1978): 39–45; and "Investigators Agree New York Blackout of 1977 Could Have Been Avoided," *Science,* (15 September 1978): 994–96.
11. James Fallows, *National Defense* (New York: Random House, 1981).
12. James Thompson, *Organizations in Action* (New York: McGraw-Hill, 1967).
13. Karl Weick, "Educational Organizations as Loosely Coupled Systems," *Administrative Science Quarterly,* 21:1 (March 1976), 1–19. For an even more stunning analysis see John Meyer and Brian Rowan, "The Structure of Educational Organizations" in Marshall Meyer and Associates, *Environment and Organizations* (San Francisco: Jossey-Bass, 1978) 78–109.
14. Melville Dalton, *Men Who Manage* (New York: John Wiley, 1959).

Chapter 4

1. J. D. Atwood, "How Hot is Too Hot?", *Ammonia Plant Safety* 18 (1976): 109–11; J. A. Davenport, "A Survey of Vapor Cloud Incidents," *Chemical Engineering Progress,* 73:9 (September 1977): 54–63; T. A. Kletz, "A Decade of Safety Lessons," *Hydrocarbon*

Processing, 58:6 (June 1979); T. A. Kletz, "The Flixborough Cyclohexane Disaster," *Loss Prevention* 9 (1975): 106–110; and Charles H. Vervalin, "Fire Losses Reported by NFPA," *Hydrocarbon Processing,* 56:2 (February 1977): 166–67.

2. Nicholas A. Ashford, *Crisis in the Workplace: Occupational Disease and Injury* (Cambridge, Mass.: MIT Press, 1976), 59–60; and David P. McCaffrey, *OSHA and the Politics of Health Regulation* (New York: Plenum Press, 1982), 22.

3. Thomas Whiteside, *The Pendulum and the Toxic Cloud* (New Haven: Yale University Press, 1979).

4. Lee Clarke, "Risk and Interorganizational Regulations," unpublished manuscript, SUNY, Stony Brook, Sociology Department, 1983.

5. Michael H. Brown, *Laying Waste* (New York: Pantheon, 1979).

6. Joyce Egginton, *The Poisoning of Michigan* (New York: W. W. Norton, 1980).

7. American Petroleum Institute, *Review of Fatal Injuries in the Petroleum Industry for 1980* (Washington, D.C.: API, September 1981); and *Summary of Occupational Injuries and Illnesses in the Petroleum Industry* (Washington, D.C.: API, September 1981).

8. Vervalin, "Fire Losses Reported by NFPA," 166–67.

9. American Petroleum Institute, *Reported Fire Losses in the Petroleum Industry for 1980* (with errata, January 18 1982) (Washington, D.C.: API, December 1981).

10. G. P. Williams. "Causes of Ammonia Plant Shutdowns," *Ammonia Plant Safety,* 20 (1978): 123–30.

11. *Catastrophe! When Man Loses Control.* Prepared by the editors of Encyclopaedia Britannica. New York: Bantam, 1979.

12. H. C. Jarvis, "Butadiene Explosion at Texas City—1," *Loss Prevention,* 5 (1971): 58.

13. Ibid.

14. R. G. Keister, B. I. Pesetsky, and S. W. Clark, "Butadiene Explosion at Texas City—3," *Loss Prevention,* 5 (1971): 67–75.

15. Ibid.

16. Department of Employment, *The Flixborough Disaster,* Report of the Court of Inquiry, London: Her Majesty's Stationery Office, 1975, 3.

17. Ibid., 81.

18. Ibid., 10.

19. C. Sadee, D. E. Samuels, and T. P. O'Brien, "The Characteristics of the Explosion of Cyclohexane at the Nypro (U.K.) Flixborough Plant on 1st June 1974," *Journal of Occupational Accidents,* 1 (1976-77): 203–35.

20. Kletz, "The Flixborough Cyclohexane Disaster," *Loss Prevention,* 9 (1975): 109.

21. Ibid.

22. Davenport, "A Survey of Vapor Cloud Incidents," 54.

23. Bruce H. Winegar, "Partial Collapse of an Atmospheric Ammonia Storage Tank," *Ammonia Plant Safety,* 22 (1980): 226–30.

24. F. G. Kokemor, "Synthesis Start-Up Heater Failure," *Ammonia Plant Safety,* 22 (1980): 160.

25. Ibid., 159.

26. Ibid., 160.

27. Ibid., 161.

28. Ibid., 162

29. M. Voros and Gy Honti, "Explosion of a Liquid CO_2 Storage Vessel in a Carbon Dioxide Plant," *Loss Prevention and Safety Promotion in the Process Industries,* ed. C. H. Buschmann (New York: Elsevier Scientific Publishing Co., 1974), 337–46.

30. Atwood, "How Hot Is Too Hot?" 109.

31. Ibid.

32. Ibid., 110–11.

33. Ibid., 111.

34. Charles Perrow, *The Organizational Context of Human Factors,* Technical Report, (DTIC number ADA 123435), U.S. Navy, Office of Naval Research, Washington, D.C., November 1982.

Notes

Chapter 5

1. These and other examples come from reports to the Aviation Safety Reporting System. See NASA, Aviation Safety Reporting System, NASA Ames/LM.
2. In this historical section I am drawing upon Jerome Lederer, *Aviation Safety Perspectives: Hindsight, Insight, Foresight* (New York: The Wings Club, 1982).
3. Ibid., 14.
4. Ibid., 25.
5. Paul Slovic, Baruch Fischhoff, and Sara Lichtenstein, "Accident Probabilities and Seat Belt Usage," *Accident Analysis and Prevention,* 10 (1978): 281–85.
6. Tom Wolfe, *The Right Stuff* (New York: Bantam Books, 1979), 17.
7. Elwyn Edwards, "Automation in Civil Transport Aircraft," *Applied Ergonomics,* 8:4 (December 1977), 194–98.
8. Thomas Mahon, *Report of the Royal Commission to Inquire into the Crash on Mount Erebus, Antarctica of a DC-10 Aircraft Operated by Air New Zealand Limited* (Wellington, New Zealand: P. D. Hasselberg, 1981).
9. Earl L. Wiener, "Controlled Flight into Terrain Accidents: System-Induced Errors," *Human Factors,* 22:5 (1980): 176.
10. NASA, Aviation Safety Reporting System Staff, "Human Factors Associated with Altitude Alert Systems," in Sixth Quarterly Report, NASA TM-78511, Washington, D.C., July 1978, 25–37.
11. This study, perhaps understandably, has not been published; it is referred to in a NASA Technical Memorandum. See Smith, *A Simulator Study of the Interaction of Pilot Workload with Errors, Vigilance, and Decisions* NASA TM 78482, Ames Research Center, Moffett Field, California, 1979. Smith's study of errors and workload is consistent with the findings of the European study.
12. Santilli, *Critical Interface Between Environment and Organisms in Class A Mishaps: A Retrospective Analysis,* Report SAM-TR-80-3, USAF School of Aerospace Medicine, Brooks Air Force Base, Texas, June 1980, 7.
13. Mahon, *Report of the Royal Commission.*
14. William B. Mackley, "Aftermath of Mount Erebus," *Flight Safety Digest* (September 1982): 1–5.
15. National Transportation Safety Board Safety Recommendation, A-81-9, 26 August 1981.
16. As of this writing, the NTSB has not issued its analysis; I have drawn upon a newspaper story by Jim Wood in the *San Francisco Examiner,* 19 December 1981.
17. *Aviation Week and Space Technology,* 9, 17 March 1980, 61.
18. Douglas B. Feaver, *Washington Post,* 7 March 1982.
19. NTSB, AAR-81-10, 7 July 1981, 12.
20. NTSB, AAR-82-3, 6 April 1982, 14–17.
21. Richard Witkin, *New York Times,* 28 November 1981.
22. Godson, *The Rise and Fall of the DC-10* (New York: David McKay, 1975), 41, 46, 62–63, 91–92, 123, 125, 235–37.
23. NTSB, AAR-81-15, 15 September 1981.
24. Ibid., 69.
25. NTSB, AAR-81-13, 19 August 1981, 21; see also NTSB Safety Recommendation A-81-92, 26 August 1981.
26. NTSB, AAR-81-15, 15 September 1981.
27. Ibid.
28. NTSB, AAR-81-13, 19 August 1981, 7.
29. Ibid., 1–34.
30. Ibid.
31. NTSB, AAR-81-18, 17 December 1981.
32. Billings et al., *A Study of Near Midair Collisions in U.S. Terminal Airspace,* NASA-TM-81225, August 1980, 21–22.
33. NTSB, SIR-81-6, 24 September 1981.

34. NASA, *Third Quarterly Report,* TM X-3546, Washington, D.C., May 1977, 63–65.

35. See Ralph L. Grayson and Charles E. Billings, "Information Transfer Between Air Traffic Control and Aircraft: Communication Problems in Flight Operations," in *Information Transfer Problems in the Aviation System,* NASA, Technical Paper 1875, Moffett Field, California, September 1981, 52.

36. NASA, *Third Quarterly Report,* p. 67.

37. Ibid.

38. See Earl L. Wiener and Renwick E. Curry, "Flight-Deck Automation: Promises and Problems," *Ergonomics,* 23:10 (1980): 997; see also M. Feazel, "Fuel Pivotal in Trunks' Earnings Slump," *Aviation Week and Space Technology,* 113 (18 February 1980): 31–32.

39. Throughout this section on air traffic control I will be relying heavily on the unpublished work of Todd LaPorte, who has studied the system intensively and compared its nearly error-free operation with that of other high-risk systems, in Todd R. LaPorte, "In Search of Nearly Error-Free Management: Lessons from U.S. Air Traffic Control for the Future of Nuclear Energy." (Unpublished manuscript, Institute of Governmental Studies, University of California, Berkeley, 1980).

40. John G. Kreifeldt, "Cockpit Displayed Traffic Information and Distributed Management in Air Traffic Control," *Human Factors,* 22:6 (1980).

41. NTSB, SIR-81-6, 9 September 1981.

42. Ibid.

43. Ibid., 30, 32.

44. Jeffrey R. Smith, "FAA Is Cool to Cabin Safety Improvements," *Science* (6 February 1981): 557.

45. Ibid.

46. Ibid., 558.

47. NASA, *Third Quarterly Report,* TM X-3546, Washington, D.C., May 1977.

48. William P. Monan, "Distractions—A Human Factor in Air Carrier Hazard Events," *Ninth Quarterly Report,* NASA TM 78608, Washington, D.C., June 1979, 22.

49. D. W. Hall and A. W. Hecht, "Summary of the Characteristics of the ASRS Database," *Ninth Quarterly Report,* NASA TM 78608, Washington, D.C., June 1979, 24–34.

Chapter 6

1. I. C. Clingan, "Safety at Sea," *Interdisciplinary Science Reviews* 6:1 (1981): 42.

2. See Patrick Lagadec, *Major Technological Risk,* (New York: Pergamon Press, 1982).

3. Willard W. Perry and William P. Articola, *Study to Modify the Vulnerability Model of the Risk Management System,* Technical Report C6-D-22-80, U.S. Department of Transportation, Washington, D.C., February 1980.

4. "Catastrophe," *Encyclopedia Britannica.*

5. S. Peltzman, "The Effects of Automobile Safety Regulation," *Journal of Political Economy,* 83 (1975).

6. L. S. Robertson, "A Critical Analysis of Peltzman's 'The Effects of Automobile Safety Regulation,'" *Journal of Economic Studies,* 11 (1977).

7. Captain A. F. Dickson, "Navigation Problems (Tankers)," *International Tanker Safety Conference* (London: International Chamber of Shipping, 1971), 2.

8. Maritime Transportation Research Board, *Human Error in Merchant Marine Safety,* AD/A-028 371 (WAshington, D.C.: National Technical Information Service, June 1976), 77.

9. Ibid.

10. Ibid., 77–78.

11. Edward Cowan, *Oil and Water* (Philadelphia: J. B. Lippincott Co., 1968), 42.

12. Ibid.

13. *Lloyd's List,* London, 21 May 1981, 1.

14. MTRB, *Human Error,* 36. Much of this is thought to be due to the restrictions our

government placed upon shipping U.S. goods in "foreign bottoms." Since a new ship built in the U.S. will cost almost three times as much as one built in Japan, we continue to use dreadfully unsafe World War II ships and have far and away the oldest fleet in the world. The restriction was meant to encourage our outmoded shipbuilding industry, but it didn't; it encouraged the continued use of rusty forty-year-old ships that the Coast Guard barely inspects. This is another aspect of this error-inducing system.

15. *Lloyd's List,* 21 May 1981, 1.
16. Ibid., 4.
17. Luther J. Carter, "*AMOCO Cadiz* Incident Points up the Elusive Goal of Tanker Safety," *Science,* 200 (5 May 1978): 514.
18. MTRB, *Human Error,* 43.
19. Noel Mostert, *Supership* (New York: Alfred A. Knopf, 1974), 28.
20. Carter, "*AMOCO Cadiz* Incident."
21. *Lloyd's List,* 21 May 1981, 2.
22. NTSB, MAR-80-5, 28 March 1980.
23. NTSB, MAR-79-16, 27 September 1979.
24. MTRB, *Human Error,* 29.
25. Dickson, "Navigation Problems (Tankers)," 6.
26. NTSB, MAR-80-16, 29 September 1980.
27. USCG/NTSB, 28 August 1973.
28. Mostert, *Supership,* 22–23.
29. Ibid., 36.
30. Ibid., 43–44, 52.
31. Ibid., 61–62.
32. Ibid., 63–64.
33. Ibid., 96.
34. Ibid., 170–71.
35. Ibid., 175.
36. Ibid., 133.
37. Ibid., 137
38. Ibid., 139.
39. NTSB, MAR-81-14, 9 December 1981.
40. NTSB, MAR-80-16, 29 September 1981.
41. NTSB, MAR-80-7, 12 May 1980.
42. NTSB, Safety Recommendation M-82-1, 18 February 1982.
43. MTRB, *Human Error,* 19.
44. USCG/NTSB, MAR-77-1, 12 May 1977.
45. Clingan, "Safety at Sea," 41.
46. Chamber of Shipping of the United Kingdom, *Marine Casualty Report Scheme* (London, October 1972).
47. John S. Gardenier, "Ship Navigational Failure Detection and Diagnosis," in *Human Detection and Diagnosis of System Failures,* ed. Jens Rasmussen and William B. Rouse (New York: Plenum Publishing Corp., 1981), 59.
48. USCG/NTSB, *Marine Casualty Report,* No. 16732/92368, 31 July 1979, 37, 39.
49. Ibid.
50. NTSB, MAR-82-3, 9 February 1982.
51. USCG/NTSB, *Marine Casualty Report—SS* Transhuron *on 24 September and Grounding on 26 September 1974, Arabian Sea* (Washington, D.C.: Government Printing office, 16 September 1976).
52. Ibid., 24.
53. Ibid., 30.
54. NTSB, S.S. *Badger State,* 7 December 1971.
55. Ibid., 32.
56. Michael E. Gaffney, "Bridge Simulation: Trends and Comparisons" (unpublished manuscript, Maritime Transportation Research Board, National Academy of Sciences, Washington, D.C., 1982).
57. See Charles Perrow, *The Organizational Context of Human Factors* Technical Report, DTIC Number ADA 123435 (Washington, D.C.: U.S. Office of Naval Research, November 1982).

Chapter 7

1. Gregory B. Baecher, M. Elizabeth Pate, and Richard de Neufville, "Risk of Dam Failure in Benefit-Cost Analysis," *Water Resources Research,* 16:3 (June 1980).
2. Committee on Government Operations, *Teton Dam Disaster* (Washington, D.C.: Government Printing Office, September 1976.), 31; and Asit K. Biswas and Samar Chatterjee, "Dam Disasters: An Assessment," *Engineering Journal,* 54:3 (March 1971): 3.
3. Kai Erikson, *Everything in its Path: Destruction of Community in the Buffalo Creek Flood* (New York: Simon and Schuster, 1976).
4. Committee on Government Operations, *Teton Dam Disaster,* 6, 7.
5. Nigel Calder, *The Restless Earth,* (New York: Viking Press, 1972), 133.
6. Committee on Government Operations, *Teton Dam Disaster,* 171.
7. Ibid., 19.
8. Ibid., 21.
9. Ibid., 25.
10. Ibid.
11. Ibid., 28–29.
12. Ibid., 31.
13. Biswas and Chatterjee, "Dam Disasters: An Assessment," 7.
14. Committee on Government Operations, *Teton Dam Disaster,* 10–11.
15. Ibid., 12.
16. Mary Ellen Hynes and Erick H. Vanmarcke, "Reliability of Embankment Performance Predictions," Proceedings of the ASCE Engineerings Mechanics Division Specialty Conference May 1976 (Waterloo, Canada: University of Waterloo Press, 1976). (I am indebted to Paul Slovik and Baruch Fischhoff for alerting me to this study.)
17. U.S. Congress House Interior Committee, *Mill Tailings Dam Break at Church Rock, New Mexico* (Washington, D.C.: Government Printing Office, 1980), 9.
18. Ibid., 47, 229–31.
19. Ibid., 22.
20. Ibid., 3, 34, 39, 42, 47, 227.
21. Calder, *The Restless Earth,* 136.
22. Ibid., 137–38.
23. "Could These Deaths Have Been Averted?" *Mine Safety and Health,* 1979.
24. Ibid.
25. Ibid.
26. E. D. Seals and R. A. Speirer, "Analysis of Accidents Related to Falls of Ground in Metal and Nonmetal Mines, 1972-1973," U.S. Bureau of Mines, Mining Enforcement and Safety Administration, Pittsburgh Information Report 1009, 1.
27. "Catastrophe," *Encyclopaedia Britannica.*
28. "The Belle Isle Explosion," *Mine Safety and Health,* 1980, 5.
29. Michael Gold, "Who Pulled the Plug on Lake Peigneur?" *Science* 81, (November 1981).
30. Biswas and Chatterjee, "Dam Disasters: An Assessment," 7.

Chapter 8

1. William Hines, "NASA: The Image Misfires," *The Nation* (24 April 1967).
2. John W. Finney, "Project Mercury Defects Laid to Private Industry," *New York Times,* 4 October 1963.
3. *Aviation Week,* 17 April 1967.
4. W. W. Weaver, "Pitfalls in Current Design Requirements," *Nuclear Safety,* 22:3 (May-June 1981).

5. William J. Broad, "Fallout from Nuclear Power in Space," *Science* (20 July 1979): 281–86.
6. P. W. Krey, "Atmospheric Burnup of a Plutonium-238 Generator," *Science,* (10 November 1967): 769.
7. *Time,* 20 December 1982, 68.
8. Tom Wolfe, *The Right Stuff* (New York: Bantam Books, 1979).
9. Ibid., 151.
10. Ibid., 30.
11. Ibid., 356–61.
12. Ibid., 309–14.
13. Ibid., 235–43.
14. Ibid., 186.
15. Ibid., 106, 254.
16. Ibid., 274.
17. William K. Stevens, "Man Has Yet to Master Shuttle's Sophistication," *New York Times,* 15 November 1981.
18. Henry S. F. Cooper, *Thirteen: The Flight that Failed* (New York: Dial Press, 1973).
19. Ibid., 16–18.
20. Ibid., 23–24.
21. Ibid., 24.
22. Ibid., 41.
23. Ibid., 62–63.
24. Graham Allison, *Essence of Decision* (New York: Little Brown, 1971).
25. See Stephen Talbot, "The H-Bombs Next Door," *The Nation* (7 February 1981) for this and other references to "broken arrows."
26. Gary Hart and Barry Goldwater, "Recent False Alerts from the Nation's Missile Attack Warning System," Report to the Senate Committee on Armed Services (Washington, D.C.: Government Printing Office, 9 October 1980).
27. *New York Times,* 16 December 1979.
28. Hart and Goldwater, "Recent False Alerts."
29. *Washington Post,* 28 April 1982.
30. Nicholas Wade, "Recombinant DNA: Warming Up for the Big Payoff," *Science,* 206 (29 November 1979).
31. Martin Kenney, et al., "Genetic Engineering and Agriculture," *Bulletin No. 125,* Cornell Rural Sociology Bulletin Series (July 1982): 2.
32. Wade, "Recombinant DNA."
33. Kenney, "Genetic Engineering and Agriculture," 1.
34. Quoted in Nancy Pfund, "Recombinant DNA: Miracles and Menace in *Do No Harm: Health Risks and Public Choices,* ed. Diana Sutton (Berkeley, California: University of California Press, 1984).
35. I am drawing upon the accounts of Nicholas Wade, *The Ultimate Experiment* (New York: Walker, 1977); John Lear, *Recombinant DNA: The Untold Story* (New York: Grown, 1978); and Sheldon Krimsky and D. Ozonoff, *Genetic Alchemy: The Social History of the Recombinant DNA Controversy* (Cambridge: M.I.T. Press, in press).
36. Krimsky and Ozonoff, Chapter 5.
37. Barbara J. Cullington, "Recombinant DNA Bills Derailed: Congress Still Trying to Pass a Law," *Science,* 199 (20 January 1978).
38. Eliot Marshall, "Gene Splicers Simulate a 'Disaster,' Find No Risk," *Science,* 203 (23 March 1979): 1223.
39. W. H. Thomasson, "Recombinant DNA and Regulating Uncertainty," *Bulletin of the Atomic Scientists of Chicago,* 35:10 (December 1979).
40. Marshall, "Gene Splicers," 1223.
41. This scenario was developed independently by biologist Jonathan Beckwith and by biologist Jonathan King. The description here is from Krimsky and Ozonoff, Chapters 6–14.
42. Charles Weiner, "Relations of Science, Government, and Industry: The Case of Recombinant DNA" in American Association for the Advancement of Science, Policy Outlook: Science, Technology and the Issues of the Eighties, Washington, D.C.: AAAS, 1981, 109–56; and Pfund, "Recombinant DNA: Miracles and Menace."

43. Barbara Goldoftas, "Recombinant DNA: The Ups and Downs of Regulation," *Technology Review* (May-June 1982): 32.
44. Kenney, "Genetic Engineering and Agriculture."
45. Cited in Kenny, ibid. 58.
46. Cited in Pfund, "Recombinant DNA: Miracles and Menace."

Chapter 9

1. See, for example, Roger Kasperson, C. Hohenemser, and J. X. Kasperson, "Institutional Response to Different Perceptions of Risk" in *Accident at Three Mile Island: The Human Dimensions,* ed. David L. Sills, Charles P. Wolfe, and Vivian B. Shelanski (Boulder, Colorado: Westview Press, 1982), 39–48.
2. Stephen Talbot, "The H-Bombs Next Door," *Nation* (7 February 1981).
3. See, for example, the papers in *Societal Risk Assessment: How Safe Is Safe Enough?,* ed. Richard C. Schwing and Walter A. Albers Jr. (New York: Plenum Press, 1980), 129–41; and Lester B. Lave, "Introduction" in *Quantitative Risk Assessment in Regulation,* ed. Lester B. Lave (Washington, D.C.: Brookings Institution, 1983), 8.
4. Schwing and Albers, *Societal Risk Assessment.*
5. See John D. Graham and James W. Vaupel, "Value of a Life, What Difference Does It Make?" *Risk Analysis,* 1:1 (1981): 89–95, for discussion.
6. See the body-count mentality in David Okrent, "Comment on Societal Risk," *Science,* 208 (25 April, 1980): 372–75.
7. Barbara Combs and Paul Slovic, "Newspaper Coverage of Causes of Death," *Journalism Quarterly,* 56:4 (Winter 1979): 837.
8. Paul Slovic, Baruch Fischhoff, and Sara Lichtenstein "Facts and Fears: Understanding Perceived Risk," in *Societal Risk Assessment,* ed. Schwing and Albers.
9. See, for example, Bernard L. Cohen and I-Sing Lee, "A Catalog of Risks," *Health Physics,* 36 (June 1979).
10. This is one of the main distinctions of the first body-count analysis by Chauncey Starr in 1969. See Starr, "Social Benefit versus Technological Risk," *Science,* 165 (1969): 1232–38.
11. See Richard Wilson, "The Costs of Safety," *New Scientist* (30 October 1975): 274–75, for estimates of $750 million spent per life saved in nuclear plants.
12. Bernard L. Cohen, "Society's Evaluation of Lifesaving and Radiation Protection and Other Contexts," *Health Physics,* 38 (January 1980): 33–51.
13. J. Patrick Wright, *On a Clear Day You Can See General Motors* (New York: Avon Books, 1980), 65–67.
14. Baruch Fischhoff, "Cost-Benefit Analysis and the Art of Motorcycle Maintenance," *Policy Sciences,* 8 (1977), 177–202.
15. For an argument that wage earners in the periphery labor market are subject to more risks and not compensated for them, see Julie Graham and Don Shakow, "Risks and Rewards: Hazard Pay for Workers," *Environment,* 23: 8 (October 1981): 14–45.
16. See, for example, sociologist Robert Nisbet, "Quintessential Liberal," *Commentary,* 72 (September 1981); 61–64; political scientist Aaron Wildavsky, "No Risk Is the Highest Risk of All," *American Scientist,* 67 (January-February 1979): 32–37; and anthropologist Mary Douglas and Wildavsky, *Risk and Culture* (Berkeley, California: University of California Press, 1982).
17. See, for example, the assortment in Schwing and Albers, *Societal Risk Assessment.*
18. Ronald A. Howard, "On Making Life and Death Decisions," in *Societal Risk Assessment,* ed. Schwing and Albers, 89–106.
19. Chauncey Starr and Chris Whippel, "The Risk of Risk Decisions," *Science,* 208 (6 June 1980): 1114–19.
20. See review of risk assessors' attitudes in William J. Broad, "Public Attitudes to Technological Progress," *Science,* 205 (20 July 1979): 281–86.

21. Howard Raffia, "Concluding Remarks," in *Societal Risk Assessment,* ed. Schwing and Albers, 340.
22. William C. Clark, "Witch, Floods and Wonder Drugs: Historical Perspectives on Risk Management," in *Societal Risk Assessment,* ed. Schwing and Albers, 305.
23. Starr and Whippel, "The Risk of Risk Decisions"; and Schwing, "Trade Offs," in *Societal Risk Assessment,* ed. Schwing and Albers, 137.
24. Kasperson, Hohenemser, and Kasperson, "Institutional Response," 40, 43.
25. Lester B. Lave, "Quantitative Risk Assessment in Regulation," (Washington, D.C.: The Brookings Institution, 1982), 8.
26. Amos Tversky and Daniel Kahneman, "Availability: A Heuristic for Judging Frequency and Probability," *Cognitive Psychology,* 5, (1973): 207–232.
27. See, for example, the critical review of cognitive psychology by cognitive psychologists Hillel J. Einhorn and Robin M. Hogarth, "Behavioral Decision Theory: Processes of Judgment and Choice, Annual Review of Psychology, 32 (1981): 53–58; and Robin M. Hogarth, "Beyond Discrete Biases: Functional and Dysfunctional Aspects of Judgmental Heuristics," *Psychological Bulletin,* 90:2 (197–217). See also the whole fascinating September 1981 issue of the Journal *Behavioral and Brain Sciences.*
28. See David M. Eddy, "Probabilistic Reasoning in Clinical Medicine: Problems and Opportunities," in *Judgment under Uncertainty: Heuristics and Biases,* ed. Daniel Kahneman, Paul Slovic, and Amos Tversky (Cambridge, Massachusetts: Cambridge University Press, 1982), 249–67.
29. Daniel Kahneman and Amos Tversky, "On the Psychology of Prediction," *Psychological Review,* 80 (1973): 237–51.
30. Baruch Fischhoff, "Behavioural Aspects of Cost-Benefit Analysis," in *Energy Risk Management,* ed. G. Goodman, W. D. Rowe (London: Academic Press, 1979).
31. Slovic, Fischhoff, and Lichtenstein "Facts and Fears."
32. See Robert Cameron Mitchell, "Public Response to a Major Failure of a Controversial Technology," in *Accident at Three Mile Island.*
33. See James G. March, "Bounded Rationality, Ambiguity, and the Engineering of Choice," *Bell Journal of Economics* (Autumn, 1978): 587–608, for an elegant statement of the assumptions behind what is called the "garbage can" theory, and see James C. March and Johan Olsen, eds., *Ambiguity in Choice in Organizations* (Vergen: Universitets Forlaget, 1976), for applications.
34. Paul Slovic, Baruch Fischhoff, and Sarah Lichtenstein, "Perceived Risk: Psychological Factors and Social Implications," Proceedings of the Royal Society of London, A 376, 1981, 17–34.
35. Slovic, Fischhoff, and Lichtenstein, "Perceived Risk," 25.
36. Clifford Geertz, *The Interpretation of Cultures* (New York: Basic Books, 1973), 3–32.
37. See Ralph Blumenthal, "Illegal Dumping of Toxins Laid to Organized Crime," *New York Times,* 5 June 1983.
38. Transcript from Kemeny Commission, closed session 19 September 1979. Transcript available in Nuclear Regulatory Commission's Three Mile Island Reading Room.
39. Charles Perrow, "The President's Commission and the Normal Accident," in *The Accident at Three Mile Island: The Human Dimensions,* eds. Vivian Shelanski, David Sills, and Charles Wolf (Boulder, Colorado: Westview Press, 1981), 173–184.

BIBLIOGRAPHY

Three Mile Island References

The Three Mile Island accident material is drawn from the following bibliography. If the references are limited to Chapter 1, they do not appear again in the general list of references for the whole book.

Babcock and Wilcox. Press Conference, June 5, 1979, 82–3, 90. Comments of J. H. McMillan

Bird and Prial, *New York Times,* 1, 2, 3, 9, and 25 November 1982.

———, *New York Times,* 7 December 1982.

———, *New York Times,* 25 January 1983.

Comey, David Dinsmore. "The Incident at Browns Ferry." In *The Silent Bomb,* edited by Peter Faulkner, 3–22. New York: Random House, 1977.

Essex Corporation. *Human Factors Evaluation of Control Room Design and Operator Performance at Three Mile Island-2.* NUREG/ CR-1270, 1, January 1980, v.

Faulkner, John. *We Almost Lost Detroit.* New York: Reader's Digest Press, 1975.

Kemeny, John, et al. *The Need for Change: The Legacy of TMI.* Report of the President's Commission on the Accident at Three Mile Island. Washington, D. C.: Government Printing Office, 1979.

Mason, John F. "The Technical Blow-by-Blow." *IEEE Spectrum,* 16 November 1979, 33–42.

Perrow, Charles. "The President's Commission and the Normal Accident." In *The Accident at Three Mile Island: The Human Dimensions.* Edited by Vivian Shelanski, David Sills, and Charles Wolf, 173–184. Boulder, Colorado: Westview Press, 1981.

President's Commission on the Accident at Three Mile Island. *Hearings.* 30, 31 May, 1 June, 18 July, 1979.

Rubenstein, Ellis. "The Accident That Shouldn't Have Happened," *IEEE Spectrum,* 16 November 1979, 33–42.

Science, 206, 19 October 1979, 308.

Scott, R. L., Jr. "Fuel-Melting Accident at the Fermi Reactor on October 5, 1966." *Nuclear Safety* 12, 1971. 122–134.

Union of Concerned Scientists. *The Risk of Nuclear Power Reactors.* Cambridge, Mass.: UCS, 1977, 10–16.

REFERENCES

Abraham, P., D. Pattnaik, and S. D. Soman. "Safety Experiences in the Operation of a BWR Station in India." In *Principles and Standards of Reactor Safety,* edited by the International Atomic Energy Agency. Proceedings of a symposium. IAEA, Vienna, 1973, 459–71.

Allison, Graham. *Essence of Decision,* New York; Little Brown, 1971.

American Petroleum Institute. *Reported Fire Losses in the Petroleum Industry for 1980* (with errata, January 18, 1982), Washington, D.C.: API, December 1981.

American Petroleum Institute. *Review of Fatal Injuries in the Petroleum Industry for 1980,* Washington, D.C., API, September 1981.

American Petroleum Institute. *Summary of Occupational Injuries and Illnesses in the Petroleum Industry,* Washington, D.C.: API, September 1981.

Ashford, Nicholas A. *Crisis in the Workplace: Occupational Disease and Injury.* Cambridge, Mass.: MIT Press, 1976, 59–60.

Atwood, J. D. "How Hot Is Too Hot?" *Ammonia Plant Safety* 18, 1976, 109–11.

Aviation Week, 17 April 1967.

Aviation Week and Space Technology 17 March 1980, 61.

Baecher, Gregory B., M. Elisabeth Pate, and Richard de Neufville. "Risk of Dam Failure in Benefit-Cost Analysis." *Water Resources Research,* 16:3, June (1980), 449–56.

Bagne, Paul. "The Glow Boys," *Mother Jones,* November 1982, 24–27.

"The Belle Isle Explosion." *Mine Safety and Health,* 5:4, 1980, 2–7, 28.

Billings, Charles, Ralph Grayson, William Hecht, and Renwick Curry. *A Study of Near Midair Collisions in U. S. Terminal Airspace.* NASA TM-81225, August 1980.

Biswas, Asit K., and Samar Chatterjee. "Dam Disasters: An Assessment." *Engineering Journal,* 54:3, March 1971.

Blumenthal, Ralph. "Illegal Dumping of Toxins Laid to Organized Crime," *New York Times,* 5 June 1983.

Braverman, Harry. *Labor and Monopoly Capital.* New York: Monthly Review Press, 1974.

Broad, William J. "Fallout from Nuclear Power in Space." *Science,* 219, 7 January 1983, 38–9.

Broad, William J. "Public Attitudes to Technological Progress." *Science,* 205, 20 July 1979, 281–286.

Brown, Michael H. *Laying Waste,* New York. Pantheon, 1979.

Bupp, Irvin C., and Jean-Claude Derian. *Light Water: How the Nuclear Dream Dissolved.* New York: Basic Books, 1978.

Burns, Tom, and G. M. Stalker. *The Management of Innovation.* New York: Barnes and Noble, 1961.

Calder, Nigel. *The Restless Earth.* New York: Viking Press, 1972.

Carson, Rachel. *Silent Spring.* New York: Houghton-Mifflin, 1962.

Carter, Luther J. "AMOCO Cadiz Incident Points Up the Elusive Goal of Tanker Safety." *Science,* 200, 5 May 1978, 514.

Castro, W. R. "Safety-Related Occurrences Reported in October-November 1970." *Nuclear Safety,* 12:2, March-April 1971, 145.

———. "Operating Experiences." *Nuclear Safety,* 12:3, May-June 1971, 249.

———. "Operating Experiences." *Nuclear Safety,* 12:4, July-August 1971.

———. "Operating Experiences." *Nuclear Safety,* 13:3, May-June 1972, 236.
———. "Operating Experiences." *Nuclear Safety,* 16:2, March-April 1975, 233.
Catastrophe! When Man Loses Control. Prepared by the editors of Encyclopaedia Britannica. New York: Bantam, 1979.
Chamber of Shipping of the United Kingdom. London: *Marine Casualty Report Scheme.* October 1972.
Clark, William C. "Witches, Floods and Wonder Drugs: Historical Perspectives on Risk Management." In *Societal Risk Assessment: How Safe Is Safe Enough?,* edited by Richard C. Schwing and Walter A. Albers, pp. 287–311. New York: Plenum Press, 1980.
Clarke, Lee. "Risk and Interorganizational Relations." Unpublished manuscript, SUNY, Stony Brook, Sociology Department, 1983.
Clawson, Dan. *Bureaucracy and the Labor Process.* New York: Monthly Review Press, 1980.
Clingan. I. C. "Safety at Sea." *Interdisciplinary Science Reviews,* 6:1, 1981, 36–48.
Cohen, Bernard L. "Society's Evaluation of Lifesaving and Radiation Protection and Other Contexts." *Health Physics,* 38, January 1980, 33–51.
Cohen, Bernard L., and I-Sing Lee. "A Catalog of Risks." *Health Physics,* 36, June 1979, 707–22.
Combs, Barbara, and Paul Slovic. "Newspaper Coverage of Causes of Death." *Journalism Quarterly,* 56:4, Winter 1979, 837–43.
Committee on Government Operations. *Teton Dam Disaster.* Washington, D.C.: Government Printing Office, September 1976.
Cooper, Henry S. F. *Thirteen: The Flight That Failed.* New York: Dial Press, 1973.
"Could These Deaths Have Been Averted?" *Mine Safety and Health* 1979, 4, 5, 8.
Cowan, Edward. *Oil and Water.* Philadelphia: J. B. Lippincott Co., 1968, 38–45.
Cullington, Barbara J. "Recombinant DNA Bills Derailed: Congress Still Trying to Pass a Law." *Science,* 199, 20 January 1978, 274–77.
Dalton, Melville. *Men Who Manage.* New York: John Wiley, 1959.
Davenport, J. A. "A Survey of Vapor Cloud Incidents." *Chemical Engineering Progress,* 73:9, September 1977, 54–63.
Department of Employment. *The Flixborough Disaster,* Report of the Court of Inquiry. London: Her Majesty's Stationery Office, 1975.
Dickson, Captain A. F. "Navigation Problems (Tankers)." *International Tanker Safety Conference,* 1971. International Chamber of Shipping, London 1971, 1–23.
Douglas, Mary, and Aaron Wildavsky. *Risk and Culture.* Berkeley, Calif. University of California Press, 1982.
Eddy, David M. "Probabilistic Reasoning in Clinical Medicine: Problems and Opportunities." In *Judgement Under Uncertainty: Heuristics and Biases,* edited by Daniel Kahneman, Paul Slovic, and Amos Tversky, Cambridge, Mass.: Cambridge University Press, 1982, 249–67.
Edwards, Elwyn. "Automation in Civil Transport Aircraft." *Applied Ergonomics,* 8:4, December 1977, 194–98.
Egginton, Joyce. *The Poisoning of Michigan.* New York: W. W. Norton, 1980.
Einhorn, Hillel J., and Robin M. Hogarth, "Behavioral Decision Theory: Processes of Judgment and Choice." *Annual Review of Psychology,* 32, 1981, 53–88.
Emshwiller, John R. "Construction Halt at Nuclear Plant Raises Questions," *Wall Street Journal,* 24 October 1979.
Erikson, Kai. *Everything in its Path: Destruction of Community in the Buffalo Creek Flood.* New York: Simon and Schuster, 1976.
Fallows, James. *National Defense.* New York: Random House, 1981.
Feaver, Douglas B. *Washington Post,* 7 March 1982, C4.
Feazel, M. "Fuel Pivotal in Trunks' Earnings Slump." *Aviation Week and Space Technology,* 113, 18 February 1980: 31–2.
Finney, John W. "Project Mercury Defects Laid to Private Industry." *New York Times,* 4 October 1963.
Fischhoff, Baruch. "Behavioural Aspects of Cost-Benefit Analysis." In *Energy Risk Management,* edited by G. Goodman and W. T. D. Rowe. London: Academic Press, 1979.
Fischhoff, Baruch. "Cost-Benefit Analysis and the Art of Motorcycle Maintenance." *Policy Sciences,* 8, 1977, 177–202.

References

Fitts, Paul M., and R. E. Jones. "Analysis of Factors Contributing to 460 'Pilot Error' Experiences in Operating Aircraft Controls." In *Selected Papers on Human Factors in the Design and Use of Control Systems,* edited by H. Wallace. Sinaiko, N. Y.: Dover Publications, Inc., 1961.

Franklin, Ben A. "Toxic Wastes Turned Area of Non-Profit." *New York Times,* 25 February 1983, B8.

Fuller, John G. "We Almost Lost Detroit." In *The Silent Bomb,* edited by Peter Faulkner, New York: Random House, 1977, 45–59.

Gaffney, Michael E. "Bridge Simulation: Trends and Comparisons." Unpublished manuscript, Maritime Transportation Research Board, National Academy of Sciences, Washington, D. C., 1982.

Gardenier, John S. "Ship Navigational Failure Detection and Diagnosis." *Human Detection and Diagnosis of System Failures,* ed. Jens Rasmussen and William B. Rouse, 49–74, New York: Plenum Publishing Corp., 1981.

Geertz, Clifford. *The Interpretation of Cultures.* New York: Basic Books, 1973, 3–32.

Gilinsky, Victor. "Full Ahead for Nuclear Power?" *Technology Review,* February/March 1982, 10.

Godson, John, *The Rise and Fall of the DC-10.* New York: David McKay, 1975.

Gold, Michael. "Who Pulled the Plug on Lake Peigneur?" *Science 81,* November 1981, 56–63.

Goldoftas, Barbara. "Recombinant DNA: The Ups and Downs of Regulation." *Technology Review,* May/June 1982, 29–32.

Goller, O. "Report on Three Serious Accidents in Oxygen Plants." In C. H. Buschmann, ed. *Loss Prevention and Safety Promotion in the Process Industries.* New York: Elsevier, 1974, 325–30.

Graham, John D., and James W. Waupel. "Value of a Life, What Difference Does It Make?" *Risk Analysis,* 1:1, 1981, 89–95.

Graham, Julie, and Don Shakow. "Risks and Rewards: Hazard Pay for Workers." *Environment,* 23:8, October 1981, 14–45.

Graham, Julie, Don Shakow, and Christopher Cyr. "Risk Compensation—In Theory and Practice." *Environment* 25:1, January/February 1983, 14–40.

Grayson, Ralph L., and Charles E. Billings. "Information Transfer Between Air Traffic Control and Aircraft: Communication Problems in Flight Operations" in *Information Transfer Problems in the Aviation System,* NASA, Technical Paper 1875, Moffett Field, Calif. September 1981, 47–62.

Greenberg, Daniel. *The Politics of Pure Science.* New York: New American Library, 1967.

Gulbransen, Earl A. "Not Safe Enough." *Bulletin of the Atomic Scientist,* June 1975, 5.

Gyorgy, Anna, and friends. *No Nukes: Everyone's Guide to Nuclear Power.* Montreal: Black Rose Books, 1979.

Hagen, E. W. "Common-Mode/Common Cause Failure: A Review." *Nuclear Safety,* 21:2, March-April 1980, 184–92.

Hall, D. W., and A. W. Hecht. "Summary of the Characteristics of the ASRS Database." *Ninth Quarterly Report,* NASA, 78608, Washington, D. C. June 1979, 24–34.

Hart, Gary, and Barry Goldwater. "Recent False Alerts from the Nation's Missile Attack Warning System." *Report to the Senate Committee on Armed Services.* Washington, D.C.: Government Printing Office, 9 October 1980.

Hines, William. "NASA: The Image Misfires." *The Nation,* 24 April 1967, 517–19.

Hirschhorn, Larry. "The Soul of a New Worker." *Working Papers,* January/February 1982, 42–7.

Hogarth, Robin M. "Beyond Discrete Biases: Functional and Dysfunctional Aspects of Judgmental Heuristics." *Psychological Bulletin,* 90:2, 197–217.

Howard, Ronald A. "On Making Life and Death Decisions." In *Societal Risk Assessment: How Safe Is Safe Enough?* edited by Richard C. Schwing and Walter A. Albers, 89–106.

Hoy-Petersen, R. "Fire Prevention in Solvent Extraction Plants." In C. H. Buschmann, ed., *Loss Prevention and Safety Promotion in the Process Industries.* New York, Elsevier, 1974, 325–30.

Hunt, D. C. "Restricted Release of Plutonium—Part 1. Observational Data." *Nuclear Safety,* 12:2, March/April 1971, 85–9.

Hynes, Mary Ellen, and Erick H. Vanmarcke. "Reliability of Embankment Performance

Predictions." Proceedings of the ASCE Engineering Mechanics Division Specialty Conference, May, 1976. Waterloo, Canada: University of Waterloo Press, 1976.

Jarvis, H. C. "Butadiene Explosion at Texas City—1." *Loss Prevention,* 5, 1971, 57–60.

Jervis, Robert. *Perception and Misperception in International Politics.* Princeton, N.J.: Princeton University Press, 1976.

Kahneman, Daniel, and Amos Tversky. "On the Psychology of Prediction." *Psychological Review,* 80, 1973, 237–51.

Kasperson, Roger, C. Hohenemser, and J. X. Kasperson. Institutional Response to Different Perceptions of Risk." In *Accident at Three Mile Island: The Human Dimensions,* edited by David L. Sills, C. P. Wolfe, and Vivian B. Shelanski, 39–48, Westview Press, 1982.

Keister, R. G., B. I. Pesetsky, and S. W. Clark. "Butadiene Explosion at Texas City—3." *Loss Prevention,* 5, 1971, 67–75.

Kemeny, John, et al. *The Need for Change: The Legacy of TMI.* Report of the President's Commission on the Accident at Three Mile Island. Washington, D.C.: Government Printing Office, October 1979.

Kenney, Martin, and Frederick Buttel, J. Tadlock Cowan, and Jack Kloppenburg, Jr. "Genetic Engineering and Agriculture." *Bulletin No. 125,* Cornell Rural Sociology Bulletin Series, July 1982.

Kletz, T. A. "A Decade of Safety Lessons." *Hydrocarbon Processing,* 58:6, June 1979, 202.

Kletz, T. A. "The Flixborough Cyclohexane Disaster." *Loss Prevention* 9, 1975, 106–10.

Kletz, T. A. "Seek Intrinsically Safe Plants." *Hydrocarbon Processing,* 59:8, August 1980, 137–51.

Kokemor, F. G. "Synthesis Start-Up Heater Failure." *Ammonia Plant Safety,* 22, 1980, 159–69.

Kompass, E. J. "A Long Perspective on Integrated Process Control Systems." *Control Engineering,* August 1981, 4–9.

Kreifeldt, John G. "Cockpit Displayed Traffic Information and Distribution Management in Air Traffic Control." *Human Factors,* 22:6, 1980, 671–91.

Krey, P. W. "Atmospheric Burnup of a Plutonium-238 Generator." *Science,* 10 November 1967, 769–71.

Krimsky, Sheldon, and D. Ozonoff. *Genetic Alchemy: The Social History of the Recombinant DNA Controversy.* Cambridge, Mass.: M.I.T. Press, in press.

Lagadec, Patrick. *Major Technological Risk.* New York: Pergamon Press, 1982.

LaPorte, Todd R. "In Search of Nearly Error-Free Management: Lessons from U.S. Air Traffic Control for the Future of Nuclear Energy." Unpublished manuscript, Institute of Governmental Studies, University of California, Berkeley, 1980.

LaPorte, Todd R. "On the Design and Management of Nearly Error-Free Control Systems." *Social Science Aspects of the Accident at Three Mile Island,* edited by David Sills, et al. Boulder, Colorado: Westview Press, 1982, 185–202.

Lave, Lester B., "Introduction." In *Quantitative Risk Assessment and Regulation.* Edited by Lester Lave, Washington, D.C.: Brookings Institute, 1983, 8.

Lave, Lester B., ed. *Quantitative Risk Assessment in Regulation.* Washington, D.C.: The Brookings Institute, 1983.

Lawrence, Paul, and Jay Lorsch. *Organizations and Environment.* Cambridge, Mass.: Harvard University Press, 1967.

Lear, John. *Recombinant DNA: The Untold Story.* New York: Crown, 1978.

Lederer, Jerome. *Aviation Safety Perspectives: Hindsight, Insight, Foresight.* New York: The Wings Club, 1982.

Lloyd's List. London, 21 May, 1981, 1–4.

Lyons, Richard D. "Crews at Reactor Criticize Cleanup." *New York Times,* 28 March 1983.

McCaffrey, David P. *OSHA and The Politics of Health Regulation.* New York: Plenum Press, 1982.

McKinley, Olson. *Unacceptable Risk: The Nuclear Power Controversy,* New York: Bantam Books, 1976, 22.

Mackley, William B. "Aftermath of Mount Erebus." *Flight Safety Digest,* September 1982, 1–5.

Mahon, Thomas. *Report of the Royal Commission to Inquire Into the Crash on Mount*

Erebus, Antarctica of a DC-10 Aircraft Operated by Air *New Zealand Limited.* Wellington, New Zealand: P. D. Hasselberg, 1981.

March, James G. "Bounded Rationality, Ambiguity, and the Engineering of Choice." *Bell Journal of Economics,* Autumn 1978, 587–608.

March, James, and Johan Olsen, eds. *Ambiguity and Choice in Organizations.* Bergen: Universitets Forlaget, 1976.

March, James, and Herbert Simon. *Organizations.* New York: John Wiley, 1958.

Maritime Transportation Research Board. *Human Error in Merchant Marine Safety,* AD/A-028 371. Washington, D.C.: National Technical Information Service, June 1976.

Marshall, Eliot. "Gene Splicers Simulate a 'Disaster,' Find No Risk." Science, 203, 23 March 1979, 1223.

Marshall, Eliot. "NRC Takes a Second Look at Reactor Design." *Science,* 207, 28 March 1980, 1445–48.

Metcalf, Lee, and Vic Reinemer. *Overcharge.* New York: D. McKay Co., 1967.

Meyer, John, and Brian Rowen. "The Structure of Educational Organizations," *Environment and Organizations.* San Francisco: Jossey-Bass, 1978, 78–109.

Mitchell, Robert Cameron. "Public Response to a Major Failure of a Controversial Technology." In *Accident at Three Mile Island: The Human Dimensions,* edited by Sills, et al., pp. 21–38. Boulder, Colorado: Westview Press, 1982.

Monan, William P. "Distraction—A Human Factor in Air Carrier Hazard Events." *Ninth Quarterly Report,* NASA Technical Memorandum 78608, Washington, D.C., June 1979, 2–22.

Morris, P. A., and R. H. Engelken. "Safety Experience in the Operation of Nuclear Power Plants." In *International Atomic Energy Agency Principles and Standards of Reactor Safety.* Proceedings of a Symposium, IAEA, Vienna, 1973, 429–46.

Mostert, Noel. *Supership.* New York: Alfred A. Knopf, 1974.

NASA, Aviation Safety Reporting System Staff. "Human Factors Associated wtih Altitude Alert Systems." In *Sixth Quarterly Report,* NASA TM-78511, Washington, D.C., July 1978, 25–37.

NASA, Aviation Safety Reporting System Staff. NASA-Ames L/M: 239–3. Perrow. This is a special report prepared by the Aviation Safety Reporting System for Perrow, 1982.

NASA, Third Quarterly Report, TM X–3546. Washington, D.C., May 1977.

NASA, TM 81225. Ames Research Center, Moffett Field, Calif. August 1980.

National Research Council. *Computer-Aided Manufacturing: An International Comparison.* Prepared by Hiroyuki Yoshikawa, Keith Rathmill, and Jozsef Hatvany, for the Assembly of Engineering, National Research Council. Washington, D.C.: National Academy Press, 1981.

NBC, *Nightly News,* 24 July 1982.

New Indicator. "You Have One Hour to Evacuate . . . " University of California, San Diego, 7:6, 2 December–4 January 1982, 1–2.

New York Times, 14 May 1980.

New York Times, 17 September, 1980.

New York Times, 26 September 1981.

New York Times, 28, 29, 30 October; 12 December, 1981. Nisbet, Robert. "Quintessential Liberal." *Commentary,* 72, September 1981, 61–4.

National Transportation Safety Board—Aircraft Accident Reports:

NTSB, AAR-81-10, 7 July 1981.

NTSB, AAR-81-11, 21 July 1981.

NTSB, AAR-81-12, 19 August 1981.

NTSB, AAR-81-13, 19 August 1981.

NTSB, AAR-81-15, 15 September 1981.

NTSB, AAR-81-17, 17 December 1981.

NTSB, AAR-81-18, 17 December 1981.

NTSB, AAR-82-3, 6 April 1982.

NTSB SS *Badger State,* 7 December 1971.

Marine Accident Reports:

NTSB, MAR-75-5, 6 July 1978.

NTSB, MAR-79-16, 27 September 1979.

NTSB, MAR-80-5, 28 March 1980.

NTSB, MAR-80-7, 12 May 1980.
NTSB, MAR-80-11, 28 August 1980.
NTSB, MAR-80-16, 29 September 1980.
NTSB, MAR-81-3, 10 April 1981.
NTSB, MAR-81-14, 9 December 1981.
NTSB, MAR-82-3, 9 February 1982
NTSB, Safety Recommendation(s) A-81-69, 29 June 1981.
NTSB, Safety Recommendation A-81-92, 26 August 1981.
NTSB, Safety Recommendation A-81-93, 26 August 1981.
NTSB, Safety Recommendation A-81-150, 9 November 1981.
NTSB, Safety Recommendation A-82-17, 5 March 1982.
NTSB, Safety Recommendation M-82-1, 18 February 1982.
NTSB, SIR-81-6, 9 September 1981.
NTSB, SIR-81-6, 24 September 1981.
NTSB. *Special Study: Major Marine Collisions and Effects of Preventive Recommendations,* MSS-81-1, 9 September 1981.
Okrent, David. "Comment on Societal Risk." *Science,* 208, 25 April 1980, 372–75.
Palladino, N. J. "Defends Zirconium." *Bulletin of the Atomic Scientist.* March 1976, 5.
Peltzman, S. "The Effects of Automobile Safety Regulation." *Journal of Political Economy,* 83, 1975, 677–725.
Perlman, David. *San Francisco Chronicle,* 6 November 1981.
Perrow, Charles. "The Bureaucratic Paradox: The Efficient Organization Centralizes in Order to Decentralize." *Organizational Dynamics,* Spring 1977, 2–14.
Perrow, Charles. *Complex Organizations: A Critical Essay,* rev. ed. Glenview, Ill.: Scott-Foresman, 1979.
Perrow, Charles. "A Framework for the Comparative Analysis of Organizations." *American Sociological Review,* 32:2, April 1967, 194–208.
Perrow, Charles. "Hospitals: Technology, Goals and Structure." In *Handbook of Organizations,* edited by James March, pp. 910–71. Chicago, Ill.: Rand McNally, 1965.
Perrow, Charles. "Normal Accident at Three Mile Island." *Society,* 18:5, July/August 1981, 17–26.
Perrow, Charles. *Organizational Analysis: A Sociological View.* Belmont, Calif.: Brooks/Cole, 1970.
Perrow, Charles. *The Organizational Context of Human Factors.* Technical Report, DTIC number ADA 123435, U.S. Navy, Office of Naval Research, Washington, D.C., November 1982.
Perrow, Charles. "The Organizational Context of Human Factors Engineering." *Administrative Science Quarterly,* December 1983.
Perrow, Charles. "The President's Commission and the Normal Accident." In *The Accident at Three Mile Island: The Human Dimensions,* edited by David Sills, Charles Wolf, and Vivian Shelanski, 173–84. Boulder, Colo.: Westview Press, 1981.
Perry, Willard W., and William P. Articola. *Study to Modify the Vulnerability Model of the Risk Management System.* Technical Report CG-D-22-80, U.S. Department of Transportation, Washington, D.C., February 1980.
Pfund, Nancy. "Recombinant DNA: Miracles and Menaces." In *Do No Harm: Health Risks and Public Choice,* edited by Diana Dutton. University of California Press, forthcoming.
President's Commission on the Accident at Three Mile Island. *Closed Hearings.* 15 September 1979.
Pryde, Philip R. *The Soviet Energy System.* New York: John Wiley, 1981.
Raiffa, Howard. "Concluding Remarks." In *Societal Risk Assessment: How Safe Is Safe Enough?,* edited by Richard C. Schwing and Walter A. Albers, 339–420. New York: Plenum Press, 1980.
Robertson, L. S. "A Critical Analysis of Peltzman's 'The Effects of Automobile Safety Regulation.' " *Journal of Economic Studies,* 11, 1977, 587–600.
Sadee, C., D. E. Samuels, and T. P. O'Brien. "The Characteristics of the Explosion of Cyclohexane at the Nypro (U.K.) Flixborough Plant on 1st June 1974." *Journal of Occupational Accidents,* 1, 1976/77, 203–35.

References

Saia, S. A. "Vapor Clouds and Fires in a Light Hydrocarbon Plant." *Chemical Engineering Progress,* 72:11, November 1976, 56–61.

Santilli, Major Stan R. *Critical Interface Between Environment and Organisms in Class A Mishaps: A Retrospective Analysis.* Report SAM-TR-80-3. USAF School of Aerospace Medicine, Brooks Air Force Base, Texas, June 1980.

Schwing, Richard C. and Walter A. Albers, Jr. *Societal Risk Assessment: How Safe Is Safe Enough?* New York: Plenum Press, 1980.

Schwing, Richard C. "Trade Offs." In *Societal Risk Assessment: How Safe Is Safe Enough?,* edited by Richard C. Schwing and Walter A. Albers, New York: Plenum Press, 1980, p. 129–41.

Science. "Investigators Agree New York Blackout of 1977 Could Have Been Avoided," 201, 15 September 1978, 994–96.

Science. "UCSD Gene Splicing Incident Ends Unresolved," 209, 26 September 1980.

Scott, R. L. Jr., "Fuel Melting Incident at the Fermi Reactor on October 5, 1966." *Nuclear Safety,* 12:2, March-April 1971, 123–34.

Seale, Robert L. "Consequences of Criticality Accidents." In *Nuclear Criticality Safety,* edited by R. Douglas O'Dell, 16–24. USAEC, Technical Information Center, 1974.

Seals, E. D., and R. A. Speirer. "Analysis of Accidents Related to Falls of Ground in Metal and Nonmetal Mines, 1972–1973." Pittsburgh, U.S. Bureau of Mines Mining Enforcement and Safety Administration. Pittsburgh, Information Report 1009, 1.

Slovic, Paul, Baruch Fischhoff, and Sara Lichtenstein. "Accident Probabilities and Seat Belt Usage." *Accident Analysis and Prevention* 10, 1978, 281–85.

Slovic, Paul, Baruch Fischhoff, and Sara Lichtenstein. "Facts and Fears: Understanding Perceived Risk." In *Societal Risk Assessment: How Safe Is Safe Enough?* edited by Richard C. Schwing, and Walter A. Albers, 181–212. New York: Plenum Press, 1980.

Slovic, Paul, Baruch Fischhoff, and Sarah Lichtenstein. "Facts versus Fears: Understanding Perceived Risk." In *Judgement Under Uncertainty: Heuristics and Biases,* edited by Daniel Kahneman, Paul Slovic, and Amos Tversky, 463–92. Cambridge, Mass.: Cambridge University Press, 1982.

Slovic, Paul, Baruch Fischhoff, and Sarah Lichtenstein. "Perceived Risk: Psychological Factors and Social Implications." *Proceedings of the Royal Society of London,* A 376, 1981, 17–34.

Smith, H. P. Ruffell. *A Simulator Study of the Interaction of Pilot Workload with Errors, Vigilance, and Decisions.* NASA TM 78482, Ames Research Center, Moffett Field, Calif.: 1979.

Smith, R. Jeffrey. "FAA Is Cool to Cabin Safety Improvements." *Science,* 211, 6 February 1981, 557–60.

Spahn, Mark J. "Analysis of the System Effectiveness Information System (SEIS) Data Base." In *The Human Element in Air Traffic Control,* edited by Glenn C. Kinney. McLean, Va.: The Mitre Corporation, December, 1977.

Starr, Chauncey, "Social Benefit versus Technological Risk." *Science,* 165, 1969, 1232–38.

Starr, Chauncey, and Chris Whipple. "The Risk of Risk Decisions." *Science,* 208, 6 June 1980, 1114–19.

Stevens, William K. "Man Has Yet to Master Shuttle's Sophistication." *New York Times,* 15 November 1981.

Stevens, William K. *New York Times,* 15 November 1982.

Street, David, Robert Vinter, and Charles Perrow. *Organization for Treatment.* New York: The Free Press, 1966.

Talbot, Stephen. "The H-Bombs Next Door." *Nation,* 7 February 1981.

Thomasson, W. A. "Recombinant DNA and Regulating Uncertainty." *Bulletin of the Atomic Scientists* 35:10, December 1979, 26–32.

Thompson, James. *Organizations in Action.* New York: McGraw-Hill, 1967.

Time, 20 December 1982, 68.

Turner, Walker. *New York Times,* 2 December 1981.

Tversky, Amos, and Daniel Kahneman. "Availability: A Heuristic for Judging Frequency and Probability." *Cognitive Psychology,* 5, 1973, 207–32.

Union of Concerned Scientists. *The Risks of Nuclear Power Reactors.* Cambridge, Mass.: USC, August 1977.

"Unsettling Questions." *Wall Street Journal,* 15 December 1982.
U.S. Atomic Energy Commission. WASH-1192: *Operational Accidents,* 1943–1970. Washington, D.C.: 1972.
U.S. Coast Guard/National Transportation Safety Board. *Marine Casualty Report* No. 16732/92368, 31 July 1979.
U.S. Coast Guard/National Transportation Safety Board. MAR-73-1, 28 August 1973.
U.S. Coast Guard/National Transportation Safety Board. *Marine Casualty Report—SS"Transhuron" Fire on 24 September and Grounding on 26 September 1974, Arabian Sea,* Washington, D.C.: Government Printing Office, September 1976.
U.S. Coast Guard/National Transportation Safety Board. MAR-77-1, 12 May 1977.
U.S. Congress House Interior Committee. *Mill Tailings Dam Break at Church Rock, New Mexico.* Washington, D.C.: Government Printing Office, 1980.
U.S. Nuclear Regulatory Commission. *NRC Licensee Assessments.* NUREG-0834, USNRC, Washington, D.C., August 1981.
U.S. Nuclear Regulatory Commission. "Safety Goals for Nuclear Power Plants: A Discussion Paper." *Nureg* 0880, USNRC, Washington, D.C., February 1982.
van Eijnatten, A. L. M. "Explosion in a Naphtha Cracking Unit." *Loss Prevention,* 11 September 1977, 11–14.
Vervalin, Charles H. "Fire Losses Reported by NFPA." *Hydrocarbon Processing,* 56:2, February 1977, 166–67.
Voros, M. and Gy Honti, "Explosion of a Liquid CO_2 Storage Vessel in a Carbon Dioxide Plant." In *Loss Prevention and Safety Promotion in the Process Industries,* edited by C. H. Buschmann, 337–46. New York: Elsevier Scientific Publishing Co., 1974.
Wade, Nicholas. "Recombinant DNA: Warming Up for the Big Payoff." *Science,* 206, 29 November 1979, 663–65.
Wade, Nicholas. *The Ultimate Experiment.* New York: Walker, 1977.
Wade, Nicholas. "UCSD Gene Splicing Incident Ends Unresolved." *Science,* 209, 26 September 1980, 1494–95.
Wald, Matthew L., *New York Times,* 21 September 1981.
Washington Post, 29 February 1980.
Washington Post, 28 April 1982, A14.
Weaver, W. W. "Pitfalls in Current Design Requirements." *Nuclear Safety,* 22:3, May–June 1981, 328–29.
Webb, Richard E. *The Accident Hazards of Nuclear Power Plants,* Amherst, Mass.: University of Massachusetts Press, 1976.
Weick, Karl. "Educational Organizations as Loosely Coupled Systems." *Administrative Science Quarterly* 21:1, March 1976, 1–19.
Weiner, Charles. "Relations of Science, Government, and Industry: The Case of Recombinant DNA" in American Association for the Advancement of Science, Policy Outlook: Science, Technology and the Issues of the Eighties. Washington, D.C: AAAS, 1981, 109–56.
Whiteside, Thomas. *The Pendulum and the Toxic Cloud.* New Haven: Yale University Press, 1979.
Wiener, Earl L. "Controlled Flight into Terrain Accidents: System-Induced Errors." *Human Factors,* 19, 1977, 171–81.
Wiener, Earl L. "Midair Collisions: The Accidents, the Systems, and the Realpolitick." *Human Factors,* 22:5, 1980, 521–33.
Wiener, Earl L., and Renwick E. Curry. "Flight-Deck Automation: Promises and Problems." *Ergonomics,* 23:10, 1980, 995–1011.
Wildavsky, Aaron. "No Risk Is the Highest Risk of All." *American Scientist,* 67, January-February 1979, 32–7.
Williams, G. P. "Causes of Ammonia Plant Shutdowns." *Ammonia Plant Safety,* 20, 1978, 123–30.
Wilson, G. L. and P. Zarakas. "Anatomy of a Blackout." *Spectrum,* 15 February 1978, 39–45.
Wilson, Richard. "The Costs of Safety." *New Scientist,* 30 October 1975, 274–75.
Winegar, Bruce H. "Partial Collapse of an Atmospheric Ammonia Storage Tank." *Ammonia Plant Safety,* 22, 1980, 226–30.

References

Witkin, Richard. *New York Times,* 28 November 1981, 9.
Wolfe, Tom. *The Right Stuff.* New York: Bantam Books, 1979.
Wood, Jim. *San Francisco Examiner,* 19 December 1981.
Woodward, Joan. *Industrial Organization: Theory and Practice.* London: Oxford University Press, 1965.
Wright, J. Patrick. *On a Clear Day You Can See General Motors.* New York: Avon Books, 1980.

INDEX

Index

Index

Index

Index

Index